D1759856

LIVERPOOL JMU LIBRARY

3 1111 01411 6154

VOLUME SIXTY FOUR

Advances in
MARINE BIOLOGY

The Ecology and Biology of *Nephrops norvegicus*

Advances in MARINE BIOLOGY

Series Editor

MICHAEL LESSER
Department of Molecular, Cellular and Biomedical Sciences
University of New Hampshire, Durham, USA

Editors Emeritus

LEE A. FUIMAN
University of Texas at Austin

CRAIG M. YOUNG
Oregon Institute of Marine Biology

Advisory Editorial Board

ANDREW J. GOODAY
Southampton Oceanography Centre

SANDRA E. SHUMWAY
University of Connecticut

VOLUME SIXTY FOUR

ADVANCES IN
MARINE BIOLOGY
The Ecology and Biology of *Nephrops norvegicus*

Edited by

MAGNUS L. JOHNSON
Centre for Environmental and Marine Sciences,
University of Hull,
Scarborough, United Kingdom

MARK P. JOHNSON
Ryan Institute for Environmental,
Marine and Energy Research,
School of Natural Sciences,
National University of Ireland, Galway,
Ireland

AMSTERDAM • BOSTON • HEIDELBERG • LONDON
NEW YORK • OXFORD • PARIS • SAN DIEGO
SAN FRANCISCO • SINGAPORE • SYDNEY • TOKYO
Academic Press is an imprint of Elsevier

Academic Press is an imprint of Elsevier
32 Jamestown Road, London NW1 7BY, UK
Radarweg 29, PO Box 211, 1000 AE Amsterdam, The Netherlands
The Boulevard, Langford Lane, Kidlington, Oxford, OX5 1GB, UK
225 Wyman Street, Waltham, MA 02451, USA
525 B Street, Suite 1800, San Diego, CA 92101-4495, USA

First edition 2013

Copyright © 2013 Elsevier Ltd. All rights reserved.

No part of this publication may be reproduced, stored in a retrieval system or transmitted in
any form or by any means electronic, mechanical, photocopying, recording or otherwise
without the prior written permission of the publisher.

Permissions may be sought directly from Elsevier's Science & Technology Rights
Department in Oxford, UK: phone (+44) (0) 1865 843830; fax (+44) (0) 1865 853333;
email: permissions@elsevier.com. Alternatively you can submit your request online by
visiting the Elsevier web site at http://www.elsevier.com/locate/permissions, and selecting
Obtaining permission to use Elsevier material

Notice
No responsibility is assumed by the publisher for any injury and/or damage to persons or
property as a matter of products liability, negligence or otherwise, or from any use or
operation of any methods, products, instructions or ideas contained in the material herein.
Because of rapid advances in the medical sciences, in particular, independent verification of
diagnoses and drug dosages should be made.

ISBN: 978-0-12-410466-2
ISSN: 0065-2881

For information on all Academic Press publications
visit our website at store.elsevier.com

Printed and bound in UK

13 14 15 16   11 10 9 8 7 6 5 4 3 2 1

Working together
to grow libraries in
developing countries

www.elsevier.com • www.bookaid.org

CONTRIBUTORS TO VOLUME 64

Jacopo Aguzzi
Marine Science Institute (ICM-CSIC), Passeig Marítim de la Barceloneta, Barcelona, Spain

Susanne P. Baden
Department of Biological and Environmental Sciences-Kristineberg, University of Gothenburg, SE-45178 Fiskebäckskil, Sweden

Ewen Bell
Centre for Environment, Fisheries and Aquaculture Science (CEFAS) Lowestoft, United Kingdom

Thomas Breithaupt
School of Biological, Biomedical and Environmental Sciences, University of Hull, Hull, United Kingdom

Ralf Bublitz
Centre for Environmental and Marine Sciences, University of Hull, Scarborough, United Kingdom

John Wedgwood Clarke
Centre for Environmental and Marine Sciences, University of Hull, Scarborough, United Kingdom

Daniel Cowing
Centre for Environmental and Marine Sciences, University of Hull, Scarborough, United Kingdom

Nicola C. Dobson
Centre for Environmental and Marine Sciences, University of Hull, Scarborough, United Kingdom

Susanne P. Eriksson
Department of Biological and Environmental Sciences-Kristineberg, University of Gothenburg, SE-45178 Fiskebäckskil, Sweden

Edward Gaten
Biology Department, University of Leicester, Leicester, United Kingdom

Bodil Hernroth
The Royal Swedish Academy of Sciences-Kristineberg, SE-45178 Fiskebäckskil, Sweden, and Department of Natural Science, Kristianstad University, Kristianstad, Sweden

Magnus L. Johnson
Centre for Environmental and Marine Sciences, University of Hull, Scarborough, United Kingdom

Mark P. Johnson
Ryan Institute for Environmental, Marine and Energy Research, School of Natural Sciences, National University of Ireland, Galway, Ireland

Emi Katoh
School of Biological, Biomedical and Environmental Sciences, University of Hull, Hull, United Kingdom

Colm Lordan
Marine Institute, Rinville, Oranmore, Co. Galway, Ireland

Steve Moss
School of Biological, Biomedical and Environmental Sciences, University of Hull, Hull, United Kingdom

Adam Powell
Centre for Sustainable Aquatic Research, Swansea University, Singleton Park, Swansea, United Kingdom

Anne Marie Power
Ryan Institute for Environmental, Marine and Energy Research, School of Natural Sciences, National University of Ireland, Galway, Ireland

Jane Sandell
Scottish Fishermen's Organisation Ltd., Peterhead, United Kingdom

Valerio Sbragaglia
Marine Science Institute (ICM-CSIC), Passeig Marítim de la Barceloneta, Barcelona, Spain

Dale Tshudy
Department of Geosciences, Edinboro University of Pennsylvania, Edinboro, Pennsylvania, USA

Anette Ungfors
Department of Biological and Environmental Science Kristineberg, Gothenburg University, Fiskebäckskil, Sweden

CONTENTS

SCAMPI

Night flowers opening from burrows
or are you lice in the folds of the sea bed—

I love your armour, your *fauld o' lames*,
your gauntlet-greaves, your being none of these.

Dark eggs in the sea-dark foam like cuckoo spit
from joints, seethes of parasites,

and none of these. Pollen of dark
hunched and sprung silences shot slithering

across the deck of names in a final moult—
Langoustine, Dublin Bay Prawn, Scampi

in a basket, *Nephrops norvegicus*—silences
the buttered tongue. Just this side of night

your black kidney-eyes filter the moon,
bioluminescent shows—pictures no one inhabits.

John Wedgwood Clarke
University of Hull Leverhulme Poet in Residence
Centre for Environmental and Marine Sciences

NEPHROPS NORVEGICUS: INTRODUCTION TO THE ISSUE

M.L. Johnson*, M.P. Johnson†

*Centre for Environmental and Marine Sciences, University of Hull, Scarborough, United Kingdom
†Ryan Institute for Environmental, Marine and Energy Research, School of Natural Sciences, National University of Ireland, Galway, Ireland

The opportunity to produce this collection of papers on *Nephrops norvegicus* (L.) arose through a gathering of researchers from varied backgrounds within a European Union-funded FP7 project called, appropriately, NEPHROPS. This project will examine the potential for the development of hatcheries, ranching and enhanced survival of animals discarded from fishing. In order to meet these aims, the project will need to explore the development, physiology, ecology and exploitation of *Nephrops*.

There are a few marine invertebrate animals where quite so much of our understanding of them is a product of knowledge gained through their exploitation. The impact of the fishing industry on our understanding of *Nephrops* is a thread that connects most of the chapters throughout this volume. As with tuna, *Nephrops* service a range of consumers: from food for mass consumption, 'pub grub' in the form of scampi, to haute cuisine as langoustine.

It is not clear when *Nephrops* first became a target for human consumption. Despite the exoskeleton of *Nephrops* potentially preserving well, the species has not featured in Mesolithic shell middens. In the North Atlantic, such middens are often dominated by intertidal species such as the limpet *Patella vulgata*; where crustaceans have been identified, these are crabs generally available by shore-gathering (Pickard and Bonsall, 2009). That consideration of *Nephrops* as a valuable species is a relatively recent phenomenon is reflected in anecdotes about the species previously being used for fertilizer or simply discarded as part of the bycatch for other species.

The singularity of *Nephrops* is reflected in the review of systematics given by Dale Tshudy (Chapter 1). *Nephrops* is a monospecific genus, but it is not clear yet where the *species N. norvegicus* came from. Morphologically, *N. norvegicus* is similar to *Metanephrops* species found in the Caribbean, South Atlantic and Pacific, which contrasts with genetic evidence for closer affinity to North Atlantic lobsters. The mystery is only made more intriguing by the almost complete absence of a fossil record—with only one recognized fossil, a claw from the Miocene of Denmark.

While the fishery for *Nephrops* may be a relatively recent phenomenon (Chapter 7), landings may be in a state of flux, both due to gear and management changes and reflecting processes such as climate change. *Nephrops* live in a habitat that may be particularly susceptible to deoxygenation events. Eriksson et al. review the typical stresses that occur from living in the sediment (Chapter 5). Oxygen is a key environmental variable, but the level of oxygen is linked to the availability and concentration of other potential stressors, including sulphide and reduced metals mobilized from anoxic sediments. With deoxygenation events potentially being stimulated by eutrophication and climate change (Diaz and Rosenberg 2008), some *Nephrops* grounds may become less suitable for the continued support of populations.

To deal with the challenges posed by global change and the economic and social pressures on *Nephrops* fisheries, we need to better understand processes at both the individual and the ecosystem level. Johnson et al. emphasize that we do not really know much about the distribution of the species, except where it is fished (Chapter 2). Resource dynamics seem to shape *Nephrops* populations, in terms of the sediment composition and the potential for competition to reduce growth. However, the data are far from complete. As long as the fishing pressure is not too intense, *Nephrops* may be facilitated by trawling, and extensive *Nephrops* grounds, in association with fishing, may influence biodiversity, nutrient and carbon flows at large scales.

The details of individual behaviour and growth are central to any efforts to rear or ranch *Nephrops* and can help understand their interactions with other species in the ecosystem. Reproduction is described by Powell and Eriksson, who review attempts to raise adults from eggs and point out priority areas for research (Chapter 6). The social interactions and the impacts of environment on behaviour are covered in depth by Katoh et al. who reveal some of the secrets behind the enigmatic crepuscular behaviour of this animal (Chapter 3). As they point out, crustaceans are covered in chemical sensory organs that determine and guide much of their behaviour. However, Gaten et al. discuss in detail one of the most obvious features of *Nephrops*, their large and complex kidney-shaped superposition compound eyes (Chapter 4). These delicate eyes are tuned anatomically and physiologically to the range of depths that *Nephrops* inhabit.

While we enjoy this treatise on a single animal, we should also perhaps reflect that it is possible only because there are funds available to study it, driven by the economic importance of *Nephrops*. Recent advances and policies in fisheries resource management have emphasized the need to

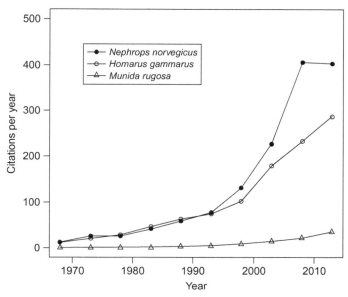

Figure 1 Numbers of papers published each year on three European species of long-bodied benthic decapod (numbers generated from searches of species names for 5-year periods using Google Scholar).

understand whole ecosystems, and in this volume, Ungfors et al. allude to changes that are occurring because of indirect effects from more successful management of whitefish in the North Sea. If the ecosystem approach implies a closer working relationship between academia, state executive agencies and fisher's organizations, then the chapter involving such a collaboration is a step forward in this respect (Chapter 7). Economical importance, however, does not always equate to ecological importance. From a cursory examination of the numbers of papers published on three well-known long-bodied decapods species found in northern Europe (Fig. 1), it is clear that there are huge discrepancies in the attention we pay to them. In order to understand ecosystems, we must develop an understanding of all of the constituents of the system, not just the attractive or economically important ones. Hopefully some future volumes of this wonderful series will also consist of thorough examinations of the roles of unfished and lesser-known species.

ACKNOWLEDGEMENTS

The authors thank all reviewers, both formal and informal, who helped produce this volume. The NEPHROPS project which stimulated this volume received funding from the European Union's Seventh Framework Programme managed by REA Research

Executive Agency http://ec.europa.eu/research/rea (FP7/2007–2013) under grant agreement number 286903.

REFERENCES

Diaz, R.J., Rosenberg, R., 2008. Spreading dead zones and consequences for marine eco-systems. Science 321, 926–929.

Pickard, C., Bonsall, C., 2009. Some observations on the Mesolithic crustacean assemblage from Ulva Cave, Inner Hebrides, Scotland. In: Burdukiewicz, J.M., Cyrek, K., Dyczek, P., Szymczak, K. (Eds.), Understanding the Past: Papers Offered to Stefan K. Kozłowski. Center for Research on the Antiquity of Southeastern Europe, University of Warsaw, Warsaw, pp. 305–313.

SERIES CONTENTS FOR LAST FIFTEEN YEARS*

*The full list of contents for volumes 1–37 can be found in volume 38

CHAPTER ONE

Systematics and Position of *Nephrops* Among the Lobsters

Dale Tshudy[1]

Department of Geosciences, Edinboro University of Pennsylvania, Edinboro, Pennsylvania, USA
[1]Corresponding author: e-mail address: dtshudy@edinboro.edu

Contents

Abstract

This chapter presents and explains the position of *Nephrops norvegicus* in the classification of lobsters. Covered, in order, are systematic classification of *Nephrops*, taxonomic history of *Nephrops*, and analysis of *Nephrops* in nephropid phylogeny.

The genus *Nephrops* was erected by Leach in 1814 and has a long and interesting taxonomic history. Prior to 1972, *Nephrops* was known by 14 Recent species. All but one of these, *N. norvegicus*, were removed to a new genus, *Metanephrops*, by Jenkins (1972). Today, *N. norvegicus* is still the only known living representative of the genus.

Similarly, *Nephrops* is known by only one fossil species, the Miocene *Nephrops kvistgaardae*, although several other fossil species have been previously referred to this genus.

Nephrops, along with the other familiar and commercially important marine clawed lobsters, is referred to Family Nephropidae, one of 17 marine clawed lobster families arrayed in 3 infraorders, 6 families each in the Astacidea and Glypheidea and 5 in the Polychelida. Infraorder Astacidea includes the Superfamily Nephropoidea, as well as the lesser known 'reef lobsters' of the Superfamily Enoplometopoidea, and the

Advances in Marine Biology, Volume 64
ISSN 0065-2881
http://dx.doi.org/10.1016/B978-0-12-410466-2.00001-7

© 2013 Elsevier Ltd.
All rights reserved.

1

freshwater crayfish, Superfamily Astacoidea. In phylogenetic analyses, the freshwater crayfish form a sister group to the Nephropoidea. It is interpreted that freshwater crayfish evolved from nephropoid lobsters, but from which lobster group is uncertain.

The taxonomic placement of *N. norvegicus* is stable at all levels, from species on up. Despite that, the phylogenetic relationships of *Nephrops* to other nephropid genera are unsettled due to conflicting results in morphological and molecular analyses. Currently, new morphological characters and new genes are being analysed in the hope of elucidating nephropid phylogeny.

Keywords: *Nephrops norvegicus*, Nephropidae, Norway lobster, *Metanephrops*, *Paraclytia*, *Palaeonephrops*

1. INTRODUCTION

The purpose of this chapter is to present and explain the position of *Nephrops norvegicus* (L.) in the classification of lobsters. It includes, in order, systematic classification of *Nephrops*, taxonomic history of *Nephrops* and analysis of *Nephrops* in nephropid phylogeny.

The genus *Nephrops* was erected by Leach in 1814. *Nephrops* means kidney (shaped) eye (Bell et al., 2006, p. 412). *Nephrops* is known by one Recent species, the Norway Lobster, *N. norvegicus* (Linnaeus, 1758), also known as the Dublin Bay Prawn, Langoustine (France) and Cigala (Spain) (Holthuis, 1991; Figure 1.1). *Nephrops* is also known by one fossil species, *Nephrops kvistgaardae* Fraaije et al., 2005, from the Miocene of Denmark.

Figure 1.1 *Nephrops norvegius* in burrow on soft bottom at depth of 40–50 m in Gullmarsfjorden, Sweden (Rune Henssel photo). (For colour version of this figure, the reader is referred to the online version of this chapter.)

Nephrops, of course, is a 'marine clawed lobster'. But what is a lobster? And a clawed lobster? 'Lobsters' are marine, reptant (crawling, rather than swimming), elongate, crustaceans with a cephalothorax (13 segments covered by an unsegmented carapace), a 6-segmented abdomen (or pleon) and 10 walking legs. 'Lobsters' include a number of different taxonomic groups, not all closely related. There are clawed lobsters, but also clawless 'spiny' lobsters, with long, whip-like antennae, and clawless 'slipper' lobsters, with antennae flattened into large plates. These different kinds of lobsters are separated at high taxonomic levels, being arrayed in four different infraorders (De Grave et al., 2009; Table 1.1; Figure 1.2). The familiar, commercially important clawed lobsters (e.g. the American lobster, *Homarus americanus*, the European lobster, *Homarus gammarus* and *N. norvegicus*) are separated from the clawless spiny and slipper lobsters at the level of infraorder, in the Astacidea and Achelata, respectively. The clawed lobsters themselves are arrayed in three different infraorders, Astacidea (including the familiar clawed lobsters), Glypheidea (mostly extinct

Table 1.1 List of the four decapod infraorders containing lobsters

Order Decapoda Latreille, 1802
Suborder Pleocyemata Burkenroad, 1963
Infraorder Astacidea Latreille, 1802
Superfamily Enoplometopoidea de Saint Laurent, 1988
Family Enoplometopidae de Saint Laurent, 1988
Family Uncinidae Beurlen, 1928
Superfamily Nephropoidea Dana, 1852
Family Chilenophoberidae Tshudy and Babcock, 1997
Family Protastacidae Albrecht, 1983
Family Stenochiridae Beurlen, 1928
Family Nephropidae Dana, 1852
Infraorder Glypheidea Winckler, 1882—mostly extinct lobsters with either claws or only 'semi-chelate' first pereiopods.
Infraorder Achelata Scholtz and Richter, 1995—non-clawed lobsters.
Infraorder Polychelida Scholtz and Richter, 1995—blind lobsters with delicate, very elongate claws.

In bold font are taxa containing *Nephrops norvegicus*.
Condensed from De Grave et al. (2009).

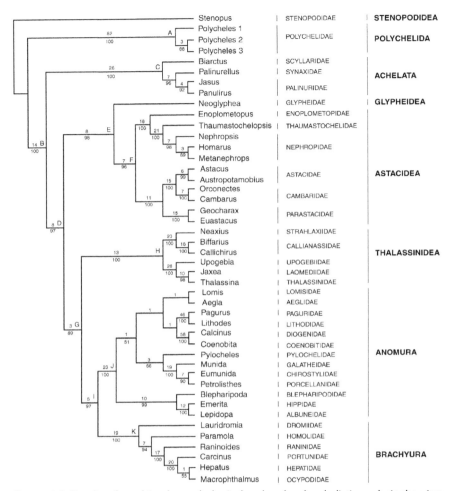

Figure 1.2 Results of combined morphological and molecular cladistic analysis showing the relative degree of derivedness of the four lobster infraorders (Polychelida [least derived], then Achelata, Glypheidea and Astacidea). *Reprinted from Ahyong and O'Meally (2004).*

and many without true claws) and Polychelida (blind, deep-sea lobsters with very elongate, delicate claws). There are 17 marine clawed lobster families in the 3 infraorders, 6 families each in the Astacidea and Glypheidea and 5 in the Polychelida. In addition, there are still other, lobster-like decapods with claws, such as the 'mud lobsters', Family Thalassinidae, which are now classified more as a kind of 'shrimp' in the infraorder Gebiidea. *Nephrops*, along with the other familiar marine clawed lobsters, is referred to the astacidean family Nephropidae.

The infraorder Astacidea includes not only the familiar marine clawed lobsters but also the freshwater crayfish. Since the Astacidea also contains the freshwater crayfish, and the Glypheidea and Polychelida also contain many clawed lobsters, 'marine clawed lobsters' is a 'paraphyletic' group rather than a natural 'monophyletic' group.

Nephrops has a long and interesting taxonomic history. Today, there is a widespread consensus among taxonomists that, in the modern ocean, this genus is monospecific. The genus once held 13 additional Recent species, but these were removed to a new genus, *Metanephrops*, by Jenkins (1972). Similarly, *Nephrops* is known by only one fossil species, although several other fossil species (or specimens) have been previously referred to this genus. This chapter summarises this taxonomic history.

2. SYSTEMATIC CLASSIFICATION

2.1. *N. norvegicus* is classified as follows

Phylum Arthropoda Latreille, 1829—Invertebrate animals with bodies segmented and covered by exoskeleton, and with jointed limbs. Includes the extinct trilobites and extant chelicerates (spiders and scorpions), myriapods (centipedes and millipedes), crustaceans (shrimps, lobsters, crabs, etc.) and insects (possibly late Precambrian; Cambrian–Recent).

Subphylum Crustacea Brünnich, 1772—Arthropods with five-segmented cephalon (head), eight-segmented thorax and (commonly) six-segmented abdomen; two pairs of antennae; biramous appendages; gills. Includes the 'insects of the sea', the shrimps, lobsters, crabs, copepods, ostracods as well as barnacles and others (Cambrian–Recent).

Class Malacostraca Latreille, 1802—Crustaceans with six-segmented abdomen and first antennae (antennules) biramous; second antennae with scale-like exopod; compound eyes on stalks; some with a carapace. Includes amphipods, krill, shrimp-like phyllocarids and hoplocarids and the Eumalacostraca, including the true shrimps, lobsters and crabs (Cambrian–Recent).

Order Decapoda Latreille, 1802—Malacostraca with five pairs of legs, the first often modified into chelipeds. Includes shrimps, lobsters and crabs (Devonian–Recent).

Class Suborder Pleocyemata Burkenroad, 1963—Fertilised eggs incubated by female. Includes some shrimp and all lobsters and crabs (Devonian–Recent).

Infraorder Astacidea Latreille, 1802—includes many marine lobsters and all freshwater crayfish, decapods with a sub-cylindrical cephalothorax and a well-developed rostrum and abdomen, with uropods of telson having a diaeresis; genital openings coxal and three pairs of claws, the first much larger than the next two pairs legs (Glaessner, 1969, p. R455) (thus excluding the spiny lobsters, Family Palinuridae, and slipper lobsters, Family Scyllaridae, of the infraorder Achelata) (Devonian–Recent). Astacidea includes 658 Recent-only species, 3 extant species with a fossil record and 111 fossil-only species (modified slightly from De Grave et al., 2009). Most (592) of the recent species are freshwater crayfish; only 66 are marine clawed lobsters. Most (101) of the fossil species are marine clawed lobsters. Even though there are many more living freshwater crayfish species than marine lobster species, fossil crayfish are relatively rare because continental environments are less likely to preserve body fossils than are marine environments (Rode and Babcock, 2003, p. 422). There are also many more marine clawed lobsters in the mostly extinct infraorder Glypheidea (256 fossil species but only 2 Recent species; many of these with only semi-chelate first pereiopods) and in the infraorder Polychelida (55 fossil species and 38 Recent species) (numbers from De Grave et al, 2009; modified slightly for Astacidea).

Superfamily Nephropoidea Dana, 1852—(according to Holthuis, 1974, p. 725; 1991, p. 19)—includes four marine lobster families with a particular carapace groove pattern (three of four families, including Chilenophoberidae, Protastacidae and Nephropidae; not Protastacidae) and claws with elongate fingers. As currently arranged (De Grave et al., 2009), this superfamily might not be a natural group. The four constituent families have not been analysed together in a cladistic analysis, and much morphological variation occurs across these families (Jurassic–Recent) (Wahle et al., 2012, p. 86).

2.1.1 Relationship of Nephropoidea to freshwater crayfish

The freshwater crayfish, superfamilies Astacoidea and Parastacoidea, form a sister group to the Nephropoidea (Chu et al., 2009; Tsang et al., 2008; Wahle et al., 2012, p. 70; Figure 1.2). It is interpreted that freshwater crayfish evolved from nephropoid lobsters, but from which lobster group is uncertain. Both the crayfish and lobster fossil records extend back into the late Paleozoic, but the crayfish fossil record is very poor.

Family Nephropidae Dana, 1852—(according to Holthuis, 1974, p. 725; 1991, p. 19) marine clawed lobsters with a particular carapace groove pattern (Lower Cretaceous–Recent [Wahle et al., 2012]). Includes the familiar, commercially important species, the American lobster, *H. americanus*, European lobster, *N. norvegicus*, and *N. norvegicus*. Diagnosis of Nephropidae Dana, 1852 in Appendix 1 (from Wahle et al., 2012, p. 87 [modified from

Tshudy and Babcock, 1997]). Nephropidae includes the formerly recognised Thaumastochelidae Bate, 1888 (Upper Cretaceous–Recent) (Ahyong, 2006; De Grave et al., 2009; Tshudy and Babcock, 1997; Tshudy et al., 2009). Nephropidae is the most diverse family (by genus and species count) of the clawed lobsters; 20 genera are recognised (10 Recent-only, 4 extant but with fossil record, 6 extinct). The total number of recognised nephropid species is 134 (54 Recent-only, 1 extant with fossil record, 79 fossils). Several times since the mid-1900s, biologists and palaeontologists have interpreted evolutionary relationships within the Nephropidae and divided the group into subfamilies. Mertin (1941), Glaessner (1960) and Holthuis (1974, 1991) all did this, based mostly on carapace groove pattern. These subfamilies have been abandoned in light of more recent cladistic analyses (Ahyong, 2006, p. 7; Tshudy and Babcock, 1997, p. 257).

2.1.2 Genus Nephrops: *defining morphologic characteristics*

Nephrops is characterised morphologically by its spiny, carinate cephalothorax, complexly sculptured abdomen and spiny, carinate, asymmetrical claws and its extremely large, pigmented eyes. *Nephrops* is unique among nephropids in exhibiting both intermediate and lateral thoracic carinae but lacking the branchial carina (Figure 1.3). *Nephrops* is also unique in exhibiting spinulose subdorsal carinae, which extend to nearly above the cervical groove. A detailed definition of the genus *Nephrops* was provided by Holthuis (1974, pp. 820–822) and is reprinted in Appendix 1 for convenience.

2.1.3 N. norvegicus: *species description*

Holthuis (1974, p. 823) wrote that descriptions of *N. norvegicus* can be found in many handbooks of European Decapoda such as Bell (1847, pp. 251–254, Figure), Lagerberg (1908, p. 44, pl. 4 Figure 1), Pesta (1918, pp. 183–189, Figure 58), Schellenberg (1928, pp. 57–59, Figures 45 and 46) and Zariquiey Alvarez (1968, 201).

3. TAXONOMIC HISTORY OF *NEPHROPS*

3.1. Species *N. norvegicus*: history, synonyms, sub-species(?)

Nephrops is known by a single Recent species; on this, there seems to be no debate among modern taxonomists. Furthermore, there are no recently proposed synonyms for *N. norvegicus*. The genus once held 13 additional Recent species, but these were removed to a new genus, *Metanephrops*, by Jenkins in

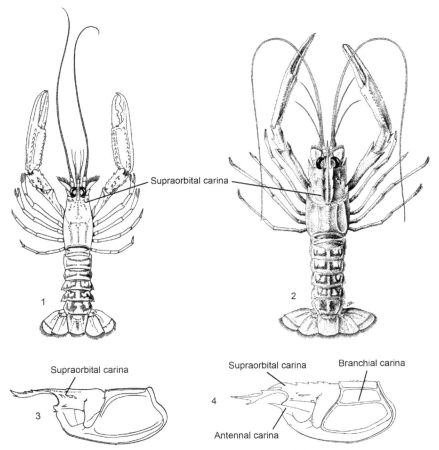

Figure 1.3 Line drawings showing locations and names of distinguishing morphologic features on *Nephrops* (1, 3) and *Metanephrops* (2, 4). *Modified from Holthuis (1991).*

1972. *Nephrops* is also known by a single fossil species, although several other species (or specimens) have been erroneously referred to the genus.

Holthuis (1974, p. 824) thoroughly explained the taxonomic history of *N. norvegicus*, detailing the early records of the species and, in the process, selected a lectotype and a type locality. The type species, *Cancer norvegicus* Linnaeus, 1758 (p. 632), was named in Linnaeus' 10th edition of *Systema Naturae*, in which modern zoological nomenclature was initiated. Linnaeus' oldest reference to *Cancer norvegicus* is that of Aldrovandus (1606, p. 113) whose specimen probably originated from the Mediterranean. Linnaeus did not designate a type specimen, but the type locality for *Cancer norvegicus* is given in Linnaeus as 'in Mari Norvegico'. Holthuis (1974, p. 824) designated as lectotype the largest of the specimens mentioned in Linnaeus (1751, p. 327) (reference according to Holthuis, p. 824), thereby establishing the type locality as the Kullen Peninsula in Southern Sweden, 56°18′N 12°28′E.

Holthuis (1974, pp. 822–823) listed several old synonymies (1761–1880) for *N. norvegicus*. After 1880, the only synonymies are typographical in nature.

Holthuis (1974, pp. 823–824) noted that a sub-species, *N. norvegicus meridionalis* Zariquey Cenarro (1935) (e.g. as used by Zariquiey Alvarez, 1968), was erected for a morphological variant in the southern part of the species' range, where variation exists in the expression of the podobranch of the second maxilliped. According to Zariquiey Cenarro (although based on just a few specimens), the podobranch is present in the southern part of its Atlantic range (e.g. Morocco and Spain) and in the Mediterranean Sea but absent in the northern part of its range (e.g. Denmark). Holthuis (1945) examined more specimens and from more localities and concluded (pp. 318–319) that 'forms with and without podobranchs can not be considered as geographical forms of one species as both occur throughout the entire range of distribution of *Nephrops norvegicus*'. Holthuis (1974) also noted that intermediate forms with a small podobranch on one side and nothing on other have also been observed. More recently, Bell et al. (2006) noted the variation in podobranch expression, stating that the validity of the subspecies is controversial (citing Holthuis, 1945 and Crnkovic, 1969), and that there is no hard evidence of spatial gradients in morphometrics of *N. norvegicus* (citing Castro et al., 1998). Molecular genetic studies, likewise, show no evidence of a geographical pattern in genetic divergence or genetic isolation (Maltagliati et al., 1998; Passamonti et al., 1997; Stamatis et al., 2004).

3.2. Genus *Nephrops*: Recent species formerly referred to *Nephrops*

In 1972, prior to a sweeping taxonomic revision by Jenkins, the genus *Nephrops* contained 14 Recent species. Jenkins' taxonomic revision was prompted by his examination of new *Nephrops*-like fossil specimens from loose cobbles on a New Zealand beach (specimens he would describe as *Metanephrops motunauensis* n. sp.). Jenkins (1972, p. 161) noted that these specimens were 'clearly related to extant lobsters which occur in the Indo-West-Pacific region, the West Indies, and off the southeast coast of South America'. Jenkins (1972, p. 162) remarked that 'It has long been recognised that extant Indo-West Pacific and American nephropsids assigned to *Nephrops* Leach differ in several important features from the European type species',

N. norvegicus. These differences cited by Jenkins include, on *Nephrops*, a spinier rostrum, a shorter supraorbital carinae extending back from the rostrum, a smaller antennal spine, absence of the branchial carina on the thoracic region, a scaphocerite that is long and lanceolate and unequal claws (Jenkins, 1972, p. 162). Jenkins removed 13 recent species from *Nephrops* to a new genus *Metanephrops*, designating *M. japonicus* Tapparone-Canefri, 1873, as the type species. The New Zealand fossil species, *M. motunauensis*, was, in 1972, the first and only fossil species of the genus.

Metanephrops is not distinguishable from other nephropids on the basis of any one morphological feature. It is, however, unique in exhibiting the combination of (1) three pairs of thoracic carinae (except *M. neptunus* [Bruce, 1965] with two pairs), (2) a prominent, spinose, supraorbital carina that extends posteriorly to near the postcervical groove, (3) a prominent, narrow, antennal carina and (4) symmetrical chelipeds that are generally long, slender, and sometimes strongly carinated.

Metanephrops bears strong morphologic similarities to *Nephrops*, although, as will be discussed below, a close relationship is disputed by DNA evidence. *Metanephrops* is most easily distinguished from *Nephrops* by its possession of four features: (1) the branchial carina (absent on *Nephrops* and all but one species of *Metanephrops*, *M. neptunus*), (2) a prominent, spinose, supraorbital carina (much shorter and lower on *Nephrops*), (3) a prominent antennal carina (much shorter and lower on *Nephrops*) and (4) symmetrical claws (Figure 1.3).

Recent species that Jenkins (1972) removed to *Metanephrops* include

Nephrops japonicus Tapparone-Canefri, 1873;

Nephrops thomsoni Spence Bate, 1888;

Nephrops andamanicus Wood-Mason, 1892;

Nephrops rubellus Moreira, 1903;

Nephrops arafurensis De Man, 1905;

Nephrops challenger Balss, 1914;

Nephrops sibogae De Man, 1916;

Nephrops sagamiensis Parisi, 1917;

Nephrops binghami Boone, 1927;

Nephrops boschmai Holthuis, 1964;

Nephrops neptunus Bruce, 1965;

Nephrops sinensis Bruce, 1966a;

Nephrops australiensis Bruce, 1966b.

Since 1972, five additional extant species (including one, *M. taiwanica* [Hu, 1983], also with a Miocene fossil record) and two more extinct fossil species of *Metanephrops* have been discovered. The latter include *M. jenkinsi*

Feldmann, 1989 from the Upper Cretaceous (Maastrichtian–Paleocene) of Antarctica, and *M. rossensis* Feldmann et al., 1993 from the Upper Cretaceous (Campanian) of Antarctica. A fifth reported fossil species, *M. elongatus* Hu and Tao, 2000, from the Pliocene of Taiwan, is poorly preserved and doubtfully *Metanephrops*. Thus, today, *Nephrops* is known by 1 extant and 1 extinct species, and *Metanephrops* is known by 18 extant species and 3 extinct species.

3.3. Fossil species of *Nephrops*, valid and otherwise

Seven fossil species of *Nephrops* have been proposed. Only one can be taken seriously as *Nephrops*. Not until 2005, after years of other specimens being mistakenly referred to the genus, did *Nephrops* have a known fossil record. *N. kvistgaardae* Fraaije et al. is reported from the Upper Miocene of Jutland, Denmark. This species is known only by a single claw, the propodus and dactylus of the left cheliped (Figure 1.4); scanty evidence for proposing a new species, but the fossil does strongly resemble modern *Nephrops*.

Followings are species previously and erroneously referred to *Nephrops*. '*Nephrops*' *aequus* Rathbun, 1920, Lower Miocene of Dominican Republic, is known by only an incomplete claw that cannot be confirmed as *Nephrops*, or even as a nephropid. Jenkins (1972) and Feldmann (1981) similarly expressed doubt about Rathbun's generic determination.

'*Nephrops*' *americanus* Rathbun, 1935, Lower Cretaceous (Upper Albian) of Texas, is known only by two claw fragments. This author agrees with previous workers (Feldmann, 1981, p. 453; Jenkins, 1972, p. 163; Stenzel, 1945, p. 429; Yaldwyn, 1954, p. 730) who have doubted the identity of these claw fragments as *Nephrops*. The palm lacks the carinae characteristic of *Nephrops* and many *Metanephrops*.

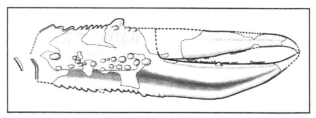

Figure 1.4 Line drawing of fossil species, *Nephrops kvistgaardae. Reprinted from Fraaije et al. (2005).*

'*Nephrops*' *buntingi* (Feldmann and Holland, 1971), Paleocene of North Dakota, is now referred to *Hoploparia* (Tshudy, 1993). Feldmann and Holland (p. 824) noted that their new species very closely resembles *Hoploparia shastensis*, also considered *Nephrops* at the time.

'*Nephrops*' *costatus* Rathbun, 1918, Pleistocene of Panama Canal Zone, is known only by four claw fingers and cannot be confirmed as *Nephrops* or even as nephropid lobster. Jenkins (1972) and Feldmann (1981) similarly expressed the doubt about the generic identity of these fragments.

'*Nephrops*' *geoffroyi* Robineau-Desvoidy, 1849, part (p. 128, pl. 5, Figure 11), Lower Cretaceous (Lower Neocomian) of Yonne, France (Glaessner, 1929; Robineau-Desvoidy, 1849) is now considered *Hoploparia edwardsi* (Robineau-Desvoidy, 1849). Robineau-Desvoidy erected several species based only on claw fragments. While it is possible that more than one nephropid species is represented in his collection, it seems best to consider these fragmentary remains as one species, *Hoploparia edwardsi*.

'*Nephrops*' *maoensis* Rathbun, 1920, Lower Miocene of Dominican Republic, is known only by two tips of claw fingers and cannot be confirmed as *Nephrops*, or even as a nephropid. Jenkins (1972) and Feldmann (1981) similarly expressed scepticism about the generic determination.

'*Nephrops*' *reedi* Carter, 1898, Pliocene of Suffolk, England, is known by only incomplete claws. Despite that these claws have been positively identified as *Nephrops* by Carter (1898) and Woods (1931), the generic identification seems uncertain. The elongate palms, in being equipped with longitudinal rows of very large tubercles, are ornamented quite unlike those of *Nephrops* or *Metanephrops*, which are spiny. Moreover, palm shape, in itself, seems an insufficient basis for giving *Nephrops* a Pliocene fossil record.

'*Nephrops*' *salviensis* Robineau-Desvoidy, 1849, Neocomian of Saint-Sauveur, Yonne, France, is not a nephropid. The three transverse grooves on the carapace, including the cervical groove that reaches the dorsomedian, suggest that the specimen is an erymid lobster.

'*Nephrops*' *shastensis* Rathbun, 1929, Upper Cretaceous (Upper Coniacian–Lower Campanian) of California; the holotype and lone specimen is well preserved and confidently referred to *Hoploparia* (Feldmann, 1981; Tshudy, 1993).

'*Nephrops*' sp. Rathbun, 1918, Pleistocene of Panama Canal Zone, known by an incomplete finger, was referred by Rathbun to *Nephrops* because its ornamentation resembled that of *Nephrops costatus* Rathbun described in the same paper, but this specimen is doubtfully a nephropid.

4. ANALYSES OF *NEPHROPS* IN NEPHROPID PHYLOGENY

4.1. Morphological analysis of phylogeny

In the 1990s, computerised cladistic analyses of nephropid phylogeny using morphological (Tshudy and Babcock, 1997) and molecular data (Tam and Kornfield, 1998) appeared in the literature. Several morphological and molecular phylogenies have followed (Morphological: Ahyong, 2006 [Figure 1.5]; Karasawa et al., 2012; Tshudy and Sorhannus, 2000a,b, 2011. Molecular: DNA: Chu et al., 2006, 2009; Tsang et al., 2008; Tshudy et al., 2005, 2009 [Figure 1.6]).

These two types of analyses, morphological and molecular, give conflicting results with regard to *Nephrops*. Morphological analyses consistently show *Nephrops* and *Metanephrops* to be closely related (usually as sister taxa), whereas molecular analyses show *Nephrops* to more closely related to *Homarus* than to *Metanephrops*.

Besides *Metanephrops* (Upper Cretaceous–Recent), two other genera, both extinct, *Palaeonephrops* (Upper Cretaceous) and *Paraclytia* (Upper Cretaceous), have been allied on morphological grounds with *Nephrops*.

Palaeonephrops Mertin (1941, p. 168) was created to accommodate a previously named lobster species (*Hoploparia browni* Whitfield, 1907) with extensive and unique carination and tuberculation of the cephalothorax (Figure 1.7). It is a monospecific genus known from the Upper Cretaceous of North America and was re-described and well illustrated by Feldmann et al. (1977). It appears in the modern literature as *Palaeonephrops browni* (Whitfield, 1907) but might be synonymous with an earlier-named species, *Hoploparia westoni* Woodward, 1900 as *P. westoni* (Tshudy, 1993, other unpublished data). The genus stands nonetheless.

Paraclytia Fritsch and Kafka, 1887 is known by five species, all Upper Cretaceous; four from Central Europe and the fifth from Iran. *Paraclytia* strongly resembles *Metanephrops* in exhibiting three pairs of thoracic carinae (intermediate, branchial and lateral), a prominent supraorbital carina extending to the post-cervical groove and a very prominent antennal carina (Figures 1.8 and 1.9). *Paraclytia* is distinguished from *Nephrops* and *Metanephrops* in exhibiting a unique sculpture of the pleon terga and pleura (Figure 1.8), and in possessing a telson with sub-median carinae converging (instead of diverging) posteriorly. Mertin (1941, pp. 169–170) provided the first detailed description of *Paraclytia*. The genus is known by *P. boetticheri* Mertin, 1941 (Upper Cretaceous of Germany), *P. nephropica* (Fritsch and

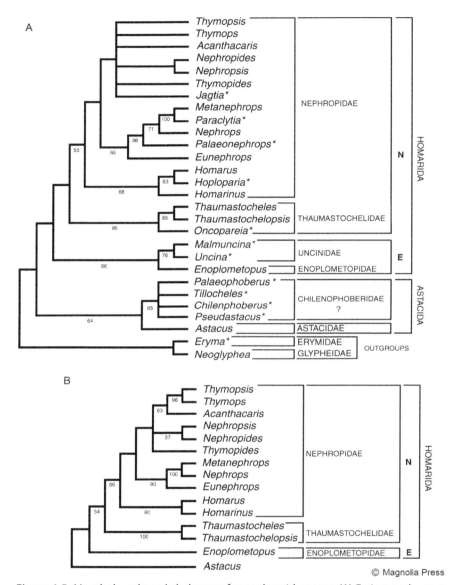

Figure 1.5 Morphology-based cladogram for nephropid genera. (A) Extinct and extant genera. (B) Extant genera only. Morphological analyses show a close relationship between the extant genera *Nephrops* and *Metanephrops*. * indicates extinct taxon. *Both reprinted from Ahyong (2006).*

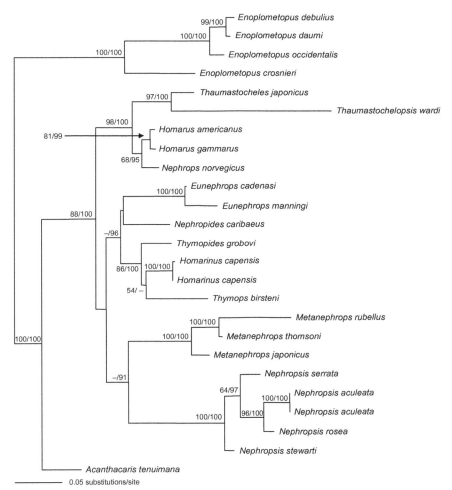

Figure 1.6 DNA-based cladogram of nephropid genera (Tshudy et al., 2009, p. 75). Based on 12S and 16S mitochondrial rRNA, *Nephrops* is a sister taxon to *Homarus*, rather than *Metanephrops*, as in morphological trees.

Kafka, 1887) (Upper Cretaceous of Czechoslovakia), *P. nephropiformis* (Schlüter, 1879) (Upper Cretaceous of Germany), *P. westfalica* Mertin, 1941 (Upper Cretaceous, Germany) and *P. valashtensis* McCobb and Hairapetian, 2009 (Upper Cretaceous of Iran).

In morphological cladistic analyses, *Paraclytia* is consistently determined to be a sister group to *Metanephrops*; these, in turn, are the sisters to *Nephrops* (Ahyong, 2006; Tshudy, 1993; Tshudy and Babcock, 1997). In these same cladistic analyses, *Palaeonephrops* forms a sister group to the

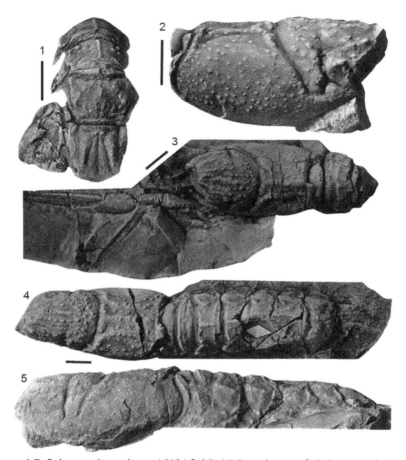

Figure 1.7 *Palaeonephrops browni* (Whitfield). (1) Dorsal view of abdomen, telson and left uropod, AMNH 32484; (2, 3) right lateral view of cephalothorax, and dorsal view of left cheliped and cephalothorax, USNM 239927; (4, 5) dorsal and left lateral views of cephalothorax and abdomen, AMNH 9572. Scale bars equal 1.0 cm. *Modified from Feldmann et al. (1977).*

Nephrops–Metanephrops–Paraclytia clade. Thus, based on morphological evidence, we can say that the closest known relatives of *Nephrops* have a fossil record beginning in the Upper Cretaceous.

4.2. Molecular analysis of phylogeny

As discussed above, it has long been held, on morphological grounds (both non-cladistic and cladistic analyses), that *Nephrops* is closely related to the extant genus *Metanephrops* as well as the extinct genera *Palaeonephrops* and

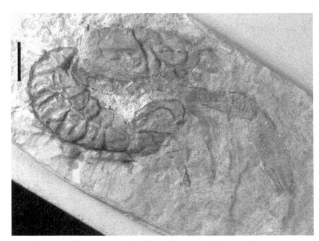

Figure 1.8 *Paraclytia nephropica*, right lateral view of complete specimen in National Museum, Prague (NMP 03459). *Modified slightly from McCobb and Hairapetian (2009).*

Paraclytia. However, recent molecular analyses on extant genera argue that *Nephrops* and *Metanephrops* are not so closely related (Figure 1.6). In fact, molecular results argue that *Nephrops* is more closely related to the plain-looking *Homarus* than to the carinated, spiny *Metanephrops*. If the molecular results are the more trustworthy, then the morphological similarities of *Nephrops* and *Metanephrops* are attributed to convergent evolution. Perhaps even more surprisingly, *Nephrops* and *Homarus*, together, are a sister group to the thaumastochelids, rather than other, more typical looking nephropids. So, we have a quandary: are the morphological or molecular results more meaningful here? This author, while a morphologist himself, has been impressed by the molecular results on lobsters. They seem more stable; that is less sensitive to selection of characters and of in-group and out-group taxa. At the time of writing this chapter, some clades are still poorly supported (e.g. bootstrap support values are not high enough), and new genes are being sequenced, but overall, the molecular trees seem more stable. Thus, this author's instinct, at this point, is to trust the molecular results more than the morphological results. With molecular results arguing that *Nephrops* and *Metanephrops* are not closely related, then, by extension, we know little about the relationship of *Nephrops* and the extinct taxa, *Palaeonephrops* and *Paraclytia*, for which there are no molecular genetic data. It still might be that *Palaeonephrops* and/or, less likely, the *Metanephrops*-like *Paraclytia* are, in fact, closely related to *Nephrops*.

    Homarus, the molecular sister group to *Nephrops*, is known from the fossil record beginning in the Lower Cretaceous (Albian age), and it looks rather

Species	Key morphological features of cephalothorax	Pleura ornament	Terga sculpture
Paraclytia nephropica			
Paraclytia boetticheri			
Paraclytia westfalica			
Paraclytia nephropiformis			
Paraclytia valashtensis			

Figure 1.9 *Paraclytia*, showing morphology of the cephalothorax, pleura and terga of the five known species. *Reprinted from McCobb and Hairapetian (2009).*

modern-looking in Cretaceous fossil form. *Homarus* is interpreted to have evolved from *Hoploparia*, the most diverse fossil clawed lobster genus (57 species known), during the early Cretaceous (Tshudy and Babcock, 1997). *Nephrops* has a fossil record extending back to the Miocene (*N. kvistgaarde*). Therefore, taking all the evidence together, it seems reasonable to hypothesise that *Nephrops* evolved from a stem nephropid, perhaps *Hoploparia*, at some time between the early Cretaceous and Miocene.

5. SUMMARY

The genus *Nephrops* Leach, 1814 has a long taxonomic history. Prior to 1972, *Nephrops* was known by 14 Recent species. All but one of these, *N. norvegicus*, was removed to a new genus, *Metanephrops*, by Jenkins (1972). Today, *N. norvegicus* is still the only known-living representative of the genus. Similarly, *Nephrops* is known by only one fossil species, *N. kvistgaardae*, from the Miocene of Denmark, despite that several other fossil species (or specimens) have been previously referred to this genus.

The taxonomic placement of *N. norvegicus* is stable at all taxonomic levels, from species on up. Despite that, the phylogenetic relationships of *Nephrops* to other nephropid genera are unsettled due to conflicting results in morphological and molecular analyses. Currently, new morphological characters and new genes are being analysed in the hope of elucidating nephropid phylogeny.

ACKNOWLEDGEMENTS

I am most grateful to Tin-Yam Chan, National Taiwan Ocean University, for many valuable discussions about *Metanephrops*; D. Mitchell, Edinboro University, for help in preparing the figures, and R. Hensell, Copenhagen, for kindly granting permission to use the unpublished photograph in Figure 1.1.

APPENDIX. SYSTEMATIC DESCRIPTIONS

A.1. Diagnosis of family Nephropidae (copied from Wahle et al., 2012, p. 87; modified from Tshudy and Babcock, 1997)

Rostrum usually large, spinose. Cervical groove extends ventrally from level of orbit to antennal groove. Postcervical groove typically curves anterodorsally from dorsal midline toward, or to, cervical groove at approximately mid-height on carapace. Branchiocardiac groove dorsally subparallel-parallel to dorsal midline, joining postcervical groove at level of orbit; combined branchiocardiac-postcervical groove extends toward, or to, hepatic groove (Tshudy and Babcock, 1997, p. 260). Incisura clavicularis (Holthuis, 1974, p. 737) in anterior margin of carapace. Pleonal pleura usually end in a point (except in thaumastochelids, which have quadrate pleura). Claws longer than carapace; equal or unequal. Fifth perieopod almost always without true chela (except in Recent thaumastochelids).

A.2. Description of genus *Nephrops* (copied from Holthuis, 1974, pp. 820–822)

The rostrum has lateral and a ventral tooth. The subdorsal carinae are distinct and spinulate, especially some spinules in the middle of its length are distinct. The supraorbital spines are well developed, but not very large, and are followed by a row of sharp spinules, there is an actual supraorbital carina. A well-developed postorbital spine is present.

There is a single additional spinule below the end of the row of spinules behind the supraorbital spine. An antennal, but no hepatic spine is present. There is a single small cervical spine. Behind the postcervical groove there is a pair of postcervical and intermediate spinules, but no others.

The postcervical, hepatic, cervical and antennal grooves are distinct. The intercervical groove is visible in its posterior part. The antennal groove anteriorly curves up behind the antennal spine, there is no forward branch. The gastro-orbital groove is not very distinct, and in the middle gives off the buccal groove, which is rather faint and extends towards the antennal groove. The inferior groove is present but faint. The marginal groove and marginal carina are well indicated. The median carina extends from the gastric tubercle back. It carries a double row of granules and is interrupted by the postcervical groove. The intermediate and lateral ridges are present behind the postcervical groove, but not the branchial ridge.

The abdominal tergites show a pattern of transverse and longitudinal carinae and grooves, the latter are wide and filled with short hairs. A median carina is present on somites 2 to 6. A broad ridge, which is interrupted posteriorly separates the tergites from the pleura. The sixth abdominal somite carries no spines on the dorsal surface, neither are any spines placed on the posterior margin.

The telson is trapezoidal, narrowing slightly posteriorly. The posterior margin is somewhat convex and the posterolateral spines are well developed. At the base of the two diverging carina of the upper surface, the telson shows a short and sharp transverse median carina but no spines.

The eyes are well developed. The cornea is large, well pigmented and kidney-shaped.

The first antennular segment is very long. The stylocerite is visible as a large lobe over the base of the segment, with an anterolateral spine, which reaches beyond the lateral margin of the segment.

The antennal peduncle bears a large and wide spine near the base of the scaphocerite. The latter is well developed, but rather narrow, with the inner margin convex, and regularly tapering towards the sharp apex. A spine is present above the opening of the antennal gland.

The epistome has a distinct ridge along the anterior margin, which bears a median forward directed tooth. The clavicular ridge is distinct. The posterior margin of the epistome is a high rounded ridge, which bears no spines or teeth. Just before the ridge there is a transverse groove and a deep median pit.

The second maxilliped has the exopod with a distinct multi-articulated flagellum. The third maxilliped has no spines or teeth on the carpus and merus.

The chelipeds of the first pair are dissimilar. One, the crushing claw, has the fingers provided with a few large, rounded teeth in addition to a few denticles, it is gaping in large specimens. The other, the cutting claw, has the cutting edges closing and provided with numerous small denticles of about the same size; a large tooth is usually present medial of the cutting edge of the fixed finger. The segments of these legs bear distinct longitudinal carinae with strong spines.

The chelae of both the second and third pereiopods have a fringe of hair along the upper and lower margin; in the third leg this fringe is inconspicuous proximally. The fingers of the second and third legs bear no row of closely placed hairs on the outer surface.

The thoracic sternites are very narrow and visible only as single or double carinae between the bases of the legs which are placed close together. In the sternite of the fourth leg of the male the ridges diverge posteriorly slightly more than in the previous somites and in the female there is a thelycum there.

The first pleopods of the male are formed into rigid copulatory stylets and consist of two immovably fused segments. There are no spines at the base of these gnathopods. The appendix masculina of the second male pleopod is elongate, more than half as long as the endopods, truncated at the apex and with short bristles there.

The sternites of the second to fifth abdominal somites of the males and young females each bear a median spine. This spine is absent or strongly reduced in adult females.

The uropods have one median ridge on the endopod and two on the exopod. The diaeresis of the exopod is distinct and bears numerous denticles on the anterior margin. The protopodite has either dorsal lobe ending in a spine.

REFERENCES

Ahyong, S.T., 2006. Phylogeny of the clawed lobsters (Crustacea: Decapoda: Homarida). Zootaxa 1109, 1–14.

Ahyong, S.T., O'Meally, D., 2004. Phylogeny of the Decapoda Reptantia: resolution using three molecular loci and morphology. Raff. Bull. Zool. 52 (2), 673–693.

Albrecht, H., 1983. Die Protastacidae n. fam., fossile Vorfahren der Flußkrebse? Neues. Jahrb. Geol. P. M. 1, 5–15.

Aldrovandus, U., 1606. "De Reliquis Animalibus exanguibus libri quatuor, editi: Nempe de Mollibus, Crustaceis, Testaceis".

Balss, H., 1914. Ostasiatiche Decapoden ii, Die Natantia und Reptantia. Abhandlungen der Bayerischen Akademie der Wissenschaften, Mathematisch-physikalische Klasse 10 (Suppl. 2), 1–101.

Bate, C.S., 1888. Report on the Crustacea Macrura collected by H.M.S. Challenger during the Years 1873–76. In: Murray, J. (Ed.), *Zoology*. Wyville Thomson, C. and Murray, J., Report on the Scientific Results of the Voyage of H.M.S. Challenger During the Years 1873–76 Under the Command of Captain George, S. Nares, R.N., F.R.S. and the Late Captain Frank Tourle Thomson, R.N., vol. 24. Neill and Company, Edinburgh.

Bell, T., 1844–1853. A History of the British Stalk-eyed Crustacea. John Van Voorst, London.

Bell, M.C., Redant, F., Tuck, I., 2006. *Nephrops* species (chapter 13). In: Phillips, B. (Ed.), Lobsters: Biology, Management, Aquaculture and Fisheries. Blackwell Publishing, Oxford, pp. 412–461.

Beurlen, K., 1928. Die Decapoden des schwäbischen Jura mit Ausnahme der aus den oberjurassischen Plattenkalken stammenden. Palaeontogr 70, 115–278.

Boone, L., 1927. Crustacea from tropical east American seas. Scientific results of the first oceanographic expedition of the "Pawnee" Bull. Bingham Ocean. Collect. 1 (2), 147.

Bruce, A.J., 1965. On a new species of *Nephrops* (Decapoda, Reptantia) from the South China Sea. Crustaceana 9 (3), 274–284.

Bruce, A.J., 1966a. *Nephrops sinensis* sp. nov., a new species of lobster from the South China Sea. Crustaceana 10, 155–166.

Bruce, A.J., 1966b. *Nephrops australiensis* sp. nov., a new species of lobster from northern Australia (Decapoda Reptantia). Crustaceana 10, 245–258.

Brünnich, M.T., 1772. Zoologiæ Fundamenta Prælectionibus Academicis Accomodata. Grunde i Dyrelaeren. Fridericus Christianus Pelt, Copenhagen.

Burkenroad, R.N., 1963. The evolution of the Eucarida, (Crustacea, Eumalacostraca), in relation to the fossil record. Tulane Stud. Geol. 2 (1), 3–17.

Carter, J., 1898. A contribution to the paleontology of the decapod Crustacea of England. Q. J. Geol. Soc. Lond. 54, 15–44.

Castro, M., Gancho, P., Henriques, P., 1998. Comparison of several populations of Norway lobster, *Nephrops norvegicus* (L.), from the Mediterranean and adjacent Atlantic—a bio-metric study. Sci. Mar. 62 (Suppl. 1), 71–79.

Chu, K.H., Li, C.P., Qi, J., 2006. Ribosomal RNA as molecular barcodes: a simple corre-lation analysis without sequence alignment. Bioinformatics 22, 1690–1701.

Chu, K. H., Tsang, L. M., Ma, K. Y., Chan, T.-Y., Ng, P. K. L., 2009. Decapod phylogeny: what can protein-coding genes tell us? In: Martin, J.W., Crandall, K.A., Felder, D.L. (Eds.), *Decapod Crustacean Phylogenetics*. Crustacean Issues (S. Koenemann, Series Ed.), vol. 18. CRC Press, Boca Raton, pp. 89–99.

Crnkovic, D., 1969. Il problema della specie *Nephrops norvegicus* (L.) in rapporto con la varietá Meridionale Zariquiey. Thalassia Jugoslavica 5, 67–68.

Dana, J.D., 1852. Conspectus Crustaceorum, Conspectus of the Crustacea of the Exploring Expedition under Capt. Wilkes, U.S.N. Macroura. Proc. Acad. Nat. Sci. Philadelphia 6, 10–28 (Preprint of 1854 publication).

De Grave, S., Pentcheff, N.D., Ahyong, S.T., Chan, T.-Y., Crandall, K.A., Dworschak, P.C., Felder, D.L., Feldmann, R.M., Fransen, C.H.J.M., Goulding, L.Y.D., Lemaitre, R., Low, M.E.Y., Martin, J.W., Ng, P.K.L., Schweitzer, C.E., Tan, S.H., Tshudy, D., Wetzer, R., 2009. A classification of living and fossil genera of decapod crustaceans. Raff. Bull. Zool. (Suppl. 21), 1–109.

De Man, J.G., 1905. Tijdschrift d. Nederlands Dierk. Vereen. (2) Dl. IX, Afl. 3.

De Man, J.G. (1916). "Families Eryonidae, Palinuridae, Scyllaridae, and Nephropsidae. The Expedition, Part III, Sibogae-Expeditie, 39".

de Saint Laurent, M., 1988. Enoplometopoidea, nouvelle superfamille de Crustacés Décapodes Astacidea. Cr. Hebd. Acad. Sci. 307, 59–62 (Série III).

Feldmann, R.M., 1981. Paleobiogeography of North American lobsters and shrimps (Crus-tacea, Decapoda). Geobios 14, 449–468.

Feldmann, R.M., 1989. *Metanephrops jenkinsi* n. sp. (Decapoda: Nephropidae) from the Cre-taceous and Paleocene of Seymour Island, Antarctica. J. Paleontol. 63 (1), 64–69.

Feldmann, R.M., Holland, F.D., 1971. A new species of lobster from the Cannonball For-mation (Paleocene) of North Dakota. J. Paleontol. 45 (5), 838–843.

Feldmann, R.M., Bishop, G.A., Kammer, T.E., 1977. Macrurous decapods from the Bearpaw Shale (Cretaceous: Campanian) of Northeastern Montana. J. Paleontol. 51 (6), 1161–1180.

Feldmann, R.M., Tshudy, D.M., Thomson, M.R.A., 1993. Late Cretaceous and Paleocene decapod crustaceans from James Ross Basin, Antarctic Peninsula. J. Paleontol. 67. Supplement to No. 1. Part II of II. The Paleontological Society Memoir 28, 1–41.

Fraaije, R., Hansen, J., Hansen, T., 2005. Late Miocene decapod faunas from Gram, Denmark. Paleontos 7, 51–61.

Fritsch, A., Kafka, J., 1887. Die Crustaceen der böhmischen Kreideformation. Selbstverlag in Commission von F. Rivnác, Prague.

Glaessner, M.F., 1929. Pars. 41. Crustacea decapoda. In: Pompeckj, J.F. (Ed.), Fossilium Catalogus, I., Animalia. Junk, Berlin, p. 464.

Glaessner, M.F., 1960. The fossil decapod Crustacea of New Zealand and the evolution of the Order Decapoda. NZ Geol. Survey Paleontol. Bull. 31, 3–63.

Glaessner, M.F., 1969. Decapoda. In: Moore, R.C. (Ed.), Part R, Arthropoda 4(2). Treatise on Invertebrate Paleontology. The University of Kansas Press, Kansas, pp. R399–R533 R626–R628.

Holthuis, L.B., 1945. Remarks on Nephrops norvegicus (L.) and its variety meridionalis Zariquiey. Zoölogische Mededelingen Leiden 25, 317–320.

Holthuis, L.B., 1964. On some species of Nephrops (Crustacea Decapoda). Zoologische Mededelingen 39, 71–78.

Holthuis, L.B., 1974. The lobsters of the Superfamily Nephropidea of the Atlantic Ocean (Crustacea: Decapoda). Bull. Mar. Sci. 24 (4), 723–884.

Holthuis, L.B., 1991. FAO Species Catalogue. Vol. 13. Marine lobsters of the world. An annotated and illustrated catalogue of species of interest to fisheries known to date. FAO Fisheries Synopsis. Vol. 125. Food and Agriculture Organization of the United Nations, Rome.

Hu, C.-H., 1983. Discovery fossil lobster from the Kuechulin Formation (Miocene), Southern Taiwan. Ann. Taiwan Mus. 26, 129–136.

Hu, C.-H., Tao, H.-J., 2000. Crustacean fossils from southern Taiwan. Petrol. Geol. Taiwan 34, 105–195.

Jenkins, R.J.F., 1972. Metanephrops, a new genus of Late Pliocene to Recent lobsters (Decapoda, Nephropidae). Crustaceana 22 (2), 161–177.

Karasawa, H., Schweitzer, C.E., Feldmann, R.M., 2012. Phylogeny and systematics of extant and extinct lobsters. J. Crust. Biol. 33 (1), 3–63.

Lagerberg, T., 1908. Sveriges Decapoder. Göteborg Vetensk. Samh. Handl., ser. 4, vol. 11 pt. 2.

Latreille, P.A., 1802. Histoire naturelle, générale et particulière des Crustacés et des Insectes. Ouvrage faisant suite à l'histoire naturelle générale et particulière, composée par Leclerc de Buffon, et rédigée par C.S. Sonnini, membre de plusieurs sociétiés savantes. Familles naturelles des genres. Vol. 3. F. DuFart, Paris.

Latreille, P.A., 1829. Crustacés, arachnids et partie des insects. In: Cuvier, G. (Ed.), Le règne animal distribué d'après son organization, poyr server de base à l'histoire naturelle des animaux et d'introduction à l'anatomie comparée. Nouvelle édition, revue et augmentée, vol. 4. Chez Déterville, Libraire and Chez Crochard, Libraire, Paris, pp. 1–584.

Leach, W.E., 1814. Crustaceology. In: Brewster, D. (Ed.), The Edinburgh Encyclopaedia, vol. 7. pp. 383–437.

Linnaeus, C., 1751. Skånska resa, på Hoga Ofwerhetens Befallning forråttad År 1749 Med Ron och Anmårkningar uti Oeconomien, Naturalier, Naturalier, Antiquitter, Seder, Lefnads-sått.

Linnaeus, C., 1758. Systema Naturae per Regna Tria Naturae, Secundum Classes, Ordines, Genera, Species, cum Characteribus, Differentiis, Synonymis, Locis. Vol. 1. Laurentii Salvii, Holmiae.

Maltagliati, F., Camilli, L., Biagi, F., Abbiati, M., 1998. Genetic structure of Norway lobster, Nephrops norvegicus (L.) (Crustacea: Nephropidae), from the Mediterranean Sea. Sci. Mar. 62 (Suppl. 1), 91–99.

McCobb, L.M.E., Hairapetian, V., 2009. A new lobster *Paraclytia valashtensis* (Crustacea, Decapoda, Nephropidae) from the late Cretaceous of the central Alborz Range, Iran. Paläontol. Zeit. 83, 419–430.

Mertin, H., 1941. Decapode Krebse aus dem Subhercynen und Braunschweiger Emscher und Untersenon sowie Bemerkungen über einige verwandte Formen in der Oberkreide. Nova Acta Leopoldina 10, 1–264.

Moreira, C., 1903. Crustaceos. Campanhas de pesca do "Annie" Archivos do Museu Nacional do Rio de Janeiro 13, 121–145.

Parisi, B., 1917. I Decapodi giapponesi del Museo di Milano. V. Galatheida e Reptantia. Atti della Società Italiana di Scienze Naturali, Pavia 56, 1–24.

Passamonti, M., Mantovani, B., Scali, V., Froglia, C., 1997. Allozyme characterization of Scottish and Aegean populations of *Nephrops norvegicus*. J. Mar. Biol. Assoc. UK 77, 727–735.

Pesta, O., 1918. "Die Decapodenfauna der Adria. Versuch einer Monographie".

Rathbun, M.J., 1918. Decapod Crustaceans from the Panama region. Bull. U.S. Nat. Mus. 103, 123–184.

Rathbun, M.J., 1920. Additions to West Indian Tertiary decapod crustaceans. Proc. U. S. Nat. Mus. 58, 381–384.

Rathbun, M.J., 1929. New species of fossil decapod crustaceans from California. J. Washington Acad. Sci. 19 (21), 469–472.

Rathbun, M.J., 1935. Fossil Crustacea of the Atlantic and Gulf Coastal Plain. Geological Society of America Special Papers 2.

Robineau-Desvoidy, M.J.B., 1849. Mémoire sur les Crustacés du terrain neocomien de Saint Sauveur-En-Puise (Yonne). Ann. Soc. Entomol. France 2 (7), 95–141.

Rode, A.L., Babcock, L.E., 2003. Phylogeny of fossil and extant freshwater crayfish and some closely related nephropid lobsters. J. Crust. Biol. 23 (2), 418–435.

Schellenberg, A., 1928. Krebstiere oder Crustacea II: Decapoda, Zehnfüsser (14. Ordnung). In: Dahl, F. (Ed.), Die Tierwelt Deutschlands und der angrenzenden Meeresteile nach ihren Merkmalen und nach ihrer Lebenswiese, vol. 10. pt. 2.

Schlüter, C., 1879. Neue und weniger gekannte Kreide- and Tertiärkrebses des nordlichen Deutschlands. Zeitschrift der deutschen Geologischen Gesellschaft 31, 586–615.

Scholtz, G., Richter, S., 1995. Phylogenetic systematics of the reptantian Decapoda (Crustacea, Malacostraca). Zool. J. Linn. Soc. 113, 289–328.

Stamatis, C., Triantafyllidis, A., Moutou, K.A., Mamuris, Z., 2004. Mitochondrial DNA variation in Northeast Atlantic and Mediterranean populations of Norway lobster, *Nephrops norvegicus*. Mol. Ecol. 13, 1377–1390.

Stenzel, H.B., 1945. Decapod crustaceans from the Cretaceous of Texas. In: Mathews, E.J. et al., (Ed.), The University of Texas Contributions to Geology 4401, pp. 400–477.

Tam, Y.K., Kornfield, I., 1998. Phylogenetic relationships of clawed lobster genera (Decapoda: Nephropidae) based on mitochondrial 16s rRNA gene sequences. J. Crust. Biol. 18 (1), 138–146.

Tapparone-Canefri, C., 1873. Intorno ad una nuova specie di *Nephrops*, genere di Crostacei Decapodi Macruri, Nota. Memorie della Reale Accademia delle Scienze di Torino 2 (18), 325–329.

Tsang, L.M., Ma, K.Y., Ahyong, S.T., Chan, T.-Y., Chu, K.H., 2008. Phylogeny of Decapoda using two nuclear protein-coding genes: origin and evolution of the Reptantia. Mol. Phylogenet. Evol. 48, 359–368.

Tshudy, D., 1993. Taxonomy and evolution of the clawed lobster families Chilenophoberidae and Nephropidae. Unpublished Ph.D. dissertation. Kent State University, 472 p.

Tshudy, D.M., Babcock, L.E., 1997. Morphology-based phylogenetic analysis of the clawed lobsters (family Nephropidae and the new family Chilenophoberidae). J. Crust. Biol. 17, 253–263.

Tshudy, D., Sorhannus, U., 2000a. *Jagtia kunradensis*, a new genus and species of clawed lobster (Decapoda: Nephropidae) from the Upper Cretaceous (Upper Maastrichtian) Maastricht Formation, The Netherlands. J. Paleontol. 74 (2), 224–229.

Tshudy, D., Sorhannus, U., 2000b. Pectinate claws in decapod crustaceans: convergence in four lineages. J. Paleontol. 74 (3), 474–486.

Tshudy, D., Sorhannus, U., 2011. A cladistic analysis of the clawed lobsters based on an expanded morphological data set. Northeastern (46th Annual) and North-Central (45th Annual) Joint Meeting (20–22 March 2011). Geol. Soc. Amer. Abst. Prog. 43 (1), 53.

Tshudy, D., Chu, K.H., Robles, R., Ho, K.C., Chan, T.-Y., Ahyong, S.T., Felder, D., 2005. Phylogeny of marine clawed lobsters based on mitochondrial RDNA. In: Sixth International Crustacean Congress (Glasgow, Scotland), Abstracts O6.

Tshudy, D., Robles, R., Chan, T.-Y., Ho, K. C., Chu, K. H., Ahyong, S. T., Felder, D., 2009. Phylogeny of marine clawed lobster families Nephropidae Dana 1852 and Thaumastochelidae Bate 1888 based on mitochondrial genes. In: Martin, J.W., Crandall, K.A., Felder, D.L. (Eds.). *Decapod Crustacean Phylogenetics*. Crustacean Issues (S. Koenemann, Series Ed.), vol. 18. CRC Press, Taylor and Francis Group, Boca Raton, pp. 357–368

Wahle, R.A., Tshudy, D., Cobb, J.S., Factor, J., Jaini, M., 2012. Eucarida: decapoda: Astacidea P.P. (Enoplometopoidea, Nephropoidea), Glypheidea, Axiidea, Gebiideam, and Anomura. In: Schram, F.R., von Vaupel Klein, J.C. (Eds.), Treatise on Zoology—Anatomy, Taxonomy, Biology. The Crustacea Complementary to Volumes of the Traite de Zoologie, vol. 9. Brill, Leiden (part B).

Whitfield, R.P., 1907. Notice of an American species of the genus *Hoploparia* McCoy, from the Cretaceous of Montana. Bull. Amer. Mus. Nat. Hist. 23, 59–461.

Winckler, T.C., 1882. Carcinological investigation on the genera *Pemphix*, *Glyphea* and *Araeosternus*. Ann. Mag. Nat. Hist. 10, 133–149. Series 5, 306–317.

Wood-Mason, J., 1892. "Crustacea, Pt. 1. Illustrations of the Zoology of the Royal Indian Marine Survey Steamer 'Investigator'".

Woods, H., 1925–1931. A Monograph of the Fossil Macrurous Crustacea of England. Paleontographical Society, London, 122 p.

Woodward, H., 1900. Further notes on podophthalmous Crustacea from the Upper Cretaceous formation of British Columbia. Geol. Mag. (Decade IV) 7 (9, 10), 392–401.

Yaldwyn, J.C., 1954. *Nephrops challenger* Balss, 1914, (Crustacea, Decapoda, Reptantia) from New Zealand and Chatham Island Waters. Trans. R. Soc. NZ 82 (3), 721–732.

Zariquey Cenarro, R., 1935. Crustáceos del Mediterráneo. *Nephrops norvegicus*, Linné, var. *meridionalis*. Butlleti de la Institucio Catalana d'historia Natural 35, 26–32.

Zariquiey Alvarez, R., 1968. Crustáceos Decápodes Ibéricos. Investigaciones Pesqueras Bercelona 32, 1–164.

CHAPTER TWO

Habitat and Ecology of *Nephrops norvegicus*

Mark P. Johnson*,1, Colm Lordan†, Anne Marie Power*

*Ryan Institute for Environmental, Marine and Energy Research, School of Natural Sciences, National University of Ireland, Galway, Ireland
†Marine Institute, Rinville, Oranmore, Co. Galway, Ireland
1Corresponding author: e-mail address: Mark.johnson@nuigalway.ie

Contents

Abstract

This review summarizes the data on habitat, population ecology and ecosystem roles of *Nephrops norvegicus*. The species has a broad range in the northeast Atlantic and Mediterranean, although it is possible that small or isolated patches of suitable habitat may not be occupied due to restrictions on larval supply. *Nephrops* densities are related to the silt–clay content of sediments, with interactions between habitat quality and density indicating competition for resources. An analysis of density–size interactions across fishery functional management units (FUs) suggests that growth is suppressed at high densities due to competition (e.g. in the western Irish Sea), although recruitment dynamics or size-selective mortality may also shape the size structure of populations. *Nephrops* biomass available across FUs may be similar, reflecting a constant yield due to the inverse relationship between individual size and population density. Gaps in the understanding of *Nephrops'* ecology reflect uncertain ageing criteria, reliance on

© 2013 Elsevier Ltd.
All rights reserved.

fisheries-dependent data and few if any undisturbed habitats in which to examine fisheries-independent interactions.

Keywords: Distribution, Ecosystem functions, Fishery interactions, Sediment, Density, Growth, Catch, Size

1. INTRODUCTION

To a great extent, the burrowing habit of *Nephrops norvegicus* (L.) defines the ecology of the species (Chapman, 1980). The existing gaps in understanding the ecology of *N. norvegicus* (referred to as *Nephrops* from this point) partially reflect the difficulties in gaining unbiased population samples with methods that are selective (trawls and pots). Furthermore, what we know of *Nephrops* has been largely driven by fisheries-related research. Ecosystem roles and interactions with species that are not fisheries targets are not particularly well characterized.

In this review, we start by defining the broad environmental conditions that characterize the geographic range of *Nephrops*. Within these environmental limits, *Nephrops* depends on reaching suitable mud habitat for building burrows. With a planktonic larval stage, there is a reliance on advection in reaching habitats and potentially advective controls on recruitment (Hill and White, 1990). The more suitable muds are fine cohesive sediments, suitable for burrow excavation. The factors that may structure the sizes of individuals and density within a population are discussed. These include predation, competition, fishing impacts, recruitment dynamics and density-dependent growth. The influence of some of these factors can be assessed, but in some cases, the role of a factor in structuring populations is not known or is difficult to separate from other processes acting simultaneously. A particular issue is related to the difficulties in estimating recruitment directly (Wahle, 2003), a problem compounded by a lack of reliable ageing methods (although the recent demonstration of consistent banding by Kilada et al. (2012) offers some hope in this respect).

The standing population of *Nephrops* and those removed from the seabed by fishing may represent a large proportion of the benthic biomass and production in the areas where the species is found. This suggests that *Nephrops* and the ubiquitously associated fishing effort may have important roles in the flows of carbon and nutrients through marine communities. In addition to trophic interactions, *Nephrops* may also interact with benthic species and sediment–water column fluxes through activities that shape the habitat: the construction of burrows and resuspension of sediment.

A number of knowledge gaps can be identified that would produce a fuller picture of the reasons for both the spatial variation in *Nephrops* population characteristics and the ecosystem roles played by the species. As well as informing the sustainability of *Nephrops* species, addressing gaps in knowledge is likely to increase the understanding of climate-related effects on populations and the effects of changes in fishing practice, such as gear changes to increase trawl selectivity.

2. DISTRIBUTION AND HABITAT ASSOCIATIONS

Nephrops are found on the continental shelf and slope throughout the northeast Atlantic (Figure 2.1). The range extends as far south as the Canary Islands (Barquín et al., 1998). The species is also found in the western Mediterranean, Adriatic and Aegean seas. In mapping the distribution of *Nephrops*, records from the Ocean Biogeographic Information System (OBIS) were augmented by information on a number of known fishery areas including the Iberian coast and Adriatic Sea. Environmental data associated with the records summarize the general environmental tolerances of *Nephrops*: the species is recorded between 4- and 754-m depth, at temperatures between 6.4 and 17.3 °C. *Nephrops* is generally restricted to marine waters (salinities for records are between 31.8 and 38.8) of relatively high oxygen concentration (between 5.9 and 9.4 mg O_2/dm^3). Not all records from OBIS have associated environmental data. *Nephrops* can certainly be found outside the variable ranges quoted here; for example, a maximum depth of 800 m is given in the FAO species catalogue (Holthuis, 1991), while there are records from areas of lower salinity (such as the entrance to the Baltic Sea, Figure 2.1). The values associated with OBIS records, however, approximate the broad environmental tolerances of the species.

The spatial extent of the environmental niche for *Nephrops* can be modelled using habitat suitability mapping (Figure 2.2). The suitability of different areas was derived using average annual sea floor temperature, salinity and oxygen, along with mean depth and mean satellite-derived chlorophyll (as a proxy for productivity). The suitability map emphasizes the continental shelf around the United Kingdom and Ireland as key habitat for *Nephrops*. The Azores and Rockall Bank are correctly identified as areas where *Nephrops* can occur; although there is no evidence yet that *Nephrops* are found on the eastern coast of Greenland. Small areas in the Black sea may be suitable, although distance from existing populations and more recent anoxia in the region (Gordina et al., 2001) makes these locations unlikely to host *Nephrops*.

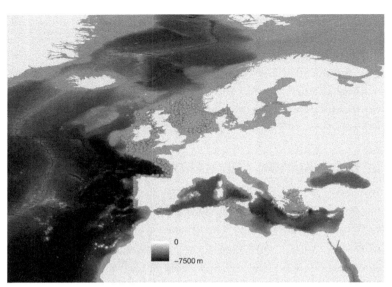

Figure 2.1 Species records for *Nephrops norvegicus* in the OBIS database. Records are drawn from the following datasets: Addinck and de Kluijver (2003), Rees (2005), BALAR (EurOBIS) (n.d.), ICES Fish predator/prey data (EurOBIS) (n.d), Picton et al. (1999), Mackie et al. (1991), MedObis (2004), UK National Biodiversity Network and Marine Biological Association—DASSH Data Archive Centre Academic Surveys (n.d..), ICES North Sea International Bottom Trawl Survey (2010), ICES Northern Ireland Survey (2010), ICES French Southern Atlantic Bottom Trawl Survey (2010), Rumohr (2006), Fockedey et al. (2004) HMAP (n.d.), ICES Baltic International Trawl Survey (2010), ICES Beam Trawl Survey (2010), ICES Contaminants and biological effects database (DOME—Biota, 2010), ICES historical plankton database (1901–1912), Olesen (2010a,b), ICES Irish ground fish survey (2010), ICES Scottish Western Coast Via Groundfish Survey (2010), Fisheries Research Service (2001), Wilkinson (2005), Olesen (2010a,b), Parr (2005a,b), Ostler (2005), Scottish Natural Heritage (2005), MEDITS (2004), Trawl Surveys (AfrOBIS) (n.d.), Whomersley (2003), Department of Invertebrate Zoology, Research and Collections Information System (n.d.), Craeymeersh et al. (1986), PANGAEA (n.d.), Rostron (2004), ICES Rockall Survey (2010), Marine Conservation Society (2005), Türkay (2006), Flanders Marine Institute (VLIZ) (2004). Additional records for known fishery areas were added using the locations of ICES functional units for the Iberian peninsula, Pampoulie et al. (2011), Sardà (1998) and Morello et al. (2009). (For colour version of this figure, the reader is referred to the online version of this chapter.)

Correlations among the predictors make it difficult to separate individual effects and hence it is not clear whether the absence of *Nephrops* from the Baltic and Black Seas is due to temperature, salinity, low oxygen or a combination of these (or other) variables.

Suitable sediment for burrow building (see below) is a pre-requisite for *Nephrops* to be found in areas that lie within the envelope of favourable

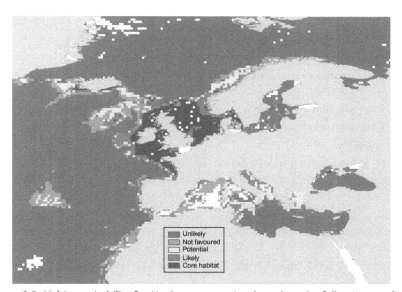

Unlikely
Not favoured
Potential
Likely
Core habitat

Figure 2.2 Habitat suitability for *Nephrops norvegicus* based on the following predictor variables: mean annual bottom temperature, salinity and oxygen, along with mean depth and mean annual surface chlorophyll. White areas had missing environmental data. The environmental data in 0.5° latitude × 0.5° longitude grid cells were downloaded from the Hexacoral database (Fautin and Buddemeier, 2008). Habitat suitability was modelled using the MAXENT programme (http://www.cs.princeton.edu/~schapire/maxent/; Phillips et al., 2004, 2006). The area under the curve (AUC) statistic for this model was 0.953, indicating a good classification of locations. Predicted suitability values were divided into five classes using Jenks natural breaks. The division of suitability values into categories is somewhat arbitrary, as Maxent works with presence-only data and therefore does not provide a robust estimate of the probability of occurrence (as would be possible if all locations had been surveyed with equal effort). (For colour version of this figure, the reader is referred to the online version of this chapter.)

environmental conditions. However, it is not clear whether all areas of suitable habitat contain *Nephrops*. If a patch of suitable sea floor is sufficiently isolated from source populations, it may not be reached by potential recruits. The pelagic phases of *Nephrops* larvae (stages I–III) can last up to approximately 50 days depending on temperature (de Figueiredo and Vilela, 1972; Dickey-Collas et al., 2000; Nichols et al., 1987). This relatively long larval duration leads to another possibility: small patches of mud may not sustain self-recruiting populations, as diffusion or advection removes larvae from the area (Hill and White, 1990; Marta-Almeida et al., 2008). For example, Tremadoc Bay is a relatively small mud patch (100 km²) off the northwest Welsh coast. Using reasonable estimates of turbulent diffusion, the larval density above this patch could be reduced by 90% in as little as two days (Hill and White, 1990). Isolated mud patches may, of course, contain viable populations if externally sourced recruits maintain the population density. This sort of metapopulation structure, where a number of patches are linked by larval dispersal, seems to be present in *Nephrops*. Modelling studies suggest that larvae can be transported between 100 and 300 km along the south and southwest coasts of the Iberian peninsula (Marta-Almeida et al., 2008).

The potential for links between adult *Nephrops* populations will clearly reflect regional variability in the size and configuration of suitable habitats and the interactions of these spatial patterns with current-associated larval transport. Direct evidence of larval transport among populations can be inferred from genetic studies showing low genetic differentiation between *Nephrops* populations (Maltagliati et al., 1998; Passamonti et al., 1997; Stamatis et al., 2004; Streiff et al., 2001). These low levels of inter-population differentiation are consistent with a metapopulation structure involving the exchange of larvae, although at the scale of an Atlantic–Mediterranean comparison, the low differentiation may also reflect a geologically recent expansion of *Nephrops* populations (Stamatis et al., 2004). Genetic evidence for the exchange of larvae between different habitat patches has also been demonstrated in the presence of differentiation in fishery variables (mean size, catch per unit effort) among isolated populations (Pampoulie et al., 2011). Unfortunately, relatively low levels of genetic differentiation, while indicating that some larval migration occurs, are insufficient to judge the demographic importance of migration between separate *Nephrops* populations.

Regardless of whether a patch of suitable habitat is large enough to support a self-sustaining population, other factors may influence whether *Nephrops* are found in a mud patch. In comparison to *Homarus* species, relatively little is known about *Nephrops* larval behaviour and appropriate cues at settlement (Cobb and Wahle, 1994). Importantly, juvenile *Nephrops* are rarely found outside the burrows of adults, so there may be a requirement for existing burrows for successful settlement (Tuck et al., 1994). Although juvenile *Nephrops* can burrow in aquaria, it may be that there is a selection to enter adult burrows before excavating a tunnel, as this avoids the risk of predation by fish that would occur while excavating at the mud surface (Chapman, 1980). The potential dependence of recruitment on existing burrows may limit the colonization of otherwise suitable habitat. Anecdotal reports from fishers have suggested that some inshore populations may have become established by the survival of discards from offshore trawling. This supports the idea

that the creation of burrows by adult *Nephrops* may make population establishment with recruitment from larvae more likely in suitable habitat.

2.1. Sediment-density response

'Mud' substrates are a known pre-requisite for the presence of *Nephrops*, but
the precise fraction of mud, how well sorted it is and a range of other factors
may help to better refine our understanding of population factors such as
density–size effects. Within a *Nephrops* inhabited area, it is clear that there
may be distinct preferences for particular grades of sediment (Tully and
Hillis, 1995). The most recent analyses by Afonso-Dias (1998) and
Campbell et al. (2009a) investigated a 'dome-shaped' relationship between
burrow density and sediment composition (specifically, they examined percentage silt plus clay which we will refer to hereafter as 'silt–clay'). The characteristic shape is explained by sandy sediments being too fluid for *Nephrops*
burrowing, so densities are low at high sand ratios. Densities are optimal at
moderately high silt–clay ratios (i.e. low sand); however, extremely high
silt–clay ratios become 'too much of a good thing' as they stimulate more
extensive burrowing, potentially leading to increased intra-specific interactions and therefore to lower densities in these sediments.

In a comparison of *Nephrops* fishing grounds, the North Minch con-
formed to the dome-shaped density-sediment response although the relationship at the Fladen grounds was more linear (Campbell et al., 2009a).
The sediments at the Fladen grounds were slightly coarser and less well
sorted than at North Minch; accordingly, neither extremely high silt–clay
sediments nor densities were observed there. The response shape could plau-
sibly be more linear in sediments where a high-density threshold is not
reached. Consulting the plot of the Aran grounds in the west of Ireland
(Figure 2.3), the *Nephrops* density slope may have an inflexion in sediments
with ~60% silt plus clay, which was a similar range as at North Minch and
Fladen grounds (Campbell et al., 2009a). But maximum density was higher
in the Aran grounds (~1.25 burrows m^{-2}) compared with ~0.75 burrows
m^{-2} at North Minch and only ~0.4 burrows m^{-2} in the Fladen grounds
(Campbell et al., 2009a). The shape of the response observed at the Aran
grounds was intermediate between the two Scottish grounds, even though
we might have expected a stronger (i.e. more dome-shaped) effect at Aran
due to the higher burrow density than the other grounds.

There are variables alongside silt–clay content that may contribute to the
relationship between sediment type and density; the silt–clay versus density

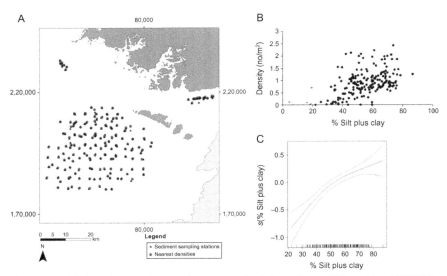

Figure 2.3 (A) Sampling stations at Aran grounds, Galway Bay and Slyne Head (FU17) in the west of Ireland. One hundred and ninety-three stations were sampled for sedimentology using a Day grab (2002–2006) and density (2002–2009) by UWTV. Burrow density estimates for each transect were centred on the point plotted in the map and burrow counts were carried out by two experienced counters. The particle size analysis of sediments was carried out with a Low Angle Laser Light Scattering method using a Malvern Instrument. (B) Sediment (% silt plus clay) versus density of *Nephrops* burrows (numbers m^{-2}). Outliers, that is, stations with >80% sand, are indicated by dashes; these were removed for analysis. Maximum densities occurred at % silt plus clay in the range of 60–70%. (C) A smoother fit for the response of *Nephrops* density to sediment % silt plus clay. The % silt plus clay and the year in which density sampling was carried out were examined by fitting a Generalized Additive Model (GAM) with gamma error distribution. Competing models were ranked using AICc weight and the model with the highest AICcw (lowest delta AICc) was selected as the most parsimonious model. This was Density $\sim s$(% silt plus clay) $+ s$(year), where 's' represents the smoother function, $k = 4$; $P_{sediment} < 0.001$; $P_{year\_density} < 0.001$, deviance $D = 43.7$ and coefficient of determination $R^2 = 0.42$. Model residuals were evaluated to check for non-normality, heteroscedasticity and autocorrelation of errors. Modelling was carried out using procedures within the mgcv library of R version 2.15 (Wood 2006). (For colour version of this figure, the reader is referred to the online version of this chapter.)

relationships also display quite a bit of scatter. Year-to-year variations in density may account for some of this, but when sampling year was taken into account, the percentage of silt–clay explained at best 42% of the variation in the data (see Figure 2.3), which was of a similar order to the model fit of Campbell et al. (2009a). In addition, deviations from a general dome-shaped pattern were seen in patches in the Scottish grounds. One of the additional variables may be the percentage (or quality) of organic matter (OM). This variable was discounted as being a major factor associated with *Nephrops* adult density in the Catalan Sea (Maynou and Sarda, 1997), nevertheless including OM could improve predictive power since density in coarser sediments may be enhanced where OM is present. The positive role of OM may reflect improvements in sediment cohesion that aid burrowing or may be associated with higher prey availability. Preliminary analyses have also shown a dome-shaped response with respect to sediment sorting variables such as skew and kurtosis at the Aran Grounds, suggesting some optimum for these in the mid range of the observations (Lordan et al., 2007).

The rationale for a dome-shaped response between density and percentage silt–clay, seen above, implies that we might further expect to see a reduction in *Nephrops* size on silt–clay sediments due to density-dependent effects on biomass. This appears to be the case. A negative relationship has been shown between the percentage of silt–clay and mean size, accounting for almost 40% of the variance in mean weight (e.g. Maynou and Sarda; 1997; Tully and Hillis, 1995). Such population level interactions are explored in the following section.

3. POPULATION LEVEL INTERACTIONS: DENSITY, GROWTH AND FISHERIES

As you would expect in a commercially important species, the best ecological information for *Nephrops* is available at the population level, particularly with regard to interactions involving size, density, biomass and exploitation.

3.1. Density–size relationships

There are extensive data relating to *Nephrops* density from trawling, almost all of it fishery dependent. But underwater video (UWTV, see Chapman, 1985) is gaining traction as a stock assessment method and it is now treated as an absolute measure of stock size by ICES (ICES, 2009). It may be used to estimate absolute (as opposed to relative) population density and biomass as long as the over-estimations inherent in the method can be corrected (e.g. Sarda and Aguzzi, 2012). Accordingly, protocols to reduce biases that arise during video analysis are being developed; these include biases in edge effect, detection of burrows, correct identification of *Nephrops* burrows and occupancy (Campbell et al., 2009b; Morello et al., 2007). Once these have been quantified, the density estimate is divided by a cumulative bias correction factor to generate more realistic estimations of stock abundance.

Figure 2.4 shows that bias-corrected burrow density from UWTV footage has an inverse relationship with *Nephrops* size (mean weight in the landings) at several functional management units (FUs). Generally speaking, the density data were more variable than the size data, although size data were also quite variable in individual FUs, for example, at the Porcupine grounds. The trend suggests that, where *Nephrops* are found at high densities, there is a trade-off in the mean size achieved. At one end of the spectrum are the low-density-large *Nephrops* (e.g. at the Porcupine Bank or Fladen grounds, FU16 and 7, respectively), and at the other extreme we see high-density-small *Nephrops*, for example, at Irish Sea West (FU15). The shift in mean weight between these extremes is large, up to 28.5 g, or 175% difference by average weight of *Nephrops* per FU. Because selectivity in the fisheries, both gear and onboard, may bias these data, efforts to examine other metrics, such as catch data, are also necessary to investigate this trend. Modal length in the *catches* (i.e. before any undersized *Nephrops* may have been discarded) is less sensitive to selectivity, particularly grading bias, than mean size data. Figure 2.5 demonstrates that the overall modal length (totalled across all years) in catch data for males was inversely related to density (data sources are listed in Table 2.1). This relationship was quite shallow and noisy compared with

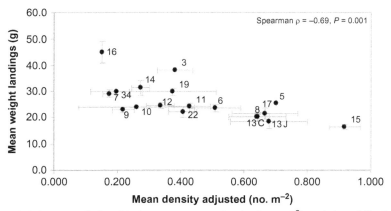

Figure 2.4 Inverse relationship between mean density (no. m^{-2}) and size of *Nephrops* (mean weight in grams) calculated from landings data ±2 standard errors of the mean. Burrow densities were calculated from UWTV and adjusted for bias (see Campbell et al., 2009b; ICES, 2009) in each functional unit (FU is indicated by data labels—see Table 2.2). For details about landings, see Table 2.1.

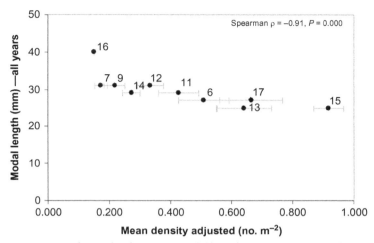

Figure 2.5 Inverse relationship between modal length (carapace, mm) in the catch versus mean density (no. m^{-2}) ±2 standard errors of the mean for *Nephrops* males. Burrow densities from UWTV have been adjusted for bias in each functional unit, as before (FU is indicated by data labels—see Table 2.2 for details about FUs). Overall modal length was extracted across all years for which catch data are available (catch details are given in Table 2.1).

Table 2.1 Functional units for which adjusted burrow densities (taken from UWTV) and length frequency distributions (LFDs) were available are listed below

Functional unit/location	Years when catch/landings data were available
6/Farn Deeps	1985–2007*
7/Fladen Grounds	1990–2008
8/Firth of Forth	1989–2008
9/Moray Firth	1989–2008
11/North Minch	1990–2007
12/South Minch	1990–2008
13/Clyde	1990–2008
14/Irish Sea East	1991–2008
15/Irish Sea West	1999–2008*
16/Porcupine Bank	1986–2007
17/Aran Grounds	1995–2008*

The sampling shown represents an unbroken annual run in most cases, except for occasional missing years where indicated *. LFDs were available for *Nephrops* both in catch and landings except at FUs 8 and 9 from 1981–1988, when landings data only were available.

the corresponding mean size, perhaps because the modal size is categorical and probably reflects several age classes.

3.2. Relationship between density and growth

Reference has frequently been made to density-dependent suppression of growth, for example, during analysis of 'stocklets' (Bailey, 1986; Bailey and Chapman, 1983). However, growth parameters such as von Bertalanffy's L_∞ (the mean size that an animal can achieve over infinite time) are rarely calculated in a density-specific manner. An exception was Tuck et al. (1997), who examined variability of L_∞/density at a small spatial scale (southern Clyde) and found these to be inversely related at this scale; that is, smaller L_∞ was observed at higher density. Figure 2.6 presents a broadscale analysis of L_∞ for males versus burrow density across all FUs for which these data were available—see Table 2.2. There was an inverse relationship between density and male L_∞ and therefore a suggestion of growth suppression at high densities over broadscales. This translates into a maximum carapace length which was 15 mm larger on the Porcupine Bank than it was in the Irish Sea West. If we could be confident that L_∞ values were representative across this scale and if we could rule out other factors, this would support the hypothesis that density-dependent suppression of growth partially explains size distributions of *Nephrops* at broad scales.

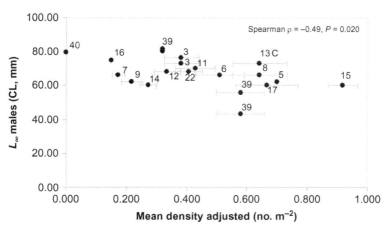

Figure 2.6 Inverse relationship between von Bertalanffy's growth parameter L_∞ (carapace length (CL), mm) for male *Nephrops* and mean density±2 standard errors of the mean. Burrow densities are adjusted for bias (see Campbell et al., 2009b; ICES, 2009). Functional units are indicated by data labels and data sources are indicated in Table 2.2.

Table 2.2 Burrow densities and von Bertalanffy's growth parameter L_∞ for functional units (FUs) in *Nephrops norvegicus* fisheries; values are based on carapace length

FU	Location	Density (no. m^{-2})	Density reference	L_∞ males (CL mm)	VBGF reference
40	Ionian Sea	0.001*	Smith and Papadopoulou (2003)	79.70	Maiorano et al. (2010)
16	Porcupine	0.151	This study	75.00	ICES (2000)
7	Fladen	0.174	This study	66.00	ICES (2000)
34	Devils Hole	0.198	This study	–	–
9	Moray Firth	0.218	This study	62.00	Bailey and Chapman (1983) and ICES (2000)
10	Noup	0.260	This study	–	–
14	Irish Sea East	0.273	This study	60.00	Hillis (1979) and ICES (2000)
39	Adriatic (off Ancona)	0.320	Morello et al. (2009) derived from Froglia, et al. (1997)	80.8 79.9 81.6 99.8[a] 81.5	Mytilineou et al. (1998) and Sarda et al. (1998)
12	South Minch	0.336	This study	68.00	ICES (2000)
19	SW and South of Ireland	0.375	This study	–	–
3	Skaggerak	0.383	This study	72.90 76.00	Ulmestrand and Eggert (2001)
22	Celtic Sea (Smalls)	0.407	This study	68.00	ICES (2000)
11	North Minch	0.428	This study	70.00	ICES (2000)
6	Farn Deeps	0.510	This study	66.00	Bailey and Chapman (1983) and ICES (2000)
39	Pomo Pit	0.580	Morello et al. (2009)	55.90 43.40	Vrgoc et al. (2004)

Continued

Table 2.2 Burrow densities and von Bertalanffy's growth parameter L_∞ for functional units (FUs) in *Nephrops norvegicus* fisheries; values are based on carapace length— cont'd

FU	Location	Density (no. m^{-2})	Density reference	L_∞ males (CL mm)	VBGF reference
8	Firth of Forth	0.642	This study	66.00	ICES (2000)
13	Clyde	0.642	This study	73.00[b]	Bailey and Chapman (1983) and ICES (2000)
17	Aran	0.665	This study	60.00	ICES (2000)
5	Botney Gut and Silver Pit	0.700	This study	62.00	ICES (2000)
15	Irish Sea West	0.918	This study	60.00	ICES (2000)
20/ 21	Labadie	–	–	–	–

[a]Omitted as outlying value.
[b]Tuck et al. (1997) suggested substantially lower L_∞ in Clyde 'stocklets'.
FUs are listed in the order of increasing density values. Density data are from UWTV which has been bias corrected, except where indicated *. Density data were collected during late spring or summer months and estimates are the most recent data available at WGCSE and WGNSSK at ICES (denoted as 'This study'), or these are taken from the literature where indicated.

The difficulties in estimating growth in *Nephrops*, for example, deriving von Bertalanffy's functions from length cohort analysis (LCA) have been discussed in detail by Mytilineou et al. (1998) and Ulmestrand and Eggert (2001). Calculation of function parameters fundamentally depends on an unambiguous designation of length/age cohorts, often from poorly structured samples. In addition, large variability in estimates may be generated due to such factors as the months included in the analysis, the precise stations in which sampling was conducted, or the calculation method. Indeed, due to the latter, several estimates of Mytilineou et al., 1998 for the Adriatic were included in our analysis. Confidence intervals are not routinely available for function parameters even though these may help judge the support or weighting for estimates. Growth in females is even more complex to model because they tend to be less well represented in samples and growth ceases during reproduction. The density–growth relationship was not significant in females in the present analysis (Spearman correlation -0.251, $P=0.259$,

$N=11$; data not shown). Von Bertalanffy's K (i.e. the rate at which it reaches L_∞) was less variable than L_∞ and did not vary with density for either males or females.

Since accurate calculation of growth depends on a myriad of factors, more context-specific estimates with confidence intervals are necessary. Calculating growth parameters based on tagging rather than LCA is an obvious route (e.g. Ulmestrand and Eggert, 2001), though this approach has been limited by low tag returns in the past. A possible area for wider exploration is the suggestion that mean carapace size and L_∞ are correlated (Mytilineou et al., 1998; Tuck et al., 1997), but only if this metric could be established as a reliable proxy for growth, and if its limitations were understood. For the purpose of the present discussion, however, we can see that the trend in L_∞ versus density is supported by empirical data because the overall modal length in the catch for males was also inversely related to density (Figure 2.5).

3.3. Factors that suppress size

The main factors that might suppress mean size are competition, which may lead to growth suppression at high densities (shown above) and recruitment: if recruitment were relatively higher inside FUs such as Irish Sea West, this could reduce mean size in the landings. This is a strong possibility because it is known that there is high larval production in this area (Briggs et al., 2002; Dickey-Collas et al., 2000). Further, the 'average' hydrological conditions at that time of year may act to retain the peak larval production and bring about high levels of recruitment in the Irish Sea (e.g. Hill, 1990, Hill et al., 1997). By contrast, larval retention (and hence recruitment) may be poor, for example, in southern Portugal (Marta-Almeida et al., 2008), or recruitment may cause periodic fluctuations in mean size, as has been suggested at the Porcupine grounds (ICES, 2011).

Size may also be restricted because of competition for food or space, or both. Evidence of nutritional limitation has been shown in high-density Clyde stocklets (Bailey, 1986; Parslow-Williams et al., 2001). However, this evidence must be tempered by the fact that energy budgets and scope for growth did not differ statistically between high- and low-density stocklets in the same area (Parslow-Williams et al., 2002). Measuring scope for growth suffers from methodological constraints including accurately calculating metabolic rates and dealing with high individual variability. Other difficulties in this area include the likelihood that *Nephrops* nutrition varies between seasons (Rosa and Nunes, 2003), as well as the influence habitats have on foraging behaviour (e.g. Aguzzi et al., 2003), even if habitat-specific emergence patterns do not necessarily translate into differences in stomach fullness (Aguzzi et al., 2004). Again, in this case, more research is required to explore whether different rhythms in feeding and digestive activity perhaps cancel out habitat-specific foraging patterns (Cristo and Castro, 2005).

Competition is evident in bouts of fighting of up to 20 min in length, which sometimes led to burrow displacements (Chapman and Rice, 1971). Burrow fidelity may be a loose arrangement, and therefore, territoriality is proposed to increase in high-density situations because of fewer unoccupied burrows (Rice and Chapman, 1971). This may lead to increased activity levels in addition to a reduced opportunity to feed as more time is spent on defending burrows and foraging expeditions are longer. Apart from easy access to prey, availability of sufficient mud with the correct silt–clay ratio to support burrow complexes is another expected limited resource. It may be possible to infer from spatial dispersion patterns whether food and/or space are limiting, but no evidence to date has been presented for progressions to over-dispersed (uniform) spatial patterns in *Nephrops* at higher densities. Instead of this, random or aggregated spacings have been seen in Biscay (Trenkel et al., 2007) and Loch Sween, Scotland (Tuck et al., 1994) even though high densities are possible in the latter site. Regular spaced burrow complexes at high-density grounds would strongly imply negative interactions from crowding, as have been seen in intertidal systems where space is severely limited. Density-dependent suppression of mean size has been seen shown to be remarkably constant with an exponent of $-3/2$ relating mass and density for intertidal barnacles and mussels (Hughes and Griffiths, 1988). For mobile animals (limpets), the distance between nearest neighbours increases as animals increase in size, leading to lower overall densities (Black, 1979). A comprehensive analysis of spatial dispersion in *Nephrops* would be very informative, especially if it were carried out along a density gradient. This should be achievable using UWTV footage. In addition, interspecific spacing and dispersion comparisons where *Nephrops* occurs with taxa such as *Munida* and *Goneplax* would greatly add to our understanding of competitive effects (e.g. Trenkel et al., 2007).

Unfortunately, density-dependent suppression of mean size due to competition and recruitment is difficult to separate and distinguishing between them is confounded because recruitment both adds numbers (strengthening competition) and exerts a downward influence on mean size. This conundrum applies to most stocks but seems particularly difficult in *Nephrops*, which lacks effective ageing criteria (but see Kilada et al., 2012) and has problematic growth parameterization (ICES, 2000). Examining pelagic larvae of *Nephrops* would require a frequency of temporal and spatial sampling which is almost certainly prohibitive. But possibilities may exist to track older recruits via analysis of UWTV burrow characteristics.

3.4. Factors that bring about losses to the population

Size-selective mortality by predation or by fisheries could reduce mean animal size if the rate involved was selective for larger *Nephrops*. These would tend to act in a density-dependent fashion if predation acted more strongly at high prey densities, for example, via a functional response (density-dependence may also apply to losses due to fisheries –see below). As discussed in the next section, *Nephrops* apparently has relatively few fish predators (Serrano et al., 2003). Pinnegar and Platts (2011) examined $n = 16,903$ predator stomachs in the Irish Sea and showed that cod predation was by far the most significant predator of *Nephrops* in this area. Nevertheless, cod predation does not explain the size distribution of *Nephrops*, since the quantity of *Nephrops* consumed by cod has declined steadily since the 1980s (Pinnegar and Platts, 2011; cf. Armstrong et al., 1991). Cod consumption represented a relatively low mortality rate in *Nephrops*: predation and fishing removed 0.61 and 8.4 thousand tonnes, respectively, from the Irish Sea in 2007 (Pinnegar and Platts, 2011). Predator influence is probably declining relative to fishing in many functional units; reductions of large predators have also taken place in recent decades in the Clyde, for example, Heath and Speirs (2011).

We can next consider the effect of fishing on size distribution and on size–density relationships. *Nephrops* are a sedentary species which means that high-density areas may be targeted very effectively by fishers. A simulation showed that areas of highest densities are targeted in the Farn Deeps, for example, Bell et al. (2005). Our analysis also suggests that, with possible exceptions at Aran (FU17) and Irish Sea West (FU15), cumulative effort is concentrated on high-density management units (Figure 2.7). But, while exploitation of Irish Sea West (FU15) does not appear particularly high from these data (Figure 2.7), the problems of measuring effort are acknowledged, including variable spans of time series providing this information, illegal and unreported effort, etc. In general, fisheries are well known to modify the size distribution in the resource, and this includes examples for *Nephrops*. Abello et al. (2002) suggested that fishery pressure controls *Nephrops* size in the Mediterranean. Dimech et al. (2012) showed a non-significant reduction in mean size of *Nephrops* between trawled and non-trawled areas in the Strait of Sicily; this equated to a difference of 1.96 mm mean carapace length between areas.

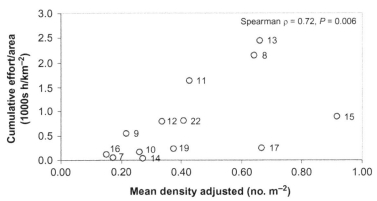

Figure 2.7 Mean density versus cumulative fishing effort/area of grounds. Fishing effort was cumulated across all years and countries in each functional unit (FU). Only years with comparable units of effort (1000s h) were included; the span of these data varied across FUs, for example, 1981–2008 at FU7, 8, 9, 10; 1965–2008 at FU11, 12, 13; 1995–2008 at FU14, 15, 16, 17, 19, 22 for Republic of Ireland's effort and in addition 1984–2008 for Northern Ireland's effort at FU15 and 1983–2008 for France's effort in FU16, 22. Burrow densities are adjusted for bias (see Campbell et al., 2009b; ICES, 2009). FUs are indicated by data labels (see Table 2.2).

To test whether a historical increase fishing pressure could explain the small size distribution in Irish Sea West (FU15), a preliminary analysis was carried out using data collected in the early 1960s. This dates back to a time before the *Nephrops* fishery developed to its current level (∼10,000 ton). The ICES Bulletin Statistique shows that catch data for the Irish and international fleet were <1500 ton/per annum until after 1960 (http://www.ices.dk/fish/CAT-CHSTATISTICS.asp). Length frequency data from catches at Irish Sea West during recent years (2002–2008) have not dramatically changed from historical levels in 1960–1962 (Cole, 1965); the only point to note is a slight decrease in the largest size categories; that is, greater than ∼37-mm carapace length (Figure 2.8). The historical data are not as spatially extensive as those of recent catch data, and so could represent 'pockets' of unusual length distributions, for example, Tully and Hillis (1995). In addition, analyses based on size distributions from fishery-dependent sampling are always prone to bias from fishery selection for large animals. Overall, we can tentatively suggest that there has not been a dramatic reduction over time in mean size caught at the Irish Sea West. We might further suggest that the differences in effort between grounds such as Irish Sea West (FU15) and Porcupine Bank (FU16—Figure 2.7) are probably not enough to explain the large difference between these grounds in size/density (Figures 2.4 and 2.5).

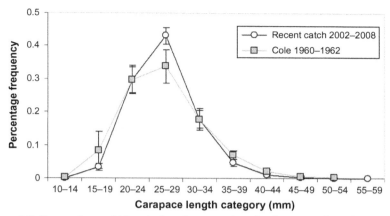

Figure 2.8 Comparison of historical and recent size distributions of *Nephrops* in the catch at Irish Sea West (FU15). 1960s data: *Nephrops* carapace length measurements were broken into 4-mm size categories and the percentage frequency recorded in each size category based on catches that took place in 1960–1962 published by Cole (1965). Recent data (2002–2008) for Ireland's catch were collapsed into comparable length categories and the mean frequency was plotted ±2 standard errors of the mean. Spatial distribution of sampling in 1960–1962 was at six locations west of Isle of Man (IoM), southwest IoM, Irish Coast and Dundalk Bay.

(Fisheries are dealt with in more detail in *Nephrops Fisheries in European Waters* by Ungfors et al., 2013).

3.5. Conclusions on density, growth and fisheries

It appears that size–density effects do occur at the scale of FUs in *Nephrops* stocks and that broad scale patterns in size distributions are at least partially explained by growth suppression due to competition and/or recruitment effects. These factors explain why *Nephrops* are smaller at high densities in areas such as Irish Sea West compared with the larger *Nephrops* found at the Fladen grounds or Porcupine Bank. As a result, when exploitation was examined, there was no relationship between mean landings per unit effort (kg/h) and mean burrow density (no. m^{-2}) (Figure 2.9). The productivity measured in landings at most of the FUs was similar. This conforms to the law of constant yield, that is, a constant biomass is available per functional unit, with density and size being regulated to achieve this limited biomass. Since other constraints including management measures will also maintain productivity below a certain level, ~40 kg/h[1] effort (see Figure 2.9) is not an absolute threshold of *Nephrops* productivity. But management constraints will tend to apply in a standard way, so common limits on

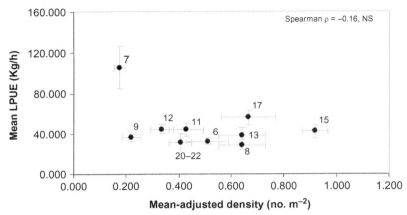

Figure 2.9 No relationship between mean landings per unit effort (kg/h) and mean burrow density (no. m^{-2}). Error bars are 2 standard errors of the mean in each case. The mean densities from UWTV have been bias-adjusted, as before. The functional units are indicated by data labels (see Table 2.2).

productivity are nevertheless possible across *Nephrops* FUs. The exception was the Fladen grounds (FU7) where the landings biomass was high even at low density. Like all exceptions, this is an interesting test of the model which requires further data. There may be a vessel effect at the Fladen grounds whereby vessels are generally larger due to the more offshore nature of this fishery; this may combine with low cumulative effort at these grounds and both factors may have resulted in unusually high LPUE.

4. ECOSYSTEM ROLES

4.1. Diet, competition and predation

Nephrops is a generalist predator and scavenger, feeding on a broad diet including crustaceans, echinoderms, polychaetes, molluscs, foraminifera and fish (Baden et al., 1990; Cristo and Cartes, 1998). *Nephrops* is also capable of removing plankton through suspension feeding (Loo et al., 1993). As might be expected in a generalist, diets may alter to reflect local variations in food availability (Parslow-Williams et al., 2002). Stable isotope studies are consistent with the view of *Nephrops* as a consumer of deposit and suspension-feeding invertebrates and of both epibenthic and burrowing crustaceans (Le Loc'h and Hily, 2005). Other crustaceans may compete for some of the same resources, including squat lobsters, *Munida rugosa* (Fabricius), angular crab, *Goneplax rhomboides* (L.), and thalassinidean crustaceans: such as *Callianassa subterranea* (Montagu), *Jaxea nocturna* (Nardo) and *Calocaris macandreae* (Bell) (Hughes and Atkinson, 1997).

The interactions of *Nephrops* with other mud-dwelling crustacea may involve competition for burrow space, exploitation competition for food resources, interference competition while foraging or direct intraguild predation (e.g. on *G. rhomboides*; Cristo, 1998). Observations from video tows suggest that some of the potential for competition may be mitigated by foraging at different times of day or by spatial segregation (for *Nephrops*— *G. rhomboides*; Trenkel et al., 2007). Aguzzi et al. (2009) also found evidence for temporal partitioning of foraging by *Nephrops* and potential prey. When species are found spatially aggregated into clumps but these clumps are independent of the clumps of other species, as found for *Munida* and *Nephrops* (Trenkel et al., 2007) and for *Nephrops* and each of *Aristeus antennatus*, *Calocaris macandreae* and *Plesionika heterocarpus* (Maynou et al., 1996), this may also promote coexistence (Hartley and Shorrocks, 2002). The relative overlap of *Nephrops* with other species has been found to vary in time and space, for example, Maynou et al. (1996) found examples of positive correlations between the densities of *Nephrops* and other crustacea (*Solenocera membranacea*, *Sergestes arcticus* and *Plesionika martia*), increasing the potential for competitive or predatory interactions.

The main predator for *Nephrops* in much of its range is cod, *Gadus morhua* (Chapman, 1980). Cod completely dominates the observations of (predominantly North Sea) gut contents containing *Nephrops* in the ICES 'Year of the Stomach' dataset (http://ecosystemdata.ices.dk/stomachdata/): of 2086 records of *Nephrops* in fish stomachs, 93% were in cod. A further 6% of *Nephrops* were in haddock (*Melanogrammus aeglefinus*) stomachs. The remaining cases were occasional records (<1% of cases) of *Nephrops* found in *Anarhichas lupus*, *Eutrigla gurnardus*, *Hippoglossoides platessoides*, *Lepidorhombus whiffiagonis*, *Merlangius merlangus* or *Psetta maxima*. Chapman also cites Thomas (1965) as finding 50% of lesser spotted dogfish, *Scyliorhinus canicula*, and thornback ray, *Raja clavata*, stomachs to contain *Nephrops*. Invertebrates may also be important predators of *Nephrops*: the modelling synthesis of Coll et al. (2006) considered benthic cephalopods to be the most important consumers of *Nephrops* (Coll et al. (2006) gave anglerfish as the second most important predator). Although it is difficult to estimate the prevalence of losses to cannibalism, *Nephrops* will eat conspecifics, both in the field and in captivity (Baden et al., 1990; Cristo and Cartes, 1998; Sarda and Valladares, 1990).

Although cod are frequent predators of *Nephrops*, consumption of *Nephrops* may be a matter of availability rather than preference. In choice tests, cod preferred capelin to *Nephrops* (Björnsson and Dombaxe, 2004). *Nephrops* are relatively low in fat and have an exoskeleton that hinders digestion; as a result, it takes twice as much *Nephrops* compared to capelin to produce a kilo of cod (Björnsson and Dombaxe, 2004). In a study of predator gut contents in the Bay of Biscay, Serrano et al. (2003) rarely found *Nephrops*, despite the species being common in the study area. These authors interpreted this absence of *Nephrops* in predator guts to arise from a combination of unavailability to predators through burrowing and a potential size-based invulnerability for larger *Nephrops*.

4.2. Ecosystem and fishery interactions

It is difficult to separate the role of *Nephrops* in ecosystems from the effects of fisheries, as there appear to be few, if any, unexploited *Nephrops* populations for comparison (although see the example of deeper Mediterranean waters (mean 556 m) in Dimech et al., 2012). Studies of the impacts of *Nephrops* fisheries have previously been constrained by a comparison of a limited number of sites, potentially introducing confounding differences between controls and impacts (e.g. Frid et al., 1999). Furthermore, using wreck sites as untrawled 'controls' risks confounding an absence of fishing disturbance with other ecological processes (Ball et al., 2000; Davis et al., 1982; Hall et al., 1993). Spatially resolved vessel monitoring system or over-flight data have the potential to produce finer resolution maps of fishing effort leading to more general evaluations of the impact of variations in fishing intensity (Gerritsen et al., 2012; Hinz et al., 2009). The importance of fishing can be so great that the removal of *Nephrops* by fishing may be of the same order as the autochthonous input of carbon passed along the food chain from the spring bloom (figures from the western Irish Sea; Hill, 2007). Hence, *Nephrops* fisheries may remove a large portion of the annual production from the fished area.

Nephrops may be relatively resilient to the effects of trawling. For example, areas in some fished regions may be trawled more than 7 times a year and yet landings have been maintained at historically high levels for 30 years (ICES, 2012a,b, Area VII). There are clearly upper limits to fishing removal and disturbance as some *Nephrops* stocks have declined (e.g. Fariña and González Herraiz, 2003). Fishing practices including trawl duration, seasonality, gear (otter or beam trawl) and net design (e.g. Drewery et al., 2010) are all likely to affect the resilience of *Nephrops* populations depending on the extent to which burrows are disturbed and/or unfished individuals damaged. For example, Sarda (1995) found variations of >100 kg in trawl door weight between regions. Heavier gear seems likely to create greater impacts on *Nephrops* burrows and individuals, reducing population resilience to trawling. The technical details of the fishery seem likely to interact with the details of the habitat, such as sediment structure and productivity, to modify the local impacts of trawling. The population and habitat-level effects of gear changes have not been researched in detail. Observations at intermediate fishing intensities, however, suggest that burrows provide some protection from the effects of trawls (Rosenberg et al., 2003). Indeed, *Nephrops* may represent an example of the sort of mobile scavenger (Bergmann et al., 2002) able to increase in relative abundance with the increase in damaged and discarded animals associated with trawling activity (Tillin et al., 2006; Vergnon and Blanchard, 2006). Queirós et al. (2006) suggested that burrowing Crustacea (*Jaxea nocturna* in their example) could dominate benthic communities in trawled areas as burrowing reduces vulnerability to trawling while sediments mobilized by trawling and bioturbation may smother filter-feeding competitors. The tendency for *Nephrops* females brooding eggs to remain in burrows during the winter (ICES, 2012c,d Area IV) also increases the resilience of stocks by protecting population reproductive capacity, although there are concerns that males may become so rare that reproduction becomes sperm limited (ICES, 2012c,d, Area IV).

Alongside a potential subsidy to *Nephrops* from carrion (although see the limitations on the value of this discussed by Kaiser and Hiddink, 2007), other ecosystem effects of fishing may also have positive effects on *Nephrops* populations. In particular, cod populations in the Irish Sea and North Sea have declined over the last 40 years (ICES, 2012a,c,d, North Sea and Irish Sea reports)—potentially reducing the predation pressure on *Nephrops* populations in these regions. The high bycatch rates with *Nephrops* fishing gear are a concern for the management of other species (Catchpole et al., 2008) and may contribute to reduced predation pressure on *Nephrops* stocks. Conversely, measures to reduce discards and bycatch of fish may reduce *Nephrops* catches due to increases in predation. A food web model of the South Catalan Sea fisheries using 40 functional groups and different scenarios for improvements in gear selectivity had complex effects on target and non-target fish, but reductions in losses of predator populations were always predicted to have negative effects on *Nephrops* biomass (Coll et al., 2008). An issue that affects all such assessments of ecosystem energy flows is that the local details of systems limit the degree of generalization. The role of species in ecosystems alters with many variables, including the level of subsidy of materials from adjacent ecosystems (Janetski et al., 2009). Responses to changes in fishing consequently vary in ways that are difficult to predict or generalize across systems. However, it does not seem likely that the changes in systems associated with different fishing practices will be understood without the investment in data gathering at appropriate scales.

The preceding comments on fishing have referred to fisheries involving the trawling of *Nephrops*. The main alternative fishing method is creeling, using baited pots on the sea floor. Creeling mostly removes the bycatch element from the fishery and the use of static gear causes less general disruption to the sea bed. Short-term experimental assessments of creel disturbance suggest that it has minimal effect on sea pen (Pennatulacea) species of conservation importance, although these experiments were potentially of lower intensity than commercial creeling (Kinnear et al., 1996). Adey et al. (2008) assessed the possible impact of 'ghost' fishing by lost or abandoned *Nephrops* pots and concluded that the impact on non-target species was minimal as, once the bait is consumed in a pot, few species are attracted to enter and most can escape if they do. Areas with static gear may have better sport-fishing catches than adjacent trawled areas, demonstrating some broader relative ecological benefits from creeling (Blyth-Skyrme et al., 2006). However, the size and other features of areas with creels may influence whether a species is able to benefit from the absence of trawling (Blyth-Skyrme et al., 2006). Economic assessments of creeling have concluded that it may be the most profitable way of fishing for *Nephrops* due to lower overheads and higher prices per kilo (Leocadio et al., 2012; Morello et al., 2009). However, sizes of individuals in the stock and the depth or distance to fishing grounds may make creeling less attractive than trawling as the benefits of creeling are reduced for populations of smaller individuals in grounds that are too deep for pots or too far from harbour for setting and revisiting traps to be practical (Morello et al., 2009). Reallocation of fishing effort from trawls to creels or additional fishing creeling activity may, however, create issues for the *Nephrops* stock. This has been seen in the Loch Torridon fishery, which lost Marine Stewardship Council certification as a sustainable fishery due to declines in landings per unit effort and uncontrolled changes in fishing effort (Bennett and Combes, 2010, second surveillance report).

4.3. Ecosystem functions and services

As a sometimes dominant species, *Nephrops* can clearly affect the flow of fixed carbon through benthic ecosystems. In addition, burrows alter the structure of the sediment, potentially affecting sediment processes such as nutrient cycling. The burrow structures of *Nephrops* create habitat heterogeneity and are known to be interconnected with other species including the echiuran worm *Maxmuelleria lankesteri*, the goby *Lesueurigobius friesii* and the thalassinidean *Jaxea nocturna* (Tuck et al., 1994). Observations of burrow use suggest that different species may temporarily use each other's burrows (Chapman and Rice, 1971). More generally, the habitat heterogeneity from structures like burrows is associated with higher benthic diversity (Thrush et al., 2001).

Burrow formers, such as *Nephrops*, flush their burrows to obtain oxygen. This, along with increases in sediment surface area, alters rates of sediment-water flux and increases oxygen transfer to the sediments (Aller and Aller, 1998; Schull et al., 2009). Burrow irrigation can be particularly important for processes such as denitrification, which occur where oxic and anoxic microenvironments are in close proximity (Gilbert et al., 2003). Similarly, fluxes of metals that change solubility with changes in sediment redox potential may be enhanced by the presence of burrowing animals. Eriksson and Baden (1998) showed that *Nephrops* from areas prone to hypoxia had higher levels of Manganese in their tissues, most likely as the anoxia promotes formation of soluble $Mn++$, while burrow irrigation increases the bioavailability of this ion. As Mn can scavenge other metals, phosphate and hazardous chemicals (PCBs) under oxic conditions, the release of $Mn++$ following anoxic reduction may also indicate increased bioavailability of other elements and compounds (Eriksson and Baden, 1998).

Attempts to understand the role of burrows in sediment-water fluxes have included modelling the system as a series of cylinder-shaped holes in the sediment (Aller, 1980). Fluxes are, however, variable and depend on interactions between the rates of reactions within the sediment and the configuration of burrows (e.g. Gilbert et al., 2003). The degree to which burrows are irrigated will also influence fluxes (Schull et al., 2009). The turnover rate of water within burrows varies between species, and this parameter does not appear to have been estimated for *Nephrops* (rates for 11 other species are summarized in Schull et al., 2009). Further complexity is added by the non-additive effects of different species on flux rates (Michaud et al., 2009), making it difficult to predict flux rates on the basis of parameters estimated for single species. Given spatial and temporal variability in sediment structure, material supply and benthic community composition, it is not surprizing that the recent review by Pillay and Branch (2011) of thalassinideans emphasizes that the effects of burrowing organisms on fluxes can be inconsistent. Considerable replication in space and time may be needed to generate a full description of benthic fluxes in areas where burrows are present (Hughes et al., 2000).

In addition to the effect of burrow irrigation on fluxes, the reworking of sediment tends to favour other burrowing species and have a negative impact on filter feeders, species that stabilize the sediment and those that are otherwise sensitive to smothering or burial (Pillay and Branch, 2011). *Nephrops* burrows are known to extend as far as 33 cm below the sediment surface. Bioturbation processes are, however, species-specific and their effects are not as well characterized for *Nephrops* as they are for thalassinidean crustaceans (Hughes and Atkinson, 1997).

In conclusion, the main ecosystem service associated with *Nephrops* is clearly related to fisheries. As a bioturbator, *Nephrops* is likely to stimulate ecosystem functions that involve sediment-water fluxes. However, quantification of these effects is complex. Not only is there a lack of specific flux measurements for *Nephrops* burrows, the details of flux will vary with sediment composition and the rate of supply of materials, along with the features of the co-occurring benthic assemblage.

5. OVERALL CONCLUSIONS

International commitments for an ecosystem approach to managing fisheries (e.g. FAO, 2001) create a need for a deeper understanding of the ecology of fished systems, including those of which *Nephrops* is a part. Fisheries data suggest that both density and fishing may have effects on size distributions. However, fishery data represent a biased sample of populations, which, along with the uncertainties around recruitment, limit the general conclusions that can be drawn. Despite the concerns over the impact of protected areas for *Nephrops* as a fisheries tool (Smith and Jensen, 2008), there may be benefits to establishing protected areas in areas with *Nephrops* to study processes such as density-dependent growth, mortality and migration in the absence of fishing. It may also be possible to gain information from spatio-temporal fishery closures such as the one in place at the Porcupine grounds since 2010 (Stokes and Lordan, 2011). Scaling up the temporal and spatial extent of video studies (Burrows et al., 2003) may also help quantify the foraging activities and interactions of *Nephrops* with conspecifics and potential predators, prey and competitors. Wahle (2003) discusses difficulties in identifying density-dependence in Crustacea and potential approaches to resolving this, including the use of field experiments.

There are a number of management processes that could be improved by better information on *Nephrops* ecology. The reform of the European Commission's Common Fisheries Policy (CFP) is intended to remove discarding of undersized and over quota catch. This is an issue for *Nephrops* trawlers as there may be a significant bycatch of depleted whitefish such as cod and whiting. Without a means of managing bycatch, fishing for *Nephrops* may have to stop as the regional quota for the other species caught is exhausted. It may be possible to address this with a suitable system of individual transferable quotas (ITQs) based on the catch for each vessel (European Commission, 2011). However, appropriate ecological and gear-related information that can improve the efficiency and selectivity of catches would reduce the pressure on any ITQ system. Further potential constraints on *Nephrops* fisheries are possible due to international commitments by the OSPAR signatory countries (covering the northeast Atlantic) to protect sea-pen and burrowing megafauna communities (OSPAR, 2010). This policy specifically mentions *Nephrops* trawling as a potential threat, but the broad scale ecology of sea pens and distributional overlap with exploited *Nephrops* stocks are not well understood. A final potential future need for detailed ecological information on *Nephrops* would arise if hatchery or ranching operations become established—creating a need for specific information about suitable sites and efficient release and ongrowing techniques (c.f. Beal et al., 2002 for lobster, *Homarus gammarus*).

This review has not covered the disease ecology of *Nephrops*. Stentiford and Neil (2011) have recently reviewed the diseases of *Nephrops*, the most well-known being the parasitic dinoflagellate *Hematodinium sp*. High prevalence of this parasite is associated with lower catches, but many knowledge gaps remain concerning the effect of fishery practices on the spread of infection and the impact on juvenile animals (Stentiford and Neil, 2011). Transport of live crustaceans for sale or aquaculture may increase the chances of parasites being found in new hosts (Bateman et al., 2012), with additional risks potentially associated with climate change (Stentiford and Neil, 2011).

The distribution and population ecology of *Nephrops* are also likely to change in response to widespread anthropogenic impacts beyond fishing. Temperature changes affect larval stage duration and mortality (Powell and Eriksson, 2013) and there may also be effects on *Nephrops* larvae and adults from ocean acidification (Hernroth et al., 2012). Climate change and eutrophication may change the level of production reaching *Nephrops* on sea floor (Hill, 2007). The stimulation of algal blooms by eutrophication has also been implicated in *Nephrops* population declines through benthic anoxia (Smith and Papadopoulou, 2003). These factors could affect *Nephrops* yields singly or in combination. With a potentially complex reliance on currents between *Nephrops* patches, changes in hydrography with climate change may also alter the recruitment to stocks. For example, the predicted strengthening of western Irish Sea gyre (e.g. Olbert et al., 2011, 2012) may have implications for the retention of larvae and potential supply to other mud patches. González Herraiz et al. (2009) concluded that the North Atlantic Oscillation had a lagged effect (6.5 years) on *Nephrops* CPUE on the Porcupine Bank. There are, however, issues on both statistical model choice and mechanism to be addressed before stock-climate links can be robustly established (Myers, 1998). Therefore, the knowledge gaps concerning *Nephrops* ecology need to be addressed if quantitative predictions are to be made about the future status of stocks and related ecosystems.

ACKNOWLEDGEMENT

With thanks to the referees who suggested additional literature and made valuable comments on the manuscript.

REFERENCES

Abello, P., Abella, A., Adamidou, A., Jukic-Peladic, S., Maiorano, P., Spedicato, M.T., 2002. Geographical patterns in abundance and population structure of *Nephrops norvegicus* and *Parapenaeus logirostris* (Crustacea: Decapoda) along the European Mediterranean coasts. Sci. Mar. 66, 125–141.
Addinck, W., de Kluijver, M., 2003. North Sea Observations of Crustacea, Polychaeta, Echinodermata, Mollusca and Some Other Groups Between 1986 and 2003. Expert Centre for Taxonomic Idenditification (ETI), The Netherlands.
Adey, J.M., Smith, I.P., Atkinson, R.J.A., Tuck, I.D., Taylor, A.C., 2008. 'Ghost fishing' of target and non-target species by Norway lobster *Nephrops norvegicus* creels. Mar. Ecol. Prog. Ser. 366, 119–127.
Afonso-Dias, M., 1998. Variability of *Nephrops norvegicus* (L.) populations in Scottish waters in relation to the sediment characteristics of the seabed. PhD thesis, University of Aberdeen, 282 pp.
Aguzzi, J., Sarda, F., Abello, P., Company, J.B., Rotllant, G., 2003. Diel and seasonal patterns of *Nephrops norvegicus* (Decapoda: Nephropidae) catchability in the western Mediterranean. Mar. Ecol. Prog. Ser. 258, 201–211.
Aguzzi, J., Company, J.B., Sarda, F., 2004. Feeding activity rhythm of *Nephrops norvegicus* of the western Mediterranean shelf and slope grounds. Mar. Biol. 144, 463–472.
Aguzzi, J., Bahamon, N., Marotta, L., 2009. The influence of light availability and predatory behaviour of the decapod crustacean *Nephrops norvegicus* on the activity rhythms of continental margin prey decapods. Mar. Ecol. 30, 366–375.
Aller, R.C., 1980. Quantifying solute distributions in the bioturbated zone of marine sediments by defining an average microenvironment. Geochim. Cosmochim. Acta 44, 1955–1965.
Aller, R.C., Aller, J.Y., 1998. The effect of biogenic irrigation intensity and solute exchange on diagenetic reaction rates in marine sediments. J. Mar. Res. 56, 905–936.
Armstrong, M.J., Smyth, D., McCurdy, W., 1991. How Much *Nephrops* is Eaten by Cod in the Western Irish Sea? ICES, Copenhagen. CM 1991/G:15.
Baden, S.P., Pihl, L., Rosenberg, R., 1990. Effects of oxygen depletion on the ecology, blood physiology and fishery of the Norway lobster *Nephrops norvegicus*. Mar. Ecol. Prog. Ser. 67, 141–155.

Bailey, N., 1986. Why does the size of *Nephrops* vary? Scott. Fish. Bull. 49, 31–36.

Bailey, N., Chapman, C.J., 1983. A Comparison of Density, Length Composition and Growth of Two *Nephrops* Populations Off the West Coast of SCOTLAND. ICES, Copenhagen. CM 1983/K: 42.

BALAR (EurOBIS). http://www.marbef.org/data/imis.php?module=dataset&dasid=3002.

Ball, B.J., Fox, G., Munday, B.W., 2000. Long- and short-term consequences of a *Nephrops* trawl fishery on the benthos and environment of the Irish Sea. ICES J. Mar. Sci. 57, 1315–1320.

Barquín, J., Brito, A., Falcon, J.M., 1998. Occurrence of the Norway lobster, *Nephrops norvegicus* (L., 1758) (Decapoda, Nephropidae), near the Canary Islands. Crustaceana 71, 344–348.

Bateman, K.S., Tew, I., French, C., Hicks, R.J., Martin, P., Munro, J., Stentiford, G.D., 2012. Susceptibility to infection and pathogenicity of White Spot Disease (WSD) in non-model crustacean host taxa from temperate regions. J. Invertebr. Pathol. 110, 340–351.

Beal, B.F., Mercer, J.P., O'Conghaile, A., 2002. Survival and growth of hatchery-reared individuals of the European lobster, *Homarus gammarus* (L.), in field-based nursery cages on the Irish west coast. Aquaculture 210, 137–157.

Bell, M.C., Elson, J.M., Addison, J.T., 2005. The effects of spatial targeting of fishing effort on the distribution of the Norway lobster, *Nephrops norvegicus*, on the Farn Deeps grounds, northeast England. N. Z. J. Mar. Freshw. Res. 39, 1023–1037.

Bennett, D., Combes, J., 2010. Surveillance Report Loch Torridon *Nephrops* Creel Fishery. Moody Marine Ltd, Derby, 24 pp. http://www.msc.org/track-a-fishery/fisheries-search/loch-torridon-nephrops-creel/ (accessed on 5.12.12).

Bergmann, M., Wieczorek, S.K., Moore, P.G., Atkinson, R.J.A., 2002. Utilisation of invertebrates discarded from the *Nephrops* fishery by variously selective benthic scavengers in the west of Scotland. Mar. Ecol. Prog. Ser. 233, 185–198.

Björnsson, B., Dombaxe, M.A.D., 2004. Quality of *Nephrops* as food for Atlantic cod (*Gadus morhua* L.) with possible implications for fisheries management. ICES J. Mar. Sci. 61, 983–991.

Black, R., 1979. Competition between intertidal limpets: an intrusive niche on a steep resource gradient. J. Anim. Ecol. 48, 401–411.

Blyth-Skyrme, R.E., Kaiser, M.J., Hiddink, J.G., Edwards-Jones, G., Hart, P.J.B., 2006. Conservation benefits of temperate marine protected areas: variation among fish species. Conserv. Biol. 20, 811–820.

Briggs, R.P., Armstrong, M.J., Dickey-Collas, M., Allen, M., McQuaid, N., Whitmore, J., 2002. The application of fecundity estimates to determine the spawning stock biomass of Irish Sea *Nephrops norvegicus* (L.) using the annual larval production method. ICES J. Mar. Sci. 59, 109–119.

Burrows, M.T., Robb, L., Nickell, L.A., Hughes, D.J., 2003. Topography as a determinant of search paths of fishes and mobile macrocrustacea on the sediment surface. J. Exp. Mar. Biol. Ecol. 285, 235–249.

Campbell, N., Allan, L., Weetman, A., Dobby, H., 2009a. Investigating the link between *Nephrops norvegicus* burrow density and sediment composition in Scottish waters. ICES J. Mar. Sci. 66, 2052–2059.

Campbell, N., Dobby, H., Bailey, N., 2009b. Investigating and mitigating uncertainties in the assessment of Scottish *Nephrops norvegicus* populations using simulated underwater television data. ICES J. Mar. Sci. 66, 646–655.

Catchpole, T., van Keeken, O., Gray, T., Piet, G., 2008. The discard problem—a comparative analysis of two fisheries: the English *Nephrops* fishery and the Dutch beam trawl fishery. Ocean Coast. Manage. 51, 772–778.

Chapman, C.J., 1980. Ecology of Juvenile and Adult Nephrops. In: Cobb, J.S., Phillips, (Eds.), The biology and management of lobsters. Ecology and Management, vol. II. Academic Press, London.

Chapman, C.J., 1985. Observing Norway lobster, *Nephrops norvegicus* (L.) by towed sledge fitted with photographic and television cameras. In: George, J.D., Lythgoe, G.I., Lythgoe, J.N. (Eds.), Underwater Photography and Television for Scientists. Clarendon Press, Oxford, UK, pp. 100–108.

Chapman, C.J., Rice, A.L., 1971. Some direct observations on the ecology and behaviour of the Norway lobster *Nephrops norvegicus*. Mar. Biol. 10, 321–329.

Cobb, J.S., Wahle, R.A., 1994. Early-life history and recruitment processes of clawed lobsters. Crustaceana 67, 1–25.

Cole, HA, 1965. The size distribution of Nephrops populations on grounds around the British coasts. Rapports et Procès-Verbaux des Rèunions, Conseil International pour l'Exploration de la Mer 156, 164–171.

Coll, M., Palomera, I., Tudela, S., Sarda, F., 2006. Trophic flows, ecosystem structure and fishing impacts in the South Catalan Sea, Northwestern Mediterranean. J. Mar. Syst. 59, 63–96.

Coll, M., Bahamon, N., Sarda, F., Palomera, I., Tudela, S., Suuronen, P., 2008. Improved trawl selectivity: effects on the ecosystem in the South Catalan Sea (NW Mediterranean). Mar. Ecol. Prog. Ser. 355, 131–147.

Craeymeersh, J, Kingston, P, Rachor, E, Duineveld, G, Heip, C, Vanden Berghe, E, 1986. North Sea Benthos Survey.

Cristo, M., 1998. Feeding ecology of *Nephrops norvegicus* (Decapoda: Nephropidae). J. Nat. Hist. 32, 1493–1498.

Cristo, M., Cartes, J.E., 1998. A comparative study of the feeding ecology of *Nephrops norvegicus* (L.), (Decapoda: Nephropidae) in the bathyal Mediterranean and the adjacent Atlantic. Sci. Mar. 62, 81–90.

Cristo, M., Castro, M., 2005. Field estimation of daily ration of Norway lobster (*Nephrops norvegicus*) in the south of Portugal. N. Z. J. Mar. Freshw. Res. 39, 485–491.

Davis, N., VanBlaricom, G.R., Dayton, P.K., 1982. Man-made structures on marine sediments: effects on adjacent benthic communities. Mar. Biol. 70, 295–303.

de Figueiredo, M.J., Vilela, M.H., 1972. On the artificial culture of *Nephrops norvegicus* reared from the egg. Aquaculture 1, 173–180.

Department of Invertebrate Zoology, Research and Collections Information System, NMNH, Smithsonian Institution. http://www.mnh.si.edu/rc/db/collection_db_policy1.html (accessed on 21.10.12).

Dickey-Collas, M., Briggs, R.P., Armstrong, M.J., Milligan, S.P., 2000. Production of *Nephrops norvegicus* larvae in the Irish Sea. Mar. Biol. 137, 973–981.

Dimech, M., Kaiser, M.J., Ragonese, S., Schembri, P.J., 2012. Ecosystem effects of fishing on the continental slope in the Central Mediterranean Sea. Mar. Ecol. Prog. Ser. 449, 41–54.

Drewery, J., Bova, D., Kynoch, R.J., Edridge, A., Fryer, R.J., O'Neill, F.G., 2010. The selectivity of the Swedish grid and 120 mm square mesh panels in the Scottish *Nephrops* trawl fishery. Fish. Res. 106, 454–459.

Eriksson, S.P., Baden, S.P., 1998. Manganese in the haemolymph and tissues of the Norway lobster, *Nephrops norvegicus* (L.), along the Swedish west coast, 1993–1995. Hydrobiologia 375 (376), 255–264.

European Commission, 2011. Impact Assessment of Discard Reducing Policies. Available at http://ec.europa.eu/fisheries/documentation/studies/discards/index_en.htm (accessed on 14.12.12).

FAO, 2001. The Reykjavik declaration on responsible fisheries in the marine ecosystem. Available at ftp://ftp.fao.org/fi/document/reykjavik/default.htm (accessed on 14.12.12).

Fariña, A.C., González Herraiz, I., 2003. Trends in catch-per-unit-effort, stock biomass and recruitment in the North and Northwest Iberian Atlantic *Nephrops* stocks. Fish. Res. 65, 351–360.

Fautin, DG, Buddemeier, RW, 2008. Biogeoinformatics of the Hexacorals. http://www.kgs.ku.edu/Hexacoral/ (accessed on 23.10.12).

Fish trawl survey: Baltic International Trawl Survey. ICES Database of trawl surveys (DATRAS). The International Council for the Exploration of the Sea, Copenhagen. 2010. http://ecosystemdata.ices.dk (accessed on 21.10.12).

Fish trawl survey: Beam Trawl Survey. ICES Database of trawl surveys (DATRAS). The International Council for the Exploration of the Sea, Copenhagen. 2010. http://ecosystemdata.ices.dk (accessed on 21.10.12).

Fish trawl survey: French Southern Atlantic Bottom Trawl Survey. ICES Database of trawl surveys (DATRAS). The International Council for the Exploration of the Sea, Copenhagen. 2010. http://ecosystemdata.ices.dk (accessed on 21.10.12).

Fish trawl survey: Irish ground fish survey. ICES Database of trawl surveys (DATRAS). The International Council for the Exploration of the Sea, Copenhagen. 2010. http://ecosystemdata.ices.dk (accessed on 21.10.12).

Fish trawl survey: North Sea International Bottom Trawl Survey. ICES Database of trawl surveys (DATRAS). The International Council for the Exploration of the Sea, Copenhagen. 2010. http://ecosystemdata.ices.dk (accessed on 21.10.12).

Fish trawl survey: Northern Ireland Survey. ICES Database of trawl surveys (DATRAS). The International Council for the Exploration of the Sea, Copenhagen. 2010. http://ecosystemdata.ices.dk (accessed on 21.10.12).

Fish trawl survey: Rockall Survey ICES Vib. ICES Database of trawl surveys (DATRAS). The International Council for the Exploration of the Sea, Copenhagen. 2010. http://ecosystemdata.ices.dk (accessed on 21.10.12).

Fish trawl survey: Scottish Western Coast Via Groundfish Survey. ICES Database of trawl surveys (DATRAS). The International Council for the Exploration of the Sea, Copenhagen. 2010. http://ecosystemdata.ices.dk (accessed on 21.10.12).

Fisheries Research Service, Marine Laboratory. Macrobenthos samples collected in the Scottish waters in 2001.

Flanders Marine Institute (VLIZ). Taxonomic Information System for the Belgian coastal area. 10 August 2004, Oostende, Belgium, Version: 1, SQL-server.

Fockedey, N., Beyst, B., Cattrijsse, A., Dewicke A., Deneudt, K., Mees J., Vincx, M., 2004. Historical hyperbenthos data (1987–2001) from the North Sea and some adjacent areas. EUROBIS, Collaboration between Ghent University (UGent), Biology Department, Marine Biology Section and Flanders Marine Institute (VLIZ). Data consulted through EurOBIS October 2012.

Frid, C.L.J., Clark, R.A., Hall, J.A., 1999. Long-term changes in the benthos on a heavily fished ground off the NE coast of England. Mar. Ecol. Prog. Ser. 188, 13–20.

Froglia, C., Atkinson, R.J., Tuck, I., Arneri, E., 1997. Underwater television survey, a tool to estimate *Nephrops* stock biomass on the Adriatic trawling grounds. In: Finka, B. (Ed.), Tisucu Godina Prvoga Spomena Ribarstva u Hrvata. Hrvatska Akademija Znanosti I Umjetnosti, Zagreb, pp. 657–667.

Gerritsen, H.D., Lordan, C., Minto, C., Kraak, S.B.M., 2012. Spatial patterns in the retained catch composition of Irish demersal otter trawlers: high-resolution fisheries data as a management tool. Fish. Res. 129, 127–136.

Gilbert, F., Aller, R.C., Hulth, S., 2003. The influence of macrofaunal burrow spacing and diffusive scaling on sedimentary nitrification and denitrification: an experimental, simulation and model approach. J. Mar. Res. 61, 101–125.

González Herraiz, I., Torres, M.A., Fariña, A.C., Freire, J., Cancelo, J.R., 2009. The NAO index and the long-term variability of *Nephrops norvegicus* population and fishery off West of Ireland. Fish. Res. 98, 1–7.

Gordina, A.D., Pavlova, E.V., Ovsyany, E.I., Wilson, J.G., Kemp, R.B., Romanov, A.S., 2001. Long-term Changes in Sevastopol Bay (the Black Sea) with Particular Reference to the Ichthyoplankton and Zooplankton. Estuar. Coast. Shelf Sci. 52, 1–13.

Hall, S.J., Robertson, M.R., Basford, D.J., Heaney, S.D., 1993. The possible effects of fishing disturbance in the northern north-sea - an analysis of spatial patterns in community structure around a wreck. Neth. J. Sea Res. 31, 201–208.

Hartley, S., Shorrocks, B., 2002. A general framework for the aggregation model of coexistence. J. Anim. Ecol. 71, 651–662.

Heath, M.R., Speirs, D.C., 2011. Changes in species diversity and size composition in the Firth of Clyde demersal fish community. Proc. R. Soc. B 279, 543–552.

Hernroth, B., Skold, H.N., Wiklander, K., Jutfelt, F., Baden, S., 2012. Simulated climate change causes immune suppression and protein damage in the crustacean *Nephrops norvegicus*. Fish Shellfish Immunol. 33, 1095–1101.

Hill, A.E., 1990. Pelagic dispersal of Norway lobster *Nephrops norvegicus* larvae examined using an advection–diffusion-mortality model. Mar. Ecol. Prog. Ser. 64, 217–226.

Hill JM (2007) Structure and flow of carbon and nitrogen to the western Irish Sea Nephrops norvegicus fishery: a stable isotope approach. PhD thesis, Queen Mary, University of London.

Hill, A.E., White, R.G., 1990. The dynamics of Norway lobster (Nephrops-norvegicus L) populations on isolated mud patches. J. Conseil 46, 167–174.

Hill, A.E., Brown, J., Fernand, L., 1997. The summer gyre in the Western Irish Sea: shelf sea paradigms and management implications. Estuar. Coast. Shelf Sci. 44, 83–95.

Hillis, J.P., 1979. Growth studies on the prawn *Nephrops norvegicus*. Rap. Proces. 175, 170–175.

Hinz, H., Prieto, V., Kaiser, M.J., 2009. Trawl disturbance on Benthic communities: chronic effects and experimental predictions. Ecol. Appl. 19, 761–773.

HMAP, n.d. http://www.hmapcoml.org/Default.asp?ID=37 (accessed on 21.10.12).

Holthuis, L.B., 1991. FAO species catalogue. In: Marine Lobsters of the World. An Annotated and Illustrated Catalogue of Species of Interest to Fisheries Known to Date. FAO Fisheries Synopsis No. 125, vol. 13. Food and Agriculture Organization of the United Nations, Rome.

Hughes, D.J., Atkinson, R.J.A., 1997. A Towed Video Survey of Megafaunal Bioturbation in the NorthEastern Irish Sea. J. Mar. Biol. Assoc. U.K. 77, 635–653.

Hughes, R.N., Griffiths, C.L., 1988. Self-thinning in barnacles and mussels: the geometry of packing. Am. Nat. 132, 484–491.

Hughes, D.J., Atkinson, R.J.A., Ansell, A.D., 2000. A field test of the effects of megafaunal burrows on benthic chamber measurements of sediment-water solute fluxes Marine. Ecol. Prog. Ser. 195, 189–199.

ICES, 2000. Report of the study group on life history of *Nephrops*. Reykjavik, Iceland 2–5 May 2000. CM 2000/G:06, 148pp.

ICES, 2009. Report of the Benchmark Workshop on Nephrops assessment (WKNEPH). ICES CM: 2009/ACOM:33. http://www.ices.dk/reports/ACOM/2009/WKNEPH/wkneph_2009.pdf (accessed on 5.12.12).

ICES, 2011. WGCSE REPORT Draft report.

ICES, 2012a. Cod in Division VIIa (Irish Sea). http://www.ices.dk/committe/acom/comwork/report/2012/2012/cod-iris.pdf (accessed on 5.12.12).

ICES, 2012b. Cod in Divisions VIIe–k (Celtic Sea cod). http://www.ices.dk/committe/acom/comwork/report/2012/2012/cod-7e-k.pdf (accessed on 5.12.12).

ICES, 2012c. Cod in Subarea IV (North Sea) and Divisions VIId (Eastern Channel) and IIIa West (Skagerrak). http://www.ices.dk/committe/acom/comwork/report/2012/2012/cod-347.pdf (accessed on 5.12.12).

ICES, 2012d. Nephrops in subarea IV (North Sea). http://www.ices.dk/committe/acom/comwork/report/2012/2012/Neph-IV.pdf (accessed on 5.12.12).

ICES Contaminants and biological effects database (DOME—Biota). The International Council for the Exploration of the Sea, Copenhagen. 2010. Online source: http://ecosystemdata.ices.dk (accessed on 5.12.12).

ICES Fish predator/prey data (EurOBIS). Retrieved (accessed on 5.12.12). from www.iobis. org.

ICES historical plankton database (1901–1912).

Janetski, D.J., Chaloner, D.T., Tiegs, S.D., Lamberti, G.A., 2009. Pacific salmon effects on stream ecosystems: a quantitative synthesis. Oecologia 159, 583–595.

Kaiser, M.J., Hiddink, J.G., 2007. Food subsidies from fisheries to continental shelf benthic scavengers. Mar. Ecol. Prog. Ser. 350, 267–276.

Kilada, R., Sainte-Marie, B., Rochette, R., Davis, N., Vanier, C., Campana, S., 2012. Direct determination of age in shrimps, crabs, and lobsters. Can. J. Fish. Aquat. Sci. 69, 1728–1733.

Kinnear, J.A.M., Barkel, P.J., Mojsiewicz, W.R., Chapman, C.J., Holbrow, A.J., Barnes, C., Greathead, C.F.F., 1996. Effect of *Nephrops* creels on environment. Fisheries Research Services Report 2/96. Scottish Office, 24 pp. http://www.scotland.gov.uk/Uploads/Documents/frsr296.pdf (accessed on 5.12.12).

Le Loc'h, F., Hily, C., 2005. Stable carbon and nitrogen isotope analysis of *Nephrops norvegicus* / *Merluccius merluccius* fishing grounds in the Bay of Biscay (Northeast Atlantic). Can. J. Fish. Aquat. Sci. 62, 123–132.

Leocadio, A.M., Whitmarsh, D., Castro, M., 2012. Comparing trawl and creel fishing for Norway lobster (*Nephrops norvegicus*): biological and economic considerations. PLoS One 7, e39567.

Loo, L.O., Baden, S.P., Ulmestrand, M., 1993. Suspension-feeding in adult Nephrops-norvegicus (L) and Homarus-gammarus (L) (decapoda). Neth. J. Sea Res. 31, 291–297.

Lordan, C., Doyle, J., Sacchetti, F., O'Driscoll, D., Heir, I., Smith, T., Allsop, C., 2007. Report of the UWTV Survey on the Aran, Galway Bay and Slyne Head *Nephrops* Grounds 2006. Working Document to WKNEPTV (2007), 36pp.

Mackie, A.S.Y., Oliver, P.G., Rees, E.I.S., 1991. Biomôr 1 Dataset. Benthic Data from the Southern Irish Sea from 1989–1991. National Museum and galleries of Wales, Cardiff, UK.

Maiorano, P., Sion, L., Carlucci, R., Capezzuto, F., Giove, A., Costantino, G., Panza, M., D'Onghia, G., Tursi, A., 2010. The demersal faunal assemblage of the north-western Ionian Sea (central Mediterranean): current knowledge and perspectives. Chem. Ecol. 26, 219–240.

Maltagliati, F., Camilli, L., Biagi, F., Abbiati, M., 1998. Genetic structure of Norway lobster, *Nephrops norvegicus* (L.) (Crustacea : Nephropidae), from the Mediterranean Sea. Sci. Mar. 62, 91–99.

Marine Conservation Society, Seasearch Marine Surveys. EUROBIS, Marine Conservation Society, 19 January 2005, Ross-on-Wye, UK, Version: 1, DiGIR Provided, http://www.marbef.org/ (accessed on 21.10.12).

Marta-Almeida, M., Dubert, J., Peliz, A., dos Santos, A., Queiroga, H., 2008. A modelling study of Norway lobster (*Nephrops norvegicus*) larval dispersal in southern Portugal: predictions of larval wastage and self-recruitment in the Algarve stock. Can. J. Fish. Aquat. Sci. 65, 2253–2268.

Maynou, F., Sarda, F., 1997. *Nephrops norvegicus* population and morphometrical characteristics in relation to substrate heterogeneity. Fish. Res. 30, 139–149.

Maynou, F., Conan, G.Y., Cartes, J.E., Company, J.B., Sarda, F., 1996. Spatial structure and seasonality of decapod crustacean populations on the northwestern Mediterranean slope. Limnol. Oceanogr. 41, 113–125.

MedOBIS. EUROBIS, Hellenic Centre for Marine Research; Institute of Marine Biology and Genetics; Biodiversity and Ecosystem Management Department, 17 December 2004, Heraklion, Greece, Version: 1, MS Access. http://www.medobis.org/ (accessed on 21.10.12).

MEDITS (EurOBIS), 2004. Retrieved (accessed on 5.12.12). from www.iobis.org.

Michaud, E., Desrosiers, G., Aller, R.C., Mermillod-Blondin, F., Sundby, B., Stora, G., 2009. Spatial interactions in the *Macoma balthica* community control biogeochemical fluxes at the sediment-water interface and microbial abundances. J. Mar. Res. 67, 43–70.

Morello, E.B., Froglia, C., Atkinson, R.J.A., 2007. Underwater television as a fishery-independent method for stock assessment of Norway lobster (*Nephrops norvegicus*) in the central Adriatic Sea (Italy). ICES J. Mar. Sci. 64, 1116–1123.

Morello, E.B., Antolini, B., Gramitto, M.E., Atkinson, R.J.A., Froglia, C., 2009. The fishery for *Nephrops norvegicus* (Linnaeus, 1758) in the central Adriatic Sea (Italy): preliminary observations comparing bottom trawl and baited creels. Fish. Res. 95, 325–331.

Myers, R.A., 1998. When do environment-recruitment correlations work? Rev. Fish Biol. Fish. 8, 285–305.

Mytilineou, C., Castro, M., Gancho, P., Fourtouni, A., 1998. Growth studies on Norway lobster, *Nephrops norvegicus* (L.), in different areas of the Mediterranean Sea and the adjacent Atlantic. Sci. Mar. 62, 43–60.

National Institute of Marine Sciences and Technologies—Trawl Surveys (AfrOBIS). Retrieved (accessed on 5.12.12) from www.iobis.org.

Nichols, J.H., Bennett, D.B., Symonds, D.J., Grainger, R., 1987. Estimation of the stock size of adult *Nephrops-norvegicus* (L) from larvae surveys in the western Irish Sea in 1982. J. Nat. Hist. 21, 1433–1450.

Olbert, A.I., Hartnett, M., Dabrowski, T., Mikolajewicz, U., 2011. Long-term inter-annual variability of a cyclonic gyre in the western Irish Sea. Cont. Shelf Res. 31, 1343–1356.

Olbert, A.I., Dabrowski, T., Nash, S., Hartnett, M., 2012. Regional modelling of the 21st century climate changes in the Irish Sea. Cont. Shelf Res. 41, 48–60.

Olesen, M., 2010a. Marine Benthic Fauna List, Island of Læsø, Denmark. The Danish Biodiversity Information Facility, DanBIF. Copenhagen. Consulted on http://www.iobis.org (accessed on 23.10.12).

Olesen M, 2010b. Marine Benthic Fauna List, Island of Læsø, Denmark. The Danish Biodiversity Information Facility, DanBIF. Copenhagen. Consulted on http://www.iobis.org (accessed on 23.10.12).

OSPAR, 2010. Recommendation 2010/11 on furthering the protection and restoration of sea-pen and burrowing megafauna communities in the OSPAR Maritime Area. OSPAR 10/23/1-E, Annex 33. http://www.ospar.org (accessed on 21.10.12).

Ostler, R., 2005. Marine Nature Conservation Review (MNCR) and associated benthic marine data held and managed by JNCC. EUROBIS, Joint Nature Conservation Committee, Centre for Ecology and hydrology, 22 November 2005, Aberdeenshire, UK, Version: 1, DiGIR Provider. http://www.marbef.org (accessed on 21.10.12).

Pampoulie, C., Skirnisdottir, S., Hauksdottir, S., Olafsson, K., Eiriksson, H., Chosson, V., Hreggvidsson, G.O., Gunnarsson, G.H., Hjorleifsdottir, S., 2011. A pilot genetic study reveals the absence of spatial genetic structure in Norway lobster (*Nephrops norvegicus*) on fishing grounds in Icelandic waters. ICES J. Mar. Sci. 68, 20–25.

PANGAEA-Publishing Network for Geoscientific and Environmental Data, AWI/MARUM. http://dx.doi.org/10.1594/pangaea.

Parr, J., 2005a. Marine Life Information Network (MarLIN) marine survey data (Professional). Marlin, Collated Marine Life Survey Datasets, Marine Biological Association of the UK, 03 December 2005, Plymouth, UK, Version: 1.0, DiGIR Provider. http://www.marbef.org/ (accessed on 21.10.12).

Parr, J., 2005b. Marine Life Survey Data (collected by volunteers) collated by MarLIN. MarLIN, collated Marine Life Survey Datasets, Marine Biological Association of the UK, 04 Dec 2005, Plymouth, UK, Version: 1, DiGIR Provider. http://www.marbef.org/ (accessed on 21.10.12).

Parslow-Williams, P.J., Atkinson, R.J.A., Taylor, A.C., 2001. Nucleic acids as indicators of nutritional condition in the Norway lobster *Nephrops norvegicus*. Mar. Ecol. Prog. Ser. 211, 235–243.

Parslow-Williams, P., Goodheir, C., Atkinson, R.J.A., Taylor, A.C., 2002. Feeding energetics of the Norway lobster, *Nephrops norvegicus* in the Firth of Clyde, Scotland. Ophelia 56, 101–120.

Passamonti, M., Mantovani, B., Scali, V., Froglia, C., 1997. Allozymic characterization of Scottish and Aegean populations of Nephrops norvegicus. J. Mar. Biol. Assoc. U.K. 77, 727–735.

Phillips, S.J., Dudik, M., Schapire, R.E., 2004. A maximum entropy approach to species distribution modeling. In: Proceedings of the 21st International Conference on Machine Learning ACM Press, New York, pp. 655–662.

Phillips, S.J., Anderson, R.P., Schapire, R.E., 2006. Maximum entropy modeling of species geographic distributions. Ecol. Model. 190, 231–259.

Picton, B.E., Emblow, C.S., Morrow, C.C., Sides, E.M., Tierney, P., McGrath, D., McGeough, G., McCrea, M., Dinneen, P., Falvey, J., Dempsey, S., Dowse, J., Costello, M.J., 1999. Marine Sites, Habitats and Species Data Collected During the Biomar Survey of Ireland. Environmental Sciences Unit, Trinity College, Dublin, Ireland.

Pillay, D., Branch, G.M., 2011. Bioengineering effects of burrowing thalassinidean shrimps on marine soft-bottom ecosystems. Oceanogr. Mar. Biol. Ann. Rev. 49, 137–191.

Pinnegar, J.K., Platts, M., 2011. DAPSTOM—An Integrated Database and Portal for Fish Stomach Records. Version 3.6. Centre for Environment, Fisheries and Aquaculture Science, Lowestoft, UK. Phase 3, Final Report, July 2011, 35pp.

Powell, A., Eriksson, S.P., 2013. Reproduction: Life cycle, larvae and larviculture of *Nephrops norvegicus*. Adv. Mar. Biol. 64, 201–247.

Queirós, A.M., Hiddink, J.G., Kaiser, M.J., Hinz, H., 2006. Effects of chronic bottom trawling disturbance on benthic biomass, production and size spectra in different habitats. J. Exp. Mar. Biol. Ecol. 335, 91–103.

Rees, H.L., et al., 2005. A comparison of benthic biodiversity in the North Sea, English Channel and Celtic Seas. EUROBIS, Centre for Environment, Fisheries and Aquaculture Science; Burnham Laboratory, 12 April 2005, Essex, UK, Version: 1, MS Excel. http://www.marbef.org/ (accessed on 21.10.12).

Rice, A.L., Chapman, C.J., 1971. Observations on the burrows and burrowing behaviour of two mud-dwelling decapod crustaceans, *Nephrops norvegicus* and *Goneplax rhomboides*. Mar. Biol. 10, 330–342.

Rosa, R., Nunes, M.L., 2003. Seasonal changes in nucleic acids, amino acids and protein content in juvenile Norway lobster (*Nephrops norvegicus*). Mar. Biol. 143, 565–572.

Rosenberg, R., Nilsson, H.C., Gremare, A., Amouroux, J.M., 2003. Effects of demersal trawling on marine sedimentary habitats analysed by sediment profile imagery. J. Exp. Mar. Biol. Ecol. 285–286, 465–477.

Rostron, D., Pembrokeshire Marine Species Atlas. EUROBIS, Countryside Council for Wales, 11 August 2004, Gwynedd, UK, Version: 1, DiGIR Provider.http://www.marbef.org/ (accessed on 21.10.12).

Rumohr, H., Historical benthos data from the North Sea and Baltic Sea from 1902–1912. Christian-Albrechts-University Kiel; Leibniz Institute of Marine Sciences; Marine Ecology Division; Benthos Ecology section, 11 June 2006, Kiel, Germany. http://www.marbef.org/ (accessed on 21.10.12).

Sarda, F., 1995. Comparative technical aspects of the *Nephrops norvegicus* (L.) fishery in the northern Mediterranean Sea. Sci. Mar. 62, 101–106.

Sardà, F., 1998. *Nephrops norvegicus* (L.): Comparative biology and fishery in the Mediterranean Sea. Introduction, conclusions and recommendations. Sci. Mar. 62, 5–15.

Sarda, F., Aguzzi, J., 2012. A review of burrow counting as an alternative to other typical methods of assessment of Norway lobster populations. Rev. Fish Biol. Fish. 22, 409–422.

Sarda, F., Valladares, F.J., 1990. Gastric evacuation of different foods by *Nephrops-norvegicus* (crustacea, decapoda) and estimation of soft-tissue ingested, maximum food-intake and cannibalism in captivity. Mar. Biol. 104, 25–30.

Sarda, F., Lleonart, J., Cartes, J.E., 1998. An analysis of the population dynamics of *Nephrops norvegicus* (L.) in the Mediterranean Sea. Sci. Mar. 62, 135–143.

Schull, D.H., Benoit, J.M., Wojcik, C., Senning, J.R., 2009. Infaunal burrow ventilation and pore-water transport in muddy sediments. Estuar. Coast. Shelf Sci. 83, 277–286.

Scottish Natural Heritage, 2005. Marine Nature Conservation Review (MNCR) and associated benthic marine data held and managed by Scottish Natural Heritage. EUROBIS, Scottish Natural Heritage, 22 April 2005, Edinburgh, UK, Version: 1, DiGIR Provider. http://www.marbef.org (accessed on 21.10.12).

Serrano, A., Velasco, F., Olaso, I., Sanchez, F., 2003. Macrobenthic crustaceans in the diet of demersal fish in the Bay of Biscay in relation to abundance in the environment. Sarsia 88, 36–48.

Smith, I.P., Jensen, A.C., 2008. Dynamics of closed areas in Norway lobster fisheries. ICES J. Mar. Sci. 65, 1600–1609.

Smith, C.J., Papadopoulou, K.N., 2003. Burrow density and stock size fluctuations of *Nephrops norvegicus* in a semi-enclosed bay. ICES J. Mar. Sci. 60, 798–805.

Stamatis, C., Triantafyllidis, A., Moutou, K.A., Mamuris, Z., 2004. Mitochondrial DNA variation in northeast atlantic and mediterranean populations of norway lobster, *Nephrops norvegicus*. Mol. Ecol. 13, 1377–1390.

Stentiford, G.D., Neil, D.M., 2011. Diseases of *Nephrops* and *Metanephrops*: a review. J. Invertebr. Pathol. 106, SI: 92–SI:109.

Stokes, D., Lordan, C., 2011. Irish fisheries-science research partnership trawl survey of the Porcupine Bank Nephrops Grounds July 2010. Ir. Fish. Bull. 39, 28pp.

Streiff, R., Guillemaud, T., Alberto, F., Magalhaes, J., Castro, M., Cancela, M.L., 2001. Isolation and characterization of microsatellite loci in the Norway lobster (*Nephrops norvegicus*). Mol. Ecol. Notes 1, 71–72.

Thomas, H.J., 1965. The white-fish communities associated with *Nephrops norvegicus* (L.) and the by-catch of white fish in the Norway lobster fishery, together with notes on Norway lobster predators. Rap. Proces. 156, 155–160.

Thrush, S.F., Hewitt, J.E., Funnell, G.A., Cummings, V.J., Ellis, J., Schultz, D., Talley, D., Norkko, A., 2001. Mar. Ecol. Prog. Ser. 221, 255–264.

Tillin, H.M., Hiddink, J.G., Jennings, S., Kaiser, M.J., 2006. Chronic bottom trawling alters the functional composition of benthic invertebrate communities on a sea-basin scale. Mar. Ecol. Prog. Ser. 318, 31–45.

Trenkel, V.M., Le Loc'h, F., Rochet, M.J., 2007. Small-scale spatial and temporal interactions among benthic crustaceans and one fish species in the Bay of Biscay. Mar. Biol. 151, 2207–2215.

Tuck, I.D., Atkinson, R.J.A., Chapman, C.J., 1994. The structure and seasonal variability in the spatial distribution of *Nephrops norvegicus* burrows. Ophelia 40, 13–25.

Tuck, I.D., Chapman, C.J., Atkinson, R.J.A., 1997. Population biology of the Norway lobster, *Nephrops norvegicus* (L.) in the Firth of Clyde, Scotland—I: growth and density. ICES J. Mar. Sci. 54, 125–135.

Tully, O., Hillis, J.P., 1995. Causes and spatial scales of variability in population structure of *Nephrops norvegicus* (L.) in the Irish Sea. Fish. Res. 21, 329–347.

Türkay, M., Senckenbergisches Sammlungsverwaltungssystem, SeSam. Senckenbergische Naturforschende Gesellschaft, 27 Oct 2006, Frankfurt, Germany, Version: 1.20, DiGIR Provider, http://sesam.senckenberg.de/ (accessed on 21.10.12).

UK National Biodiversity Network, Marine Biological Association—DASSH Data Archive Centre Academic Surveys.

Ulmestrand, U., Eggert, H., 2001. Growth of Norway lobster, *Nephrops norvegicus* (Linnaeus 1758), in the Skagerrak, estimated from tagging experiments and length frequency data. ICES J. Mar. Sci. 58, 1326–1334.

Ungfors, A., Sandell, J., Cowing, D., Dobson, N.C., Bublitz, R., Bell, E., 2013. *Nephrops* fisheries in European waters. Adv. Mar. Biol. 64, 249–316.

Vergnon, R., Blanchard, F., 2006. Evaluation of trawling disturbance on macrobenthic invertebrate communities in the Bay of Biscay, France: abundance biomass comparison (ABC method). Aquat. Living Resour. 19, 219–228.

Vrgoc, N., Arneri, E., Jukic-Peladic, S., Krstulovic Šifner, S., Mannini, .P, Marceta, B. Osmani, K., Piccinetti, C., Ungaro, N., 2004. Review of current knowledge on shared demersal stocks of the Adriatic Sea. AdriaMed Technical Documents No.12 GCP/RER/010/ITA/TD-12 Termoli (Italy), March 2004.

Wahle, R.A., 2003. Revealing stock–recruitment relationships in lobsters and crabs: is experimental ecology the key? Fish. Res. 65, 3–32.

Whomersley, P., 2003. National Marine Monitoring Programme. Benthos data of the North Sea, Irish Sea, English Channel from 2002–2003. CEFAS, Burnham On Crouch, UK.

Wilkinson, S., 2005, Marine benthic dataset (version 1) commissioned by UKOOA. EUROBIS, Joint nature Conservation Committee, 07 December 2005, Peterborough, UK, Version: 1, DiGIR Provider. http://www.marbef.org (accessed on 21.10.12).

Wood, S.N., 2006. Generalized Additive Models: An Introduction with R. Chapman and Hall/CRC, New York.

CHAPTER THREE

Sensory Biology and Behaviour of *Nephrops norvegicus*

Emi Katoh*, Valerio Sbragaglia†, Jacopo Aguzzi†, Thomas Breithaupt*,1

*School of Biological, Biomedical and Environmental Sciences, University of Hull, Hull, United Kingdom
†Marine Science Institute (ICM-CSIC), Passeig Marítim de la Barceloneta, Barcelona, Spain
1Corresponding author: e-mail address: t.breithaupt@hull.ac.uk

Contents

Abstract

The Norway lobster is one of the most important commercial crustaceans in Europe. A detailed knowledge of the behaviour of this species is crucial in order to optimize fishery yields, improve sustainability of fisheries, and identify man-made environmental threats. Due to the cryptic life-style in burrows, the great depth and low-light condition of their habitat, studies of the behaviour of this species in its natural environment are challenging. Here, we first provide an overview of the sensory modalities (vision, chemoreception, and mechanoreception) of *Nephrops norvegicus*. We focus particularly on the role of the chemical and mechanical senses in eliciting and steering spatial

© 2013 Elsevier Ltd.
All rights reserved.

orientation behaviours. We then concentrate on recent research in social behaviour and biological rhythms of *Nephrops*. A combination of laboratory approaches and newly developed tracking technologies has led to a better understanding of aggressive interactions, reproductive behaviours, activity cycles, and burrow-related behaviours. Gaps in our knowledge are identified and suggestions for future research are provided.

Keywords: Chemoreception, Mechanoreception, Vision, Rhythms, Burrowing, Actograph, Video-image analysis, Pheromone, Aggression, Mating

1. INTRODUCTION

Ethology, the study of animal behaviour, encompasses several aspects of animal biology and hence requires a multidisciplinary approach (Davies et al., 2012). Ethologists are interested in the causes of behaviour. Questions about animal behaviours can be subdivided into two major categories: questions about *proximate* (physiological) causes address 'how' an individual comes to behave in a determinate way; questions about *ultimate* (evolutionary) causes address 'why' the individual has evolved that behaviour (Tinbergen, 1963). For any putative behavioural trait, the first category of questions is usually addressed through laboratory experiments, while the second one requires the measurement of the fitness value.

Behavioural studies of the Norway lobster (*Nephrops norvegicus* L.; in the following referred to as *Nephrops*) started in the 1970s both in the field and in the laboratory, driven by fishery management issues in relation to population size, and overall exploitability of lobster stocks (Farmer, 1975). The first experiments were performed in order to relate the temporal variation in catchability with the behaviour of individuals in the laboratory (Chapman, 1980). That comparison was sustained by the assumption that the locomotor behaviour in the laboratory is a reliable proxy of burrow emergence behaviour in the field (and hence of overall catchability). Since the mid–1980s, *Nephrops* has also been used to answer fundamental neuro–ethological questions addressing the control of simple behaviours by the nervous system (e.g. Neil and Miyan, 1986; Neil and Wotherspoon, 1982; Newland and Neil, 1987). Over the past two decades, technical innovations both in monitoring *Nephrops* in the field and in studying their physiological parameters in the lab allowed insight into the proximate causes of rhythmic behaviour (e.g. Aguzzi et al., 2011a,b; Sardà and Aguzzi, 2012). These aim to provide better predictability of catch rate and more accurate population estimates. Only recently, basic biological questions about the social behaviour (aggressive interactions, reproductive behaviour) have been addressed (Katoh, 2011;

Katoh et al., 2008) targeting a better understanding of the ultimate causes of *Nephrops* behaviour.

This chapter reviews current knowledge of the behaviour of the *Nephrops*. We will start with a description of the sensory equipment of *Nephrops* and its role in triggering behaviour. We will then concentrate on two current research areas: social behaviour and biological rhythms.

2. SENSORY BIOLOGY AND THE CONTROL OF ORIENTATION RESPONSES

In common with other decapod crustaceans (Atema and Voigt, 1995), Norway lobsters possess a comprehensive set of mechanosensory, chemosensory, and visual receptors. The compound eyes are the main visual receptors (see Chapter 4) but decapods also have extraretinal light detectors such as the caudal photoreceptor (Figure 3.1). Chemoreceptors are on the antennules (olfactory receptors) and on the body appendages including second antennae, mouthparts, and walking legs (bimodal receptors; Figures 3.1 and 3.2). Mechanoreceptors include cuticular setae that are distributed over the whole exoskeleton of *Nephrops*, proprioreceptors in the joints of the appendages including the flagellae of the antennae, and the statocysts located in the basal segment of the antennules (Figure 3.2). While not all the receptors of *Nephrops* have been subject to detailed investigation, morphological similarities suggest functions that are similar to the receptors of the better investigated American lobsters *Homarus americanus*, spiny

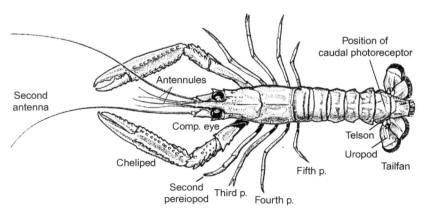

Figure 3.1 The Norway lobster *N. norvegicus*. Second to fifth pereiopod constitute the walking legs. *Drawing modified after Howard (1989) with permission from publishers.*

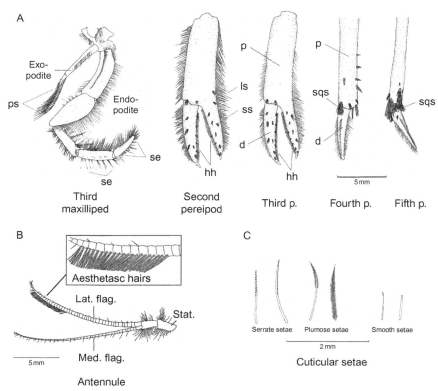

Figure 3.2 Cuticular setae of *N. norvegicus*. (A) Third maxilliped and walking legs (pereiopods 2–5) showing distribution of setae: hh, hedgehog hairs; se, serrate setae; ls, long smooth setae; ss, smooth setae; sqs, squamate setae; ps, propodite; d, dactylopodit; (B) antennule with aesthetasc hairs. Notice position of statocyst (stat.); (C) types of cuticular setae found on *N. norvegicus*. (A) Modified after Farmer (1974a)—reprinted by permission of the publisher Taylor & Francis Ltd. (B) Modified after Farmer (1973, Figure 3)—with kind permission from Springer Science + Business Media. (C) Modified after Farmer (1974a)—reprinted by permission of the publisher Taylor & Francis Ltd.

lobster *Panilurus* sp., or crayfish (e.g. *Pacifastacus leniusculus*, *Procambarus clarkii*) (see Atema and Voigt, 1995; Schmidt and Mellon, 2011).

2.1. Vision

Norway lobsters have large kidney-shaped ('reniform') compound eyes of the 'superposition type' that are typical of nocturnal crustaceans (Figure 3.1; Aréchiga and Atkinson, 1975). In contrast to other receptor organs, the compound eyes have been well investigated. They are reviewed in depth in Chapter 4. While vision can provide an accurate and instantaneous image of the external environment in daylight conditions (Dusenbery, 1992), the low light availability at the spatial and temporal activity range of Norway lobsters limits the utility of this modality. Norway lobster populations that live in shallow waters are active out of the burrow only at night; at intermediate depth they are active at dawn or dusk and in deep water they are active at daytime (Bell et al., 2006; see also Section 4.4). Long-term capture–recapture experiments in the field have shown that lobsters with light-induced eye damage are still able to survive and reproduce and do not suffer reductions in growth (Chapman et al., 2000; Shelton et al., 1985). While the compound eyes are still important sensory organs facilitating spatial orientation and biological rhythms, they do not appear to be crucial in mediating important behaviours such as foraging, predator avoidance, and courtship. Some rhythmic behaviours and burrow-related behaviours may be mediated by the caudal photoreceptor, a multimodal interneuron situated in the sixth abdominal ganglion that responds to mechanical stimuli as well as to changes in light intensity (Chapman et al., 2000; Simon and Edwards, 1990).

2.2. Chemoreception

Decapod crustaceans including *N. norvegicus* have two types of chemosensory organs: unimodal olfactory sensilla and non-olfactory bimodal sensilla (Hallberg and Skog, 2011). The olfactory sensilla named aesthetascs are located on the distal end of the lateral flagellum of the antennules (Figure 3.2; Farmer, 1973). Each aesthetasc sensillum typically contains many olfactory neurons (hundreds of olfactory neurons in some species such as spiny lobsters; Caprio and Derby, 2008) that all project into the olfactory lobe in the brain. Non-olfactory bimodal sensilla are found on almost all appendages. These sensilla contain both chemo- and mechanoreceptors. The chemoreceptors do not project into the olfactory lobe but to other centres in the brain and in the ventral nerve chord (Schmidt and Mellon, 2011). Typical bimodal sensilla that are found in decapod crustaceans including *Nephrops* are the hedgehog hairs, the smooth and squamous setae, and the serrate setae (Derby, 1982; Farmer, 1974a; Goodall, 1988; for a recent review of types of setae see also Garm and Watling, 2013). The hedgehog hairs (named 'teeth' chemoreceptors in Farmer, 1974a) are stout conical structures organized in a row lining the inner cutting edge of the chelae of the second and third walking legs (Figure 3.2A; Derby, 1982; Farmer, 1974a). The smooth setae (named 'fine simple setae' in Farmer, 1974a) and the squamous setae ('large serrate setae with scales' in Farmer, 1974a) are arranged in small groups ('tufts') rising at intervals out of small depressions along the propodit and dactylopodit of the walking legs (Figure 3.2A). Serrate setae are found on the mouthparts, including the maxillipeds (Figure 3.2A) and on the fourth and fifth pereiopods. Using electrophysiology, Derby (1982) confirmed that hedgehog hairs, smooth setae and squamous setae respond both to mechanical and chemical stimuli.

To contrast the function of olfactory unimodal and non-olfactory bimodal sensilla, the latter have been attributed as taste or gustatory receptors (Atema and Voigt, 1995; Derby and Sorensen, 2008). For terrestrial animals, olfaction and taste are clearly distinguished based on the physical medium in which they are received. Olfaction is the detection of air-borne molecules whereas taste detects water-soluble molecules. For aquatic animals, the distinction between olfaction and taste is unclear as both modalities detect water-soluble molecules. In current discussions, the distinction is based on the organization of the first order interneurons (olfactory processing is localized in the olfactory lobe, non-olfactory processing is distributed across different areas of the central nervous system; Caprio and Derby, 2008) and on functional characteristics (Schmidt and Mellon, 2011). Taste mediates simple reflexive behaviours such as grabbing, biting, and swallowing, whereas olfaction mediates more complex behaviours such as search for odour sources from a distance, courtship behaviour and learning about odours (Caprio and Derby, 2008). Ablation experiments in spiny lobsters (Reeder and Ache, 1980) and American lobsters (Devine and Atema, 1982) showed that antennular receptors are involved in detection and initial tracking of odour plumes. Weissburg (2011) suggests that in blue crabs chemoreceptors on the walking legs are involved in odour plume tracking. Once near the food source the taste receptors are important in selecting and ingesting edible food. The first and second walking legs probe the ground and use the small chelae to pick up food items detected by leg chemoreceptors (Derby and Atema, 1982). Food is transferred to the maxillipeds where it is further checked by taste receptors and passed on to the mouth (Derby and Atema, 1982). Norway lobsters are predators and scavengers feeding on polychaetes, crustaceans, molluscs, and echinoderms (Bell et al., 2006). They rely on chemoreception in their food search (Krang and Rosenqvist, 2006). First responses to food include antennule 'flicking' (a behaviour that is analogue to sniffing in mammals as it enhances sensitivity to chemical stimuli; Schmitt and Ache, 1979) and antennal sweeps, followed by olfactory tracking (Krang and Rosenqvist, 2006). Little is known about the role of bimodal receptors in food detection and selection.

Olfactory sensilla on the antennules are important in mediating social interactions. Fighting behaviour in *Nephrops* is accompanied by increased flicking (Katoh et al., 2008) and chemical signals appear to play an important role both in aggressive and in courtship interactions (Katoh et al., 2008; see below). Olfactory detection and chemical signalling are facilitated by generation of 'information currents' (Atema and Voigt, 1995). The anterior location of the urinary pores ('nephropores') in the basal segments of the second antennae allows urine-borne molecules to be introduced into the frontal water currents emanating from the gill chambers. The gill currents serve as information currents by delivering urine-borne chemical messages towards conspecific receivers (Breithaupt, 2011). A second source of information currents is generated by beating of the maxilliped-flagellae (exopodites; see Figure 3.2A). The concerted action of the exopodites of all three maxillipeds act like a fan organ producing water currents that draw odour molecules towards the antennules. This fanning behaviour enhances the perceptive range of odour detection (Denissenko et al., 2007). Olfaction is also used to sense chemicals released by predators or by injured or disturbed conspecifics (Derby and Sorensen, 2008; Hazlett, 2011). This can mediate adaptive responses to elevated predation risk mostly by reducing locomotion (Hazlett, 2011).

2.3. Mechanoreception

Mechanoreceptors can be used for tactile exploration, for the perception of water movement (hydrodynamic stimuli), and for the detection of acoustic stimuli (Breithaupt, 2002; Breithaupt and Tautz, 1990; Goodall et al., 1990). The mechanoreceptor equipment of *N. norvegicus* is similar to that of other decapod crustaceans such as crayfish and lobsters (Farmer, 1974a; Goodall, 1988). Mechanosensory structures include short cuticular setae that are distributed over the entire surface of the body, the first antennae (antennules; Figures 3.1 and 3.2), second antennae, and the statocysts located in the basal segment of the antennules (Atema and Voigt, 1995; Breithaupt and Tautz, 1990; Goodall, 1988).

Mechanosensory setae of crayfish are innervated by two mechanoreceptive neurons that respond to movements of the seta in opposite direction providing the sensory hair with a directional sensitivity (Wiese, 1976). The cuticular setae respond to tactile stimuli as well as to hydrodynamic stimuli. Some of the hairs were shown to respond to water movements as slow as 0.1 mm/s (Breithaupt and Tautz, 1990).

Mechanoreceptive interneurons in the last abdominal ganglion integrate flow information from many different setae and respond to complex flow pattern moving over the tailfan (Tautz and Plummer, 1994). This enables crayfish to recognize vortices created by fish passing behind their tailfan (Breithaupt et al., 1995). The flow stimulus elicits a turning response in crayfish that can lead to capture of small prey fish (Breithaupt et al., 1995). Norway lobsters possess mechanoreceptive setae (plumose setae, long setae; Figure 3.2C; Farmer, 1974a) on their tailfan similar to those of crayfish (Goodall, 1988).

Mechanosensory hairs also trigger escape responses of Norway lobsters. The escape response involves rapid flexion and extensions of the abdomen (i.e. tail flips; Newland and Neil, 1990). The tail flip response and its underlying neuronal circuitry has been extensively studied in crayfish and in Norway lobsters and is one of the best-known examples of a stereotyped motor pattern (Newland and Neil, 1990; Newland et al., 1992; Wine, 1984). The initial tail flip is mediated by giant nerve fibres. Mechanical stimulation of the anterior cephalothorax is mediated by the median giant fibre and leads to a tail flip that is directed backwards, away from the point of stimulation. Stimulation of the abdomen recruits the lateral giant fibre and produces an upward, forward pitching tail flip. The initial giant fibre-mediated tail flip is followed by many swimming tail flips not mediated by the giant fibres (Newland et al., 1992). The trajectories of the swimming tail flip also take into account the laterality of the stimulation. Field experiments have shown that asymmetrical stimuli to the chelipeds elicit swimming trajectories directed away from the point of stimulation (Newland and Chapman, 1989). These escape behaviours are adaptive and can be used to escape rapidly approaching predatory fish. Cod (*Gadus morhua*) has been identified as the most important predator of Norway lobsters in the Atlantic Ocean (Farmer, 1975). In the Mediterranean Sea, major predators include angler fish (*Lophius* spp.), various elasmobranchs, hake (*Merluccius merluccius*) scorpionfish (*Scorpaena* spp.), and small gadoids (e.g. *Trispoterus minutus capelanus*) (Bell et al., 2006; see also Chapter 2). Swimming behaviour also plays a role in the process of capture of *Nephrops* in trawl nets (Newland et al., 1992). Lobsters stimulated to swim by ground gear generally escape approximately parallel to the direction of tow. However, lobsters facing away from the oncoming ground gear swim up into the water column (Newland et al., 1992). In both situations, approach of predatory fish or of a trawl net, the bow wave may provide the stimulus for the mechanoreceptors that triggers the directional escape response.

Sweeping movements of the long second antennae are a common behaviour of Norway lobsters during spatial exploration (e.g. Krang and Rosenqvist, 2006). In crayfish, the second antenna is equipped with sensory setae and specialized receptors that detect when the antennal flagellum is deformed (Sandeman, 1989; Tautz et al., 1981). These receptors together with proprioreceptors in the joint of the basal segment (Taylor, 1967) can serve the animal to register movement of the flagellum and deformation due to contact with objects. Similar to other decapod crustaceans, Norway lobsters sweep the antennae to create physical contact with objects in the environment. Behavioural experiments on the crayfish *Cherax destructor* have demonstrated that in the absence of visual stimuli the antennae are used to detect and learn topographic changes in the environment (Basil and Sandeman, 2000). It was suggested that during exploration behaviour they compare the input of the bilateral antennae to make orientation decisions (McMahon et al., 2005). Crayfish can retain information about the configuration of their environment for up to 24 h (Basil and Sandeman, 2000).

The statocyst, positioned in the basal segment of the antennules (see Figure 3.2), is a fluid-filled chamber that contains a mass of sand grains that act as statolith. Hair sensilla are arranged in a crescent around the statolith mass and are in contact with the sand grains (Atema and Voigt, 1995). Due to the higher inertia of the statolith, a tilt in one of the three body axis causes a relative movement between statolith and the wall of the statocyst thereby stimulating specific groups of cuticular hairs. Statocysts are primarily involved in maintaining equilibrium by triggering righting movements of walking legs, swimmerets, and uropods (Newland and Neil, 1987; see Figure 3.1). Analogue to the fish otocyst which is used in equilibrium responses and hearing, the statocyst has been suggested as an acoustic detector in crustaceans (e.g. Popper et al., 2001). Field experiments were carried out by Goodall et al. (1990) to test hearing abilities of the Norway lobster under appropriate acoustical conditions. They found that while Norway lobsters do not respond to the pressure component of sound (i.e. the sound component that the human ear detects) they show behavioural reactions to water vibrations (hydrodynamic stimuli) in the frequency range of 20–180 Hz. Bulk movement of water, such as the bow wave created by larger fish, tidal currents, or seismic waves will cause sudden displacements of the lobster's body which, due to the statolith's higher inertia, will be detected by the statocyst organ. It may be these hydrodynamic stimuli, rather than acoustic stimuli, that the crustacean statocyst is adapted to detect in its natural environment (Breithaupt, 2002).

3. SOCIAL BEHAVIOUR

3.1. Agonistic interactions

Many animals including the Norway lobsters display aggressive behaviour and fighting occurs between individuals over limited resources such as food, shelter, and mating opportunities (Chapman and Rice, 1971; Dissanayake et al., 2009; Moore, 2007). Fighting can cause injuries such as limb loss or even death which is an issue when keeping them communally (Briffa and Sneddon, 2007; McVean, 1982; Norman and Jones, 1991; Smith and Hines, 1991). Chapman and Rice (1971) noticed that a high proportion of *Nephrops* that were caught by creels in the field have circular indentations or holes in the propodits of their great claws. Due to the nature of these wounds, they were likely inflicted by claws of conspecifics suggesting that fighting is a common event in natural populations (Chapman and Rice, 1971). The level of aggression in fights and the method of forming a dominance hierarchy are affected by intrinsic and extrinsic factors (Moore, 2007). Intrinsic factors are features inherent to the individual such as size, sex, reproductive status, and winner/loser history (Mesterton-Gibbons, 1999; Moore, 2007). The fighting ability in terms of physical prowess (i.e. its resource holding potential; Parker, 1974) is correlated with the size of an animal. Larger animals usually dominate smaller ones and the subordinate animal usually positions itself as far as possible from the dominant animal (Lee and Fielder, 1982). Extrinsic factors include chemical signals from the opponent or the value and type of the resource that is contested (McGregor and Peake, 2000; Moore, 2007). In *Nephrops*, fighting has been observed in the field by video cameras and direct observation of divers (Chapman and Rice, 1971). These recordings suggest that fighting is often caused by territorial interactions over a burrow and start with one animal approaching a burrow defended by a conspecific. In the laboratory, aggressive behaviour can be elicited by placing two animals together in a small tank. Individuals that are initially separated by a divider will start interacting as soon as the divider is lifted. Aggressive behaviours are categorized based on stereotypical agonistic behaviours (Table 3.1). Fights sometimes involve more than one bout of aggressive displays. Fight duration is analysed as the sum of bout durations. A bout ends with one animal (the loser) showing avoidance (level -1) or escape behaviour (level -2). The loser of the fight is the combatant that loses the last bout and does not show any aggression exceeding level 3 for the remaining time of the interaction. Agonistic level

Table 3.1 Definition of agonistic levels for fighting *N. norvegicus*

Level	Behaviour	Definition
−2	Fleeing	Walking backwards, walking away, or turning away, tail flipping
−1	Avoidance	Walking around but avoiding opponent, body pressed to the ground
0	Separate	No activity
L	Separate	Locomotion, cleaning
1	Approach	Animals within reach of claws, facing approaching, turning towards, following
2	Touching	Some body parts (e.g. abdomen, pereiopods) touch for extended time without any higher levels of aggression
3	Threat display	High on legs, meral spread (horizontally spread chelipeds without display physical contact)
4	Cheliped pushing	Combatants push each other face to face in meral spread position pushing
5	Wrestling	Smacking, pushing, antennal touching claw grabbing, punching

Adapted from Atema and Voigt (1995).

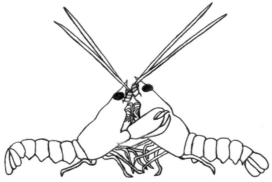

Figure 3.3 Fighting *Nephrops* using cheliped pushing behaviour (Katoh et al., 2008). Copyright 2008 by N.V. Koninklijke Brill, Leiden, The Netherlands.

4 is a *Nephrops*-specific behaviour called 'cheliped pushing' (Figure 3.3), where combatants are face-to-face making contact between the laterally outstretched chelipeds (Katoh et al., 2008).

Katoh et al. (2008) showed that the duration and intensity of dyadic contests decreases when the fight is repeated (Figure 3.4; control) suggesting the formation of a dominance relationship. However, the mechanisms

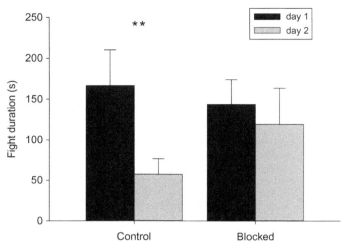

Figure 3.4 Duration (mean + SEM) of repeated fights between size-matched individuals. Animals had their urine release blocked on day 2 by catheter tubes diverting the urine to a syringe floating at the water surface (blocked) or were sham-catheterized (control, days 1 and 2; blocked, day 1) not restraining the urine output. Asterisks indicate significant difference in fight duration ($p < 0.01$). *Data from Katoh et al. (2008).*

responsible for the maintenance of dominance in *Nephrops* are unknown. There are three possible mechanisms to maintain an established dominance relationship: winner/loser effects, individual recognition, and recognition of social status (see Breithaupt, 2011, for a detailed discussion of the three mechanisms). Winner/loser effects are intrinsic factors based on the winning/losing history of an individual. Winners of previous fights tend to win again and losers are more likely to lose in future fights (Hock and Huber, 2005; Hsu and Wolf, 2001). Individual recognition has been demonstrated to mediate maintenance of dominance in some species of hermit crabs (Gherardi and Atema, 2005; Gherardi and Tiedemann, 2004; Gherardi et al., 2005; Hazlett, 1969), in mantis shrimps (Caldwell, 1979, 1985), and in lobsters (Johnson and Atema, 2005; Karavanich and Atema, 1998). Males in these species have the ability to remember the individual opponents they had fought in a previous encounter. The recognition of social status refers to sensory assessment of the winning/losing history of the opponent. A loser of a recent fight can recognize an unknown winner of a different fight. Status recognition has been demonstrated in some species of crayfish (Breithaupt and Eger, 2002; Copp, 1986; Gherardi and Daniels, 2003) and hermit crabs (Winston and Jacobson, 1978). In Norway lobsters, winner/loser effects to not appear to play an important role. Extrinsic factors (individual or status recognition), based on chemosensory assessment of the opponent, appear to be more important in dominance relationships. Experiments blocking the urine release of fighting Norway lobsters indicate that assessment of the opponent's chemical signals is crucial in the maintenance of dominance (Katoh et al., 2008). Two *Nephrops* males that had interacted in a previous fight were paired again 24 h later. In the second fight, the urinary pores located frontally in the basal segments of the second antennae were blocked by connecting catheters to the cuticle surrounding the nephropores. Urine was released into the catheters rather than into the water. In the first fight, lobsters were sham catheterized (Katoh et al., 2008). In control animals that were sham catheterized in both fights, second encounters are generally shorter than first fights indicating the formation of a dominance relationship (Figure 3.4). However, when nephropores were blocked in the second encounter, these fights were as long as the initial encounter indicating that dominance was not maintained (Figure 3.4). This finding suggested that chemical communication is necessary for maintaining dominance (Katoh et al., 2008). It is likely that it is the subordinate who recognizes the chemical identity or status of the dominant. The subordinate *Nephrops* showed significantly more antennule flicking in first and second day fights than the dominant (Katoh et al., 2008).

In order to investigate whether *Nephrops* uses individual or status recognition, two rounds of dyadic encounters were staged (Katoh, 2011). For the interactions, only males with size-matched carapace length (less than 5% difference) were used to exclude size effects influencing the outcome of interactions. Males were isolated in 3-l individual tanks for 7–10 days prior to the first fight to remove any memory of previous encounters (e.g. in lobsters; Karavanich and Atema, 1998). Fights were staged at temperatures between 10 and 12 °C in a 70-l tank with three sides darkened by a black sheet and the bottom filled with 1 cm of black sand. The tank was illuminated with dim red light (25 W). Opponents were given 30-min acclimatization time on opposite sides of a divider. The fight was started by lifting the divider. The second fight followed 24 ± 1 h after the first fight allowing for sufficient time of physical recovery and restoring of urine resources after the first fight. The first fights enabled *Nephrops* to assess the opponent. The second fights were designed as either familiar or unfamiliar treatment. In familiar treatments (14 replicates), *Nephrops* encountered the same opponent they had fought in the first fight. In the unfamiliar treatment (14 replicates), individuals fought an unknown opponent with contrasting fighting experience (i.e. dominants were paired with subordinates; Karavanich and Atema, 1998;

Katoh, 2011). In Norway lobsters, if dominance was based on individual recognition the second fights would only be shorter in the familiar treatment but not in the unfamiliar treatment.

There was a significant effect of day of fight ($F = 14$, $p < 0.001$, two-way repeated measure ANOVA) but not of the treatment ($F = 0.05$, $p = 0.824$). In both treatment groups, second fights were shorter than first fights (Figure 3.5; *post-hoc* Tukey tests, $p < 0.05$) but there was no difference between treatments. The result suggests that in *Nephrops* it does not make a difference whether the opponent is familiar or unfamiliar from previous fights. Norway lobsters appear to recognize the social status of the opponent rather than the individual identity (Katoh, 2011). This is surprising as Norway lobsters are similar to American lobsters (Atema and Voigt, 1995) in displaying some site fidelity even though they not always occupy the same burrow (Chapman and Rice, 1971; see also Section 4.2). American lobsters remember for up to 2 weeks the identity of previously fought conspecifics (Karavanich and Atema, 1998). Future field studies are necessary to understand the ecological significance of status recognition versus individual recognition in the Norway lobster.

A better understanding of the mechanisms of fighting behaviour is important to understand the relationship between population density and average body size found during the analysis of *Nephrops* stocks (see Chapter 2). Norway lobsters appear to grow less at high population densities. One

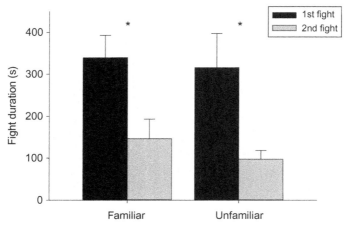

Figure 3.5 Fight duration (mean + SEM) of second encounter was lower than first day encounters, no matter if the same pair of animals fought twice (familiar) or the winner of the first fight met an unfamiliar loser (unfamiliar). Asterisks indicate significant differences ($p < 0.05$, *post-hoc* Tukey test).

explanation of this could be that higher densities lead to more encounters and more territorial aggression. This increase in territorial behaviour may reduce the time for foraging and lead to a reduced growth of individuals (Chapter 2).

Experiments on Norway lobsters with disabled urine release indicated that chemical signals are important for the formation of a dominance relationship by reducing the aggression of the subordinate (Katoh et al., 2008). Based on this finding, one could expect that adding a chemical, such as dominance odour, into a tank with fighting *Nephrops* may further reduce aggression of the competitors. In contrast, Sneddon et al. (2003) found that male shore crabs *Carcinus maenas* would increase the fighting effort when female pheromone was added but not when male water or untreated seawater was added. To test the effect of conspecific odour on male competition in *Nephrops*, experiments were conducted by adding dominant odour, female odour, or sea water, as a control, to a tank with two fighting males. In the dominant odour treatment, a dominant male of a different fight was kept for 12 h in the experimental tank to condition the water with dominant male odour. After the conditioning period, the dominant male was replaced by two unfamiliar size-matched males. The males were allowed to interact immediately for 30 min. In the second treatment, water was conditioned for 12 h by an intermoult female (IF), and in the third treatment, water was left for 12 h without an animal (Katoh, 2011). The average fight duration decreased significantly when adding dominant odour compared to the control, where no conspecific odour was added to the experiment (Figure 3.6; $p < 0.05$, one-way ANOVA; Katoh, 2011). Although the dominant male was not present, the odour alone was enough to manipulate the agonistic behaviour of the two interacting males by decreasing fight durations. Female water did not have a significant effect on fighting behaviour. However, in contrast to the study of Sneddon et al. (2003), the females used in this experiment were not receptive (i.e. they were not post-moult) and therefore may have had less impact on the males as they may not have been considered a resource to fight for. This procedure of reducing aggressive behaviour by introducing odour from dominant males can possibly become a valuable tool in aquaculture by reducing aggression in communally held *Nephrops*.

3.2. Reproductive behaviour

Crustaceans have evolved a variety of mating strategies and behaviours (Duffy and Thiel, 2007). Various ways of communication including visual, tactile or chemical signals, or a combination of them plays a role in attracting

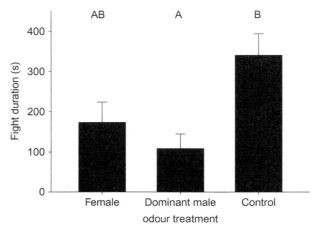

Figure 3.6 Fight durations (mean + SEM). Fight durations following the introduction of different stimuli: female odour, intermoult female conditioned seawater; male dominant odour, dominant male conditioned seawater odour; seawater, unconditioned seawater; $N = 14$. Dominant male odour significantly shortens fight duration in Norway lobster, *N. norvegicus* ($p < 0.05$; one-way ANOVA). Different letters above column indicates significant differences ($p < 0.05$).

mating partners (Duffy and Thiel, 2007). Depending on the species, the male or female initiates mating behaviour. In some species, it was observed that females initiate mating by approaching the male (Lipcius et al., 1983), visiting the males' shelter (Atema and Steinbach, 2007), or backing under the males' body (Jivoff and Hines, 1998). Females are usually selective with respect to the size of the male mate. This is seen when a female shore crab approaches and performs courtship behaviour to the largest available male (Sneddon et al., 2003). Female American lobsters (*H. americanus*) visit the males' shelters to assess the males' quality in order to choose a mating partner. Prior to their moult, *H. americanus* females enter the dominant males' shelter, cohabits with the male until it moults, after which mating occurs (Atema and Steinbach, 2007). In some species, when the female is in the pre-moult stage, she will back under the males' body to initiate his pre-copulatory mate guarding behaviour. The guarding behaviour of the male will protect the female from predators and other males during the moult stage, when the female is weak and vulnerable (Jivoff and Hines, 1998).

Ever since Farmer's classic paper on the reproduction of the Norway lobster (Farmer, 1974b), mating behaviour has not been subject to detailed investigation. Katoh's (2011) study provided the first in depth analysis of mating behaviour and the role of chemical signals in courtship of

N. norvegicus. Animals ranging in size from 31 to 45 mm carapace length were kept in individual tanks at 10°C in a 12–12-h light/dark cycle (with light dimmed down) and fed with polychaete worms once a week. Observations were made on size-matched male–female pairs ($N = 46$ pairs in total) in a 30-l tank illuminated by a 25-W red light with three sides darkened using black sheet and the bottom covered with black sand. Interactions were filmed for 12 h and recorded using time-lapse on a 3-h tape.

Mating behaviour can be divided into six stages (Table 3.2), altogether lasting between 28 s and 6.40 min. Six different behaviours were observed in 16 matings (Table 3.2). Mating generally starts with the 'male approaching' the female. The male then 'mounts' the female, 'turns' it onto her back or onto the side. This is followed by the male 'positioning' himself on top of the female so that his gonopods are close to the female receptacle. The male then 'rolls' the female to the side and deposits the spermatophores in the female receptacle using typical 'thrusting' movements (Table 3.2; Katoh, 2011). There were noticeable differences between the mating behaviours recorded in the study of Katoh (2011) compared to the study of Farmer (1974b). The males in Katoh's study did not have a specific approaching strategy, as mentioned in Farmer's study; they can approach from the front, side, or back and do not necessarily straddle the female from the rear. After 'penetration/thrusting', the pair separates and shows no interest in further interactions. In two cases, the pairs mated twice during the experiment. The observation of double mating in this experiment gives indication that females will mate more than once on occasions.

Decapod crustaceans either mate with hard shells (intermoult mating; e.g. crayfish; Berry and Breithaupt, 2008; squat lobster; Thiel and Lovrich, 2011), with soft shells shortly after moulting (post-moult mating; e.g. many Brachyuran crabs; Christy, 1987), or they sometimes mate hard-shelled, sometimes soft-shelled (e.g. *Homarus gammarus*; Skog, 2009). *N. norvegicus* has been described as a post-moult mater (Farmer, 1974b). Moreover, depending on the male guarding strategy, females can mate with more than one male. Streiff et al. (2004) discovered occurrence of multiple paternity in wild broods of Norway lobsters. Males can secure exclusive paternity only when they guard the female shortly before and after mating (post-copulatory guarding). In this situation, the male keeps other males away from the female for the short period of time during which the female is receptive. Such behaviour has been observed in other Nephropidae, such as the American lobster, *H. americanus* where the male cohabits with a female for up to 2 weeks in the mating shelter (Atema and Steinbach, 2007).

Table 3.2 Definition of courtship levels for mating *N. norvegicus*

Level	Behaviour	Definition
1	Male approach	The male approaches the female by walking towards her from the front ($N=6$), the side ($N=2$), or from the back ($N=8$)
2	Mount	The male climbs onto the females' carapace, from behind or from the side when they are parallel to each other, using his pereiopods. The female is usually passive during mounting and does not move unless she refuses to mate. In the latter case, the female tries to escape, which occurred 8 times out of 24 mounting attempts
3	Turn	The male turns the female using his walking legs (pereiopods) onto her back ($N=6$) or onto the side ($N=6$). During this procedure, the male often holds a claw or antenna of the female with one cheliped ($N=11$). In some cases, there was no reason for the male to turn the female, as in 4 out of 16 matings the female turned onto her back herself, while the male tried to climb on her
4	Positioning	The male positions himself on top of the female ($N=6$ out of 16 matings) so the ventral-to-ventral and face-to-face position can be maintained and the male gonopods are closest to the female seminal receptacle. This stage is skipped if the male has turned the female to the side directly instead of turning her onto her back
5	Rolling	The males who turned females on their backs and positioned themselves on top, will now turn with the females to the side while the hold on to the female with their pereiopods. Usually, at this point the males' claw lets go of the females' claw or antenna. The females are in a torpedo shape with outstretched chelipeds
6	Thrusting	The male moves his abdomen rapidly, while the uropods at the telson open and close (indication of spermatophore transfer and thus mating success; Skog, 2009)

Adapted from Skog (2009).

3.3. The role of pheromones in *Nephrops* mating

Post-moult mating is often triggered by sex pheromones released by the female around the time of moulting (Aggio and Derby, 2011; Hardege and Terschak, 2011). In order to explore whether chemical signals play a role in the courtship and mating of the Norway lobsters, male–female pairs were exposed to water from intermoult (IW in Figure 3.7) or from post-moult females (PW; Katoh, 2011). Freshly moulted females were used within 3 days after moult. These females were used to condition the water

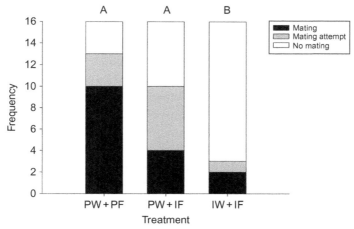

Figure 3.7 The total number of matings, mating attempts, and non-matings in *N. norvegicus* male–female pairs ($N = 16$) exposed to different treatments. PW, post-moult female water; PF, post-moult female; IW, intermoult female water; IF, intermoult female. The male was always in the intermoult stage. The letters indicate differences between the treatments. Columns with different letters are significantly different from each other with respect to the ratio between 'matings + mating attempts' and 'no-matings' ($p < 0.05$; Fisher's exact test). Different letters above column indicate significant differences ($p < 0.05$).

for 12 ± 1 h prior to staging male–female interactions in the observation tank, a glass tank ($45.5 \times 25.5 \times 25.5$ cm) filled with 30-l still water which had three sides darkened with black sheet. After the conditioning period, the female was replaced by another female which was either freshly moulted (PF in Figure 3.7; 16 trials) or intermoult (IF; 16 trials). Introduction of the second female was complemented by introduction of an intermoult male. Control experiments used an IF to condition the water (IW) which was then replaced by an IF paired with an intermoult male (16 trials; Katoh, 2011). The pairs were allowed to interact for 12 h. Behaviours were analysed by recording the presence and timing of the different elements of mating behaviours (Table 3.2). Interactions were scored as 'mating attempts' when males grabbed females from behind and tried to turn her over but did not proceed to thrusting (i.e. spermatophore transfer; Table 3.2). Interactions were scored as 'matings' when thrusting was observed.

The highest numbers of matings occurred when the water was conditioned with a freshly moulted female and interaction were staged between a post-moult female and an intermoult male (PW + PF; Figure 3.7). Out of 16 experiments, 10 pairs mated and three males attempted mating.

In the second treatment, testing intermoult mating in post-moult female conditioned water (PW + IF) 10 out of 16 males attempted to mate, with only four of these being successful. The number of matings and mating attempts was not significantly different from that of the first treatment ($p = 0.433$; Fisher's exact test). The third treatment tested whether mating takes place between an IF and male in water conditioned by an IF (IW + IF). Two of the 16 pairs mated under these conditions and there was one mating attempt. This was significantly different to the number of 'matings + mating attempts' in the first and second treatment ($p = 0.004$; Fisher's exact test; Figure 3.7). This outcome indicates that the odour of moulted female contains sex-specific substances that entice males to mate. The duration of matings was significantly longer (mean $\pm$ SEM $= 246 \pm 104$ s) when IFs were involved than between post-moult females and intermoult males (134 ± 71 s; $t = -2.6$, df $= 14$, $p = 0.022$, t-test) reflecting the observation that females were less resistant to mating when they were in post-moult than in intermoult.

Male *Nephrops* clearly were attracted to females and initiated courtship behaviour when odour from a freshly moulted female was present. Mating attempts were displayed towards post-moult female and IFs, indicating that the presence of the odour was more important than the current moult stage of the female. Odour from IFs rarely (2 out of 16 matings, 12.5% of all matings) initiated male mating attempts suggesting that the odour from post-moult females rather than from IFs contained sex pheromones. The presence of sex pheromone has been demonstrated in many other decapods crustaceans (Atema and Steinbach, 2007), including American and European lobsters and spiny lobsters (Aggio and Derby, 2011), some species of crayfish (Breithaupt, 2011), shore crabs *C. maenas* (Hardege and Terschak, 2011), blue crabs *Callinectes sapidus* (Kamio and Derby, 2011), and Caridean shrimps (Bauer, 2011). In all these examples, sex pheromones were released by the female and initiated male courtship behaviour.

It is questionable whether female *Nephrops* conduct mate choice. In shore crabs (*C. maenas*), for example, the female approaches and mates with the largest male, selecting males based on size (Sneddon et al., 2003). In blue crabs, the female is attracted by males displaying courtship 'stationary paddling' (rhythmically waving the last pair of legs; Kamio and Derby, 2011). This allows the females to assess quality of the male based on mechanical and chemical signals produced by the display (Kamio and Derby, 2011). In American lobsters and in spiny lobsters, it was shown that pre-moult females choose a mating partner by repeatedly visiting the shelter of the chosen mate (Atema and Voigt, 1995; Lipcius et al., 1983). However, similar behaviour from female *Nephrops* was not observed. In the first treatment where the female was freshly moulted, only two females managed to escape the mating attempt of the males. The vulnerable and fragile condition of the moulted females indicates that they do not have a choice whether they want to mate or not. Mating with post-moult females was significantly shorter than mating with IFs, probably due to higher resistance of the IF. Berry and Breithaupt (2010) showed that following release of female sex pheromone male signal crayfish have to overcome the resistance of the female. Female may display variation in resistance to conduct selection of particular males (Berry and Breithaupt, 2010). In *Nephrops*, it could also be that a female's willingness to mate may be higher when she is freshly moulted. Female blue crabs and female rock shrimp *Rhynchocinetes typus* demonstrated a willingness to mate by backing under the males' body to initiate pre-copulatory mate guarding (Díaz and Thiel, 2003; Jivoff and Hines, 1998). Yet, the moulted female *Nephrops* showed no such behaviour. Furthermore, males did not show any form of guarding behaviour towards the moulted females. In the experiments where the water was conditioned by a freshly moulted female, which was then replaced with an IF, only four matings were successful. Of the six males that attempted to mate, three showed repeated attempts but were not successful; in all six cases, the female managed to escape before spermatophore transfer took place (Katoh, 2011).

The source of sex pheromone in *N. norvegicus* is unknown. Where investigated in other decapods crustaceans, sex pheromones were always released in the urine (see Breithaupt and Thiel, 2011). The chemical identity of sex pheromone has been revealed as a nucleotide uridine diphosphate in the shore crab *C. maenas* (Hardege et al., 2011). Other crustacean sex pheromones have been preliminary characterized (e.g. the blue crab *C. sapidus*; Kamio and Derby, 2011). Future studies of *Nephrops* need to concentrate on identifying the source and chemical identity of the sex pheromone.

We have only begun exploring the possible existence of pheromones in *Nephrops*. It is still a long way towards a better understanding of the role that chemical signals play in the behaviour of the Norway lobster. The potential benefits of such knowledge are manifold (Thiel and Breithaupt, 2011). Chemical signals can provide efficient tools in manipulating and controlling species (e.g. by trapping invasive lampreys in the Great Lakes of the United States, Johnson et al., 2009; in insect pest management, Baker, 2011). Knowledge of the chemical identity and responses elicited by sex pheromones could benefit fisheries and aquaculture of *Nephrops*. Using concentrated female sex pheromone as bait in creels may provide a powerful and selective way of capturing males (Hardege and Terschak, 2011). Likewise, compared with other bait it would be a useful means of protecting females during the reproductive season. Sex pheromones could provide powerful tools to enhance the success of *Nephrops* aquaculture. They may be useful for selective brood-stock acquisition or to stimulate and time reproduction in captivity (Barki et al., 2011).

4. BIOLOGICAL RHYTHMS

4.1. The general concept of biological rhythms

Evolution shapes the temporal functioning of all living organisms favouring organism that adapt their behaviour to a changing world (Hochachka and Somero, 2002). Periodic habitat changes (e.g. photoperiod, temperature, tides) are driven by the rotation of the earth on its axis and in relation to its positioning respect to the sun and the moon. In response, species evolved a complex system to synchronize their biological activity within the framework of deterministic habitat changes. Such tuning is required in order to anticipate the onset of unfavourable conditions (reviewed by Naylor, 2010). As a result, all species show marked diurnal, nocturnal, crepuscular, or tide-related adaptations to their respective ecological niches (Kronfeld-Schor and Dayan, 2003).

Biological rhythms are governed by the biological clock through a three-step mechanism (Tosini and Aguzzi, 2005): (1) input pathway, (2) processing system, and (3) output pathway. The first is the sum of all the sensory processing environmental information. The second is represented by the circadian pacemaker, a group of neuronal cells that are capable to generate a self-sustained oscillation, the functioning of which can be adjusted (i.e. entrainment) to the periodic environmental fluctuations. The third compartment is represented by the rhythm itself at the level of the soma (Dunlap et al., 2004; Refinetti, 2006). The entrainment permits to distinguish endogenous (biological clock-driven) rhythms from exogenous ones (Aschoff, 1960). Endogenous rhythms persist with only slight modification of phase and period when the specimen under study is exposed to constant conditions (e.g. darkness). Such rhythms displaying a 24-h-based fluctuation are defined as circadian (from the Latin: *circa*—around and *dies*—days). Other rhythms are circatidal, when approximating the tidal periodicity (12.4 h) in their fluctuation, while we refer to ultradian or infradian when the period is shorter or longer than 24 h, respectively. In this context, the term 'masking' refers to any modification of overt rhythms that is not controlled by the pacemaker. Masking is of importance when trying to define the diurnal or nocturnal character of an animal (Mrosovsky and Hattar, 2005). In some terrestrial species (e.g. rodents), individuals show a nocturnal behaviour in the field that reverts to diurnal (i.e. locomotion peaks at subjective daytime), when animals are exposed to constant laboratory conditions.

The nature and location of the pacemaker is a current theme of investigation in many marine species. That structure is of neural nature and in vertebrates is located in the suprachiasmatic nucleus of the brain (Stephan and Zucher, 1972). In invertebrates (e.g. the fruit fly, *Drosophila melanogaster*), it is located in the brain hemispheres. In crustaceans, no master clock has been yet identified, but a model of distributed clockworks has been proposed as made by different oscillators (e.g. retinular cells, neurosecretory systems in the optic lobes; Aréchiga and Rodriguez-Sosa, 2002; Strauss and Dircksen, 2010). The circadian system of *N. norvegicus* seems to fit within this dispersed model (Aréchiga et al., 1980; Naylor, 1985).

On land, day–night cycles and seasonal photoperiod length variations are pervasive (with some exceptions such as caves). In the sea, the depth gradient and differences in water quality introduce an additional and vertical complexity to the light cycle (reviewed by Aguzzi and Company, 2010; Mercier et al., 2011). Light intensity and spectral quality strongly diminish with depth, to an extent that locally depends upon the primary productivity and overall turbidity (Herring, 2002; Jerlov, 1968; Kirk, 1996). The narrowing of the spectral diversity means that only blue light (480 nm) penetrates to deeper water. Biological clocks of marine species that inhabit a wide range of depths rely on the measurement of this wavelength, since it is the only one invariantly present all over the water column down to the twilight zone end (i.e. the theoretical depth limit of light presence). Blue light seems to be a good candidate for biological clock entrainment also in *N. norvegicus* (Aguzzi and Company, 2010) that is demonstrated to have a blue sensitive rhabdomere (Johnson et al., 2002; for more specific details on eye structure and sensibility, see Chapter 4). Furthermore, the regulation of blue light on animals' biological clocks seems to be an ancient evolutionary achievement, being present in species of several phyla (Sancar, 2003).

4.2. The burrow and the burrow emergence rhythm

Nephrops is a burrowing decapod inhabiting muddy bottoms of continental shelves and slopes of the Mediterranean and the European Atlantic (Bell et al., 2006; Sardà, 1995). Animals show a strict preference for a substratum granulometry (i.e. silt and clays) that allows the building of tunnels of precise architecture (Chapter 2). A crater-like entrance opens with an approximately 45° onto the seabed and is connected to a tubular tunnel of a diameter adjusted to the animals' body size. That tunnel ends with one or multiple ventilator shafts (Rice and Chapman, 1971). The burrow can be considered the central place during the life of *Nephrops* individuals, since animals use it as centre for the expression of a strong territorial and aggressive behaviour (Chapman and Rice, 1971). Field studies showed that some animals changed burrows, while other individuals were observed in the same burrow for several days (Chapman and Rice, 1971). The mixed scavenger/predator life habits of the species are in accordance with the strong territoriality. Animals are capable of opportunistically feeding on a wide variety of items that are found close to their burrow (Farmer, 1975; Oakley, 1979).

The mechanisms driving the daily burrow emergence rhythm of the *Nephrops* population, at different shelf and slope depths, are observed by trawling at different time of day and night. The effect of population density and territoriality on burrow emergence is still unknown (Sardà and Aguzzi, 2012). *Nephrops* seem to respond mainly to light intensity variations (Chapman et al., 1975). This would explain the depth-related temporal shift of catch rates (representing *Nephrops* activity times) from night to daytime, when moving from shallow water to greater depth, respectively (see Section 2; Bell et al., 2006). The ultimate evolutionary causes for nocturnal or low light activity of *Nephrops* are unclear and await further research. Nocturnal activity is often seen as an adaptation to escape predatory pressure from diurnal visual hunters. For example, field studies investigating predation on tethered juvenile spiny lobsters (*Panulirus argus*) showed that predation risk in open habitat increased markedly from night to daytime, while there was little difference in predation risk between day and night in sheltered lobsters (Smith and Herrnkind, 1992).

Another important parameter controlling the timing of burrow emergence is the variation in the photoperiod length throughout the year. For example, in the Western Mediterranean area, *Nephrops* catches increase in spring-summer, when animals are engaged in moulting and reproductive activities (Aguzzi et al., 2004a). Moulting takes place outside the burrow and mating (i.e. the passing of spermatophores from males to females) generally occurs when the exoskeleton of the female is still soft (Farmer, 1975; see also Section 3.2). All these processes oblige lobsters to spend more time outside the burrow, which in turn provoke an increment of their catchability (see below). Synchronization of rhythmic behaviour between individuals within a population could also be achieved by stimuli such as conspecific feeding behaviour, that in other crustaceans have been demonstrated to have an entraining effect (Naylor, 2010).

Novel scenarios of research on *Nephrops* entrainment concern the putative hydrodynamic modulation of burrow emergence. As depth increases, light fades out, especially in areas where water turbidity is elevated, so other geophysical cycles, such as periodical current flow, could modulate behavioural rhythms (Wagner et al., 2007). The Atlantic Ocean seabed is dominated by internal tides (12.4-h periodicity; Lorance and Trenkel, 2006), while in the Mediterranean Sea wind-driven inertial currents (18-h periodicity) are the only cyclic signal detected by moorings (Puig et al., 2000). Bell et al. (2008) observed an influence of tides on the catchability of *Nephrops*. In the Western Mediterranean, Aguzzi et al. (2009a) detected an 18-h patterning in the physiological activity of freshly collected animals, as indication that inertial current might control the burrow emergence behaviour. Hydrodynamic entrainment may occur via mechanoreception. The mechanoreceptors in the sensory setae, the first and second antennae, and the statocysts should be able to discriminate the direction and the strength of seabed currents (see Section 2).

There may be conflicting stimuli in those depth areas where animals are exposed to marked day–night and seabed current cycles. In the Atlantic shallow shelf areas, a tidal patterning was not reported to our best knowledge in the field or in the laboratory, as an indication that the day–night cycle may overwhelm any putative current modulation. In deeper areas, where light is almost absent and cyclical seabed currents are strong the effects on the entrainment of rhythms are still unknown. That question is currently under investigation with an appropriate actograph in the laboratory (Sbragaglia V. and Aguzzi J., unpublished data).

4.3. The laboratory-based research of locomotor activity

N. norvegicus can be considered a good model for laboratory studies of rhythmic behavioural and physiological modulation in marine species for the following reasons: it can be easily collected by trawling or creeling at any depth and a great fraction of fished animals survive capture induced stress quite well (Bergmann, 2001); it can be maintained in laboratory facilities for long periods of time (i.e. more than 1 year) if cool (12–13 °C) seawater is available; it displays a locomotor rhythmicity similarly to the diel burrow emergence pattern found in the field; finally, its burrowing behaviour is easy to track with the appropriate actographic technologies. In this sense, the species can be compared to classic model organisms in chronobiology such as the golden hamster, the mouse, and the fruit fly (Aguzzi et al., 2011a), although it is still difficult and expensive to breed Norway lobsters in the lab.

The presence of a burrow is a very important element for the structuring of *Nephrops* locomotor activity rhythm into temporally coherent bouts (Aguzzi and Company, 2010). Withdrawal into the substratum for sheltering favours the development of temporal patterns of locomotor activity. This has also been reported in the American lobster (*H. americanus*; Jury et al., 2005) and the sand-burying penaeid shrimps (e.g. *Penaeus duorarum*, *P. monodon*, and *P. semisulcatus*; Huges, 1968; Honculada-Primavera and Lebata, 1995; Moller and Jones, 1975).

Nephrops can be captured by trawl hauling only when animals have emerged (Main and Sangster, 1985; Newland and Chapman, 1989; Newland et al., 1992). They are able to immediately retire into their tunnels when residing close to the entrance (Aguzzi and Sardà, 2008). As a consequence, research has focussed on the question whether trawl data provide reliable information on their population demography. Accordingly, several laboratory trials have experimentally studied the locomotor rhythms of *Nephrops*. Standard 12–12 light and darkness photoperiod regimes have been applied using white, green (530 nm; Aréchiga and Atkinson, 1975), or blue (480 nm) monochromatic lights (Aguzzi et al., 2009b). A period of acclimation to laboratory conditions is used prior the trials to minimize the effect of any potential stress from the capture process. Specimens are fed at random times during acclimation, but they starve during experiments to avoid possible feeding entrainment (Fernández de Miguel and Aréchiga, 1994).

Results of this research provided a reliable interpretation of *Nephrops* locomotor rhythms in relation to catch fluctuations. The 24-h behavioural rhythm of *Nephrops* can be subdivided into emergence, retraction, and residence at the burrow mouth (Figure 3.8) (Aguzzi and Sardà, 2008). This latter was described as 'door-keeping', with animals guarding the tunnel entrance, with their claws projected forward (Chapter 7). The duration of each of these behavioural components depends on size and reproductive state (Aguzzi et al., 2008). Seasonal variations in sex and size structures of catches have been reported in the Atlantic and the Mediterranean (Sardà, 1995), which could be caused by variations in emergence and door-keeping between genders and size classes (reviewed by Aguzzi et al., 2004a). Apparently, berried females do not emerge with the same frequency as adults males; in fact, females are rarely captured if they carry eggs (Aguzzi and Sardà, 2008). Presently, it is still unknown if the selective capturing of larger

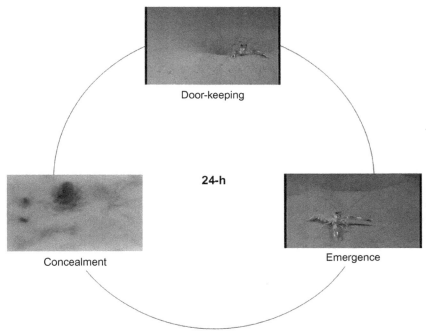

Figure 3.8 *Nephrops* images depict the behaviour of an individual portrayed at 665 m depth during the UE-funded EUROLEON survey by ROV in the Western Mediterranean. *Courtesy of Prof. M. Canals, University of Barcelona; Dr. J.B. Company, ICM-CSIC.* (For colour version of this figure, the reader is referred to the online version of this chapter.)

quantities of males in relation to females is altering the population sex-ratio and consequently behaviour, as demonstrated in other crustacean fisheries (e.g. Van Son and Thiel, 2007). Also, juveniles are captured in lower proportions than adults suggesting different emergence behaviour. Both groups (berried females and juveniles) still do not starve, since they are capable of opportunistic food retrieval (i.e. when engaged in door-keeping, and collecting what is readily available in their close proximity; Aguzzi and Sardà, 2008). Intervidual variability was recently observed in burrow emergence of *Nephrops* under simulated light cycles (Sbragaglia V. and Aguzzi J., unpublished data). Laboratory experiments results did not match with catch fluctuations in the field. Food availability and social interactions might be the major exogenous factors interplaying with the circadian modulation of burrow emergence (the latter controlled by light intensity). Moreover, the effect of predators' presence remains to be tested (see Chapter 2). Taken together, these observations add other variables to the complex scenario of endogenous and exogenous control of individuals' emergence behaviour. More systematic field studies of natural intra- and interspecific interactions using novel tracking technology are necessary to disentangle the impact of the different exogenous factors on *Nephrops*' emergence behaviour.

4.4. Our knowledge on physiological temporal patterning

Different physiological rhythms in *Nephrops* have been identified in relation to the burrow emergence behaviour. Modifications to the rate of cardiac activity, oxygen consumption, and haemolymph glucose concentration have been linked to the locomotor activity pattern (reviewed by Aguzzi and Sardà, 2008). These results pointed out the presence of two different components in the locomotor activity rhythm sustaining burrow emergence behaviour: these were door-keeping (activity at the burrow entrance related to territorial control) and active emergence (activity outside the burrow related to foraging and mating) (Chiesa et al., 2010). The light cycle could exert a masking effect on active emergence rhythm that disappears when the light stimulus is removed in laboratory constant darkness (Chiesa et al., 2010). Door-keeping and active emergence both occur during the night only in the shallower depth limits of the species distribution range (i.e. upper shelf; 10–50 m). As depth increases, the *Nephrops* locomotor pattern progressively dissociates into two components: one, door-keeping, invariantly nocturnal (as measured by constant darkness experiments in the lab) and the other, emergence, crepuscular (lower shelf; 100–200 m), or diurnal (slope; 200–400 m) (Chapman et al., 1972; Chiesa et al., 2010). Possibly, emergence is linked to light levels in order to avoid visual predators while maintaining residual light for individuals own foraging, whereas door-keeping relates to the activity of conspecifics depending on species–typical interactions.

Melatonin is an indolamine hormone that regulates the circadian cycle in animals within different phyla (Hardeland et al., 1995) by integrating the external photic information into the biology of the organism. Melatonin has been shown to be present at higher concentrations in the eyestalks of decapods, but its secretion is highly variable. In *Nephrops*, diel melatonin levels in haemolymph were studied in the laboratory, by simulating different depths (lower shelf and slope) through variable light intensity (10 and 0.1 lx, respectively) cycles. A daily increase was observed only at higher light intensity regimes (10 lx, lower shelf). Also, data indicated that levels of melatonin were two orders of magnitude higher at 10 lx than at 0.1 lx (Aguzzi et al., 2009c).

Even if melatonin is not strictly involved in the control of locomotor activity of *Nephrops* (Aguzzi et al., 2011a), its rhythmic secretion could be influenced by non-photic stimuli (periodical current flow) such as hypothesized for two species of demersal fish (Wagner et al., 2007).

4.5. The genetic control of rhythmic behaviour

The core of the biological clockwork was shown to be based on a set of genes (denominated 'clock genes') generating negative transcriptional–translational feedback loops (Dunlap et al., 2004). The clock genes have been extensively studied in the fruit fly and have been shown to regulate its circadian behaviour. Mutations in these genes cause dramatic modifications in the period and phase of activity rhythms (Peschel and Helfrich-Förster, 2011). Compared to insects very little is known about the presence and function of these genes in decapods. To date, only the gene *Clock* from the prawn *Macrobrachium rosenbergii* has been cloned (Yang et al., 2006). The characterization of the clock genes in *Nephrops* and their daily patterns of expression could speed up the research on biological rhythms in this and other marine species.

4.6. Future research insights and the new monitoring technological scenario

Studies on marine biodiversity should be linked to the concept of community functioning in the face of habitat changes through time (Smith et al., 2009). This should be carried out in a context where in the past decade, fishing effort has progressively moved to greater depths, impacting local communities in a poorly understood manner (Sheppard, 2000). The precise effects of this are unknown due to a general lack of knowledge on species distribution and their behavioural rhythms, which in turn influence our perception of local biodiversity (Aguzzi et al., 2012). The analysis of communities and biodiversity should take into account behavioural rhythms as a key parameter to fully understand the temporal dynamic.

The study of activity rhythms is of strategic importance for marine field research, especially in deep-water continental margin areas (Naylor, 2005). As the depth increases, the opportunities for conducting repeated observations decrease (Raffaelli et al., 2003). Technological limitations for direct observations are the reason for our scarce modelling capacity regarding population/stock and biodiversity assessments as well as overall ecosystem functioning in continental margin areas. The general problems with conducting repeated observations at greater depth set limitations to the study of the activities of benthic populations (Aguzzi et al., 2012). Similarly, laboratory research on activity rhythms should follow a similar concomitant technological development in order to mix information on rhythmic patterns of individuals with those of population in the field (Naylor, 2005).

In relation to the field scenario, great efforts have been devoted to *Nephrops* population assessment by video imaging (reviewed by Sardà and Aguzzi, 2012). Video surveys have been conducted as parallel fishery-independent evaluation of demography of exploited stocks, to be compared with the outcomes of trawling. The number of burrows was counted as proxy of animals' density, assuming that one burrow would account for one animal (Bell et al., 2006; Morello et al., 2007). Underwater television observations have been conducted by towed benthic sledges in middle to lower shelf areas (Chapman, 1985; Tuck and Atkinson, 1995), while stills photography with tripod cameras was carried out on the slope (Aguzzi et al., 2004b). Where populations have a more shallow distribution (e.g. in the upper Atlantic shelf), more direct observations were possible and burrows have been manually counted by scuba divers (Chapman, 1979; Chapman and Howard, 1979). To the best of our knowledge, none of these studies was performed in a temporally scheduled fashion, in order to portrait the burrow emergence in the field and to assess, at the same time, how many individuals emerged over consecutive days. The study of the behaviour of *Nephrops* in the field could benefit from the most recent technological implementations in the field of deep-water marine exploration (Sardà and Aguzzi, 2012). Remote or autonomous operated vehicles, equipped with very efficient video-imaging systems, could be used to survey at hourly frequency the same and closely parallel transects. Also, cabled video observatory technology is revealing promising applications in the long-lasting and remote study of activity rhythms in the field at several different depths of the continental margins (Aguzzi et al., 2012).

Emergence behaviour in the laboratory has usually been studied using infrared actographs (Aguzzi et al., 2008; Naylor and Atkinson, 1972) and only recently using automated video imaging (Aguzzi et al., 2009b; Menesatti et al., 2009; Figure 3.9). Also, recent technological advancements in Radio Frequency IDentification (RFID) technology allowed the expansion of locomotor studies from isolated individuals to a group of four individuals, in order to study the social modulation of burrow emergence rhythms (Aguzzi et al., 2011b). Each individual was tagged with a different geometric form on the superior part of the carapace (Figure 3.10) and tracked through video imaging. At the same time, RFID transponders were

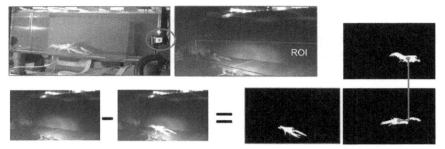

Figure 3.9 An example of actographic device used for the study of *Nephrops norvegicus* burrow emergence rhythm in the laboratory. The first two pictures in the upper row present an actograph where animals' movements are detected with a standard, low cost, mini-web camera through automated video imaging. On the left, the webcam is indicated by a grey (red) circle while on the right, the region of interest (ROI) used for the video imaging is framed by grey (red) lines. The lower row shows that subtraction of consecutive frames and binarization of images is used to extract the position of a lobster. Pictures on the right indicate displacement of the centre of the lobster body from one to the next frame. *Modified after Aguzzi et al. (2009b)—reprinted by permission of the publisher CSIC.* (For colour version of this figure, the reader is referred to the online version of this chapter.)

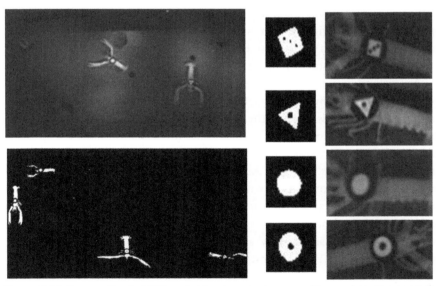

Figure 3.10 Two methods (RFID and video-imaging) are used in combination to track behavioural rhythms of four individuals of *N. norvegicus* in the same tank. On the left: (above) a video frame where dragged RFID transponders are visible as black dot behind the tails of both lobsters; (below) example of an image resulting from binarization of a video frame. On the right: the geometric tags used for the automated video imaging. *Modified after Aguzzi et al. (2011b)—reprinted by permission of the publisher MDPI.*

attached to lobsters' telson and receiving antennas at the bottom of the tank. Finally, results of both tracking methods were compared, showing similar results. In the future, these tracking technologies could be useful to study *Nephrops'* behavioural rhythms in relation to social interactions integrating different aspects of sensory biology.

5. CONCLUSION

A good understanding of the sensory biology and behaviour of an organism of such economical importance as the Norway lobster is crucial for better protection and sustainable exploitation of this species.

With respect to the sensory biology, thanks to the early work of Farmer (1973, 1974a), the morphology of the main mechano- and chemoreceptors is relatively well known and we have inferred the function of these organs by comparison with other, better studied crustaceans. This may not always be a valid comparison as the Norway lobster is adapted to its specific habitat conditions including low light, muddy sediment, and burrows with low oxygen. Our knowledge of the behavioural ecology of *Nephrops* is still very incomplete. Questions with respect to individual behaviour still await further exploration in depth studies. For example, our knowledge of the individual's activity in the field still does not go beyond anecdotal observations. Do individual defend one or several burrows? How does dominance status relate to size of territory or number of defended burrows, respectively? How far do individuals forage away from 'their' burrow(s)? How does individual behaviour change in the mating season? Do males (or females) increase activity in the mating season to find (and select) the mating partner? What are the mechanisms of mate choice in *Nephrops*? Some of the new monitoring techniques discussed above can provide insight into the behaviour of individuals by combining field and laboratory investigations. The results of such studies may eventually lead to better understanding of population parameters including sex-ratio and density/size relationships and facilitate management of natural *Nephrops* populations.

A better understanding of the chemoreceptor function and related behaviour would be important to foresee any specific threats to this species. For example, detailed investigations of the function of olfactory and gustatory receptors will be crucial to understand which specific senses *Nephrops* uses to find and select food (including bait offered in creels), to find and choose mating partners, and to avoid predators. The next step will be to identify the impact of human-caused disruption of these receptors and behaviours by ocean acidification, global warming, and pollutants including heavy metals. Olsén (2011) reviewed the effects of pollutants such as pesticides, insecticides, and heavy metals on olfactory behaviours of crustaceans. Pollutants can disrupt olfactory receptor cells, parts of the central nervous systems, and the endocrine system. Only few studies have addressed the effect of pollutants on *Nephrops* behaviour. For example, Krang and Rosenqvist (2006) showed that increased Mangan concentration interferes with food search of *Nephrops*. As a benthic predator/scavenger pollutants accumulate in the body of *Nephrops* and may cause concern for its use in the food industry. Ocean acidification may be a particular problem in the future. Recent studies revealed that a lowering of pH has sublethal effects on chemoreception and impairs olfactory discrimination, homing abilities, and predator detection in some fish (Dixson et al., 2010; Munday et al., 2009).

Nephrops rhythmic behaviours could alter our perception of population dynamic. In spite of temporal limitations of sampling repeatability, two major research lines should be developed in the next future: a suitable technology to track behaviour in the laboratory and in the field and research focussed to understand molecular biological clock functioning. Regarding tracking technology of behaviour, great progress was made in the past years. In the laboratory, video-imaging analysis has shown to be the right candidate to improve our knowledge upon different aspects of the biology and ecology of Norway lobster considering also experiment trials with more than one individual in the same tank. This methodological approach when implemented in an automated way is not time consuming and generates a great quantity of data at the same time: (1) time series of lobsters movement, (2) the possibility to reprocess data acquired (frames) focusing on different selected areas in the experimental tank, and (3) the storing of high definition videos could be of valuable help at the moment to study different behavioural traits. Using the appropriate temporal resolution will be key to successful monitoring of individuals. In the field, cabled observatories might be used to track (Aguzzi et al., 2011a) a portion of a population in the natural environment over a long period of time (many years). These techniques will help us to better understand rhythmic behaviours and to correlate this area of research to other topics such as sexual interactions, response to predators, foraging behaviour, and aggressive interactions.

As reported at the beginning of the chapter, behavioural research is a multidisciplinary field. In future studies, an integrated multidisciplinary approach will be key to elucidate still unknown aspects of *Nephrops* behaviour and ecology. Finally, as molecular sequencing technologies become less costly and more widely available (Wang et al., 2009), an important future direction will be to obtain more information upon genome and transcriptome of *Nephrops*. This will speed up our comprehension of the molecular mechanisms behind the behaviours described here.

ACKNOWLEDGEMENTS

We would like to thank the editors for inviting us to contribute to this volume; Dr. Magnus Johnson for his great support and constructive comments to the chapter; Dr. Martin Thiel for numerous valuable comments to a previous version of the manuscript that significantly improved its quality. Section 3 was part of a PhD project of E. K. and funded by Whitby Seafoods. Part of discussed findings was obtained within the framework of the RITFIM Project (CTM2010-16274; IP, J. A.). J. A. is a Postdoctoral Fellow of the *Ramón y Cajal* Program (Ministerio de Ciencia e Innovación, MICINN).

REFERENCES

Aggio, J., Derby, C.D., 2011. Chemical communication in lobsters. In: Breithaupt, T., Thiel, M. (Eds.), Chemical Communication in Crustaceans. Springer, New York, pp. 239–256.

Aguzzi, J., Company, J.B., 2010. Chronobiology of deep water continental margin decapods. Adv. Mar. Biol. Ann. Rev. 58, 155–225.

Aguzzi, J., Sardà, F., 2008. A history of recent advancements on *Nephrops norvegicus* behavioural and physiological rhythms. Rev. Fish Biol. Fish. 18, 35–48.

Aguzzi, J., Allué, R., Sardà, F., 2004a. Characterization of seasonal and diel variations in *Nephrops norvegicus* (Decapoda: Nephropidae) landings off the Catalan Coasts. Fish. Res. 69, 293–300.

Aguzzi, J., Bozzano, A., Sardà, F., 2004b. First records on *Nephrops norvegicus* (L.) burrows distributions from a deep-shelf population (100 m) in the Western Mediterranean by direct underwater survey. Crustaceana 77, 299–310.

Aguzzi, J., Sarriá, D., García, J.A., Del Rio, J., Sardà, F., Manuel, A., 2008. A new tracking system for the measurement of diel locomotor rhythms in the Norway lobster, *Nephrops norvegicus* (L.). J. Neurosci. Meth. 173, 215–224.

Aguzzi, J., Puig, P., Company, J.B., 2009a. Hydrodynamic, non-photic modulation of biorhythms in the Norway lobster, *Nephrops norvegicus* (L.). Deep-Sea Res. Pt. I 56, 366–373.

Aguzzi, J., Costa, C., Menesatti, P., García, J.A., Sardà, F., 2009b. Monochromatic blue light entrains diel activity cycles in the Norway lobster, *Nephrops norvegicus* (L.) as measured by automated video-image analysis. Sci. Mar. 73, 773–783.

Aguzzi, J., Sanchez-Pardo, J., García, J.A., Sardà, F., 2009c. Daynight and depth differences in haemolymph melatonin of the Norway lobster, *Nephrops norvegicus* (L.). Deep-Sea Res. Pt. I 56, 1894–1905.

Aguzzi, J., Company, J.B., Costa, C., Menesatti, P., Bahamon, N., Sardà, F., 2011a. Activity rhythms in the deep-sea crustacean: chronobiological challenges and potential technological scenarios. Front. Biosci. 16, 131–150.

Aguzzi, J., Sbragaglia, V., Sarriá, D., García, J.A., Costa, C., Del Río, J., Mànuel, A., Menesatti, P., Sardà, F., 2011b. A new RFID technology for the laboratory monitoring of behaviour in marine organisms. Sensors 11, 9532–9548.

Aguzzi, J., Company, J.B., Costa, C., Matabos, M., Azzurro, E., Mànuel, A., Menesatti, P., Sardà, F., Canals, M., Delory, E., Cline, D., Favali, P., Juniper, S.K., Furushima, Y., Fujiwara, Y., Chiesa, J.J., Marotta, L., Priede, I.M., 2012. Biorhythms challenge to stock and biodiversity assessments: cabled observatories video-solutions. Oceanogr. Mar. Biol. 50, 233–284.

Aréchiga, H., Atkinson, R.J.A., 1975. The eye and some effects of light on locomotor activity of Nephrops norvegicus. Mar. Biol. 32, 63–76.

Aréchiga, H., Rodriguez-Sosa, L., 2002. Distributed circadian rhythmicity in the crustacean nervous system. In: Wiese, K. (Ed.), The Crustacean Nervous System. Springer, Berlin, pp. 113–122.

Aréchiga, H., Atkinson, R.J.A., Williams, J.A., 1980. Neurohumoral basis of circadian rhythmicity in Nephrops norvegicus (L.). Mar. Behav. Physiol. 7, 185–197.

Aschoff, J., 1960. Exogenous and endogenous components in circadian rhythms. Cold Spring Harb. Symp. Quant. Biol. 25, 11–28.

Atema, J., Steinbach, M., 2007. Chemical communication and social behaviour of the lobster *Homarus americanus* and other decapods crustacean. In: Duffy, J.E., Thiel, M. (Eds.), Evolutionary Ecology of Social and Sexual Systems, Crustaceans as Model Organisms. Oxford University Press, Oxford, pp. 115–144.

Atema, J., Voigt, R., 1995. Behavior and sensory biology. In: Factor, J.R. (Ed.), Biology of the Lobster *Homarus americanus*. Academic Press, New York, pp. 313–348.

Baker, T.C., 2011. Insect pheromones: useful lessons for crustacean pheromone programs? In: Breithaupt, T., Thiel, M. (Eds.), Chemical Communication in Crustaceans. Springer, New York, pp. 531–550.

Barki, A., Jones, C., Karplus, I., 2011. Chemical communication and aquaculture of decapods crustaceans: needs, problems, and possible solutions. In: Breithaupt, T., Thiel, M. (Eds.), Chemical Communication in Crustaceans. Springer, New York, pp. 385–506.

Basil, J., Sandeman, D., 2000. Crayfish (*Cherax destructor*) use tactile cues to detect and learn topographical changes in their environment. Ethology 106, 247–259.

Bauer, R.T., 2011. Chemical communication in decapods shrimps: the influence of mating and social systems on the relative importance of olfactory and contact pheromones. In: Breithaupt, T., Thiel, M. (Eds.), Chemical Communication in Crustaceans. Springer, New York, pp. 277–296.

Bell, M.C., Redant, F., Tuck, I., 2006. Nephrops species. In: Phillips, B.F. (Ed.), Lobsters: Biology, Management, Aquaculture and Fisheries. Blackwell Publishing, Oxford, pp. 412–461.

Bell, M.C., Elson, J.M., Addison, J.T., Revill, A.S., Bevan, D., 2008. Trawl catch composition in relation to Norway lobster (*Nephrops norvegicus* L.) abundance on the farn deeps grounds, NE England. Fish. Res. 90, 128–137.

Bergmann, M., 2001. Survival of decapod crustaceans discarded in the Nephrops fishery of the Clyde Sea area, Scotland. ICES J. Mar. Sci. 58, 163–171.

Berry, F.C., Breithaupt, T., 2008. Development of behavioural and physiological assays to assess discrimination of male and female odours in crayfish, *Pacifastacus leniusculus*. Behaviour 145, 1427–1446.

Berry, F.C., Breithaupt, T., 2010. To signal or not to signal? Chemical communication by urine-borne signals mirrors sexual conflict in crayfish. BMC Biol. 8, 25.

Breithaupt, T., 2002. Sound perception in aquatic crustaceans. In: Wiese, K. (Ed.), The Crustacean Nervous System. Springer, Berlin, pp. 548–559.

Breithaupt, T., 2011. Chemical communication in crayfish. In: Breithaupt, T., Thiel, M. (Eds.), Chemical Communication in Crustaceans. Springer, New York, pp. 257–276.

Breithaupt, T., Eger, P., 2002. Urine makes the difference: chemical communication in fighting crayfish made visible. J. Exp. Biol. 205, 1221–1231.

Breithaupt, T., Tautz, J., 1990. The sensitivity of crayfish mechanoreceptors to hydrodynamic and acoustic stimuli. In: Wiese, K., Krenz, W.D., Tautz, J., Reichert, H.,

Mulloney, B. (Eds.), Frontiers in Crustacean Neurobiology. Advances in Life Sciences. Birkhäuser, Basel, pp. 114–120.

Breithaupt, T., Thiel, M., 2011. Chemical Communication in Crustaceans. Springer, New York.

Breithaupt, T., Schmitz, B., Tautz, J., 1995. Hydrodynamic orientation of crayfish (*Procambarus clarkii*) to swimming fish prey. J. Comp. Physiol. A 177, 481–491.

Briffa, M., Sneddon, L.U., 2007. Physiological constraints on contest behaviour. Funct. Ecol. 21, 627–637.

Caldwell, R.L., 1979. Cavity occupation and defensive behaviour in the stomatopod *Gonodactylus festai*: evidence for chemically mediated individual recognition. Anim. Behav. 27, 194–201.

Caldwell, R.L., 1985. A test of individual recognition in the stomatopod *Gonodactylus festae*. Anim. Behav. 33, 101–106.

Caprio, J., Derby, C.D., 2008. Aquatic animal models in the study of chemoreception. In: Basbaum, A.I., Kaneko, A., Shepherd, G.M., Westheimer, G., Firestein, S., Beauchamp, G.K. (Eds.), The senses: a comprehensive reference. Olfaction & Taste, vol. 4. Academic Press, San Diego, pp. 97–134.

Chapman, C.J., 1979. Some observations on populations of Norway lobster, *Nephrops norvegicus* (L.) using diving, television and photography. Rapports et Procés Verbaux des Reunions 175, 127–133.

Chapman, C.J., 1980. Ecology of juvenile and adult Nephrops. In: Cobb, J.S., Phillips, B.F. (Eds.), The Biology and Management of Lobsters. Academic Press, New York, pp. 143–178.

Chapman, C.J., 1985. Observing Norway lobster, Nephrops norvegicus (L.) by towed sledge fitted with photographic and television cameras. In: George, J.D., Lythgoe, G.I., Lythgoe, J.N. (Eds.), Underwater Photography and Television for Scientists. Science Publications, Oxford, pp. 100–108.

Chapman, C.J., Howard, F.G., 1979. Field observations on the emergence rhythm of the Norway lobster Nephrops norvegicus, using different methods. Mar. Biol. 51, 157–165.

Chapman, C.J., Rice, A.L., 1971. Some direct observations on the ecology and behaviour of the Norway lobster, *Nephrops norvegicus*. Mar. Biol. 10, 321–329.

Chapman, C.J., Priestley, R., Robertson, R., 1972. Observations on the diurnal activity of the Norway lobster, *Nephrops norvegicus* (L.). ICES 20, 254–269.

Chapman, C.J., Johnstone, A.D.F., Rice, A.L., 1975. The behaviour and ecology of the Norway lobster, *Nephrops norvegicus* (L.). In: Proceedings of the 9th European Marine Biology Symposium, pp. 59–74.

Chapman, C.J., Shelton, P.M.J., Shanks, A.M., Gaten, E., 2000. Survival and growth of the Norway lobster *Nephrops norvegicus* in relation to light-induced eye damage. Mar. Biol. 136, 233–241.

Chiesa, J.J., Aguzzi, J., García, J.A., Sardà, F., De la Iglesia, H., 2010. Light intensity determines temporal niche switching of behavioural activity in deep water *Nephrops norvegicus* (Crustacea: Decapoda). J. Biol. Rhythms 25, 277–287.

Christy, J.H., 1987. Competitive mating, mate choice and mating associations of Brachyuran crabs. Bull. Mar. Sci. 41, 177–191.

Copp, N.H., 1986. Dominance hierarchies in the crayfish *Procambarus clarkii* (Girard, 1852) and the question of learned individual recognition (Decapoda, Astacidea). Crustaceana 51, 9–24.

Davies, N.B., Krebs, J.R., West, S.A., 2012. An Introduction to Behavioural Ecology. Wiley-Blackwell, Oxford.

Denissenko, P., Lukaschuk, S., Breithaupt, T., 2007. Flow generated by an active olfactory system of the red swamp crayfish. J. Exp. Biol. 210, 4083–4091.

Derby, C.D., 1982. Structure and function of cuticular sensilla of the lobster *Homarus americanus*. J. Crust. Biol. 2, 1–21.

Derby, C.D., Atema, J., 1982. The function of chemo- and mechanoreceptors in lobster (*Homarus americanus*) feeding behaviour. J. Exp. Biol. 98, 317–328.

Derby, C.D., Sorensen, P.W., 2008. Neural processing, perception, and behavioural responses to natural chemical stimuli by fish and crustaceans. J. Chem. Ecol. 34, 898–914.

Devine, D.V., Atema, J., 1982. Function of chemoreceptor organs in spatial orientation of the lobster, *Homarus americanus*. Differences and overlap. Biol. Bull. 163, 144–153.

Díaz, E.R., Thiel, M., 2003. Female rock shrimp prefer dominant males. J. Mar. Biol. Assoc. U.K. 83, 941–942.

Dissanayake, A., Galloway, T.S., Jones, M.B., 2009. Physiological condition and intraspecific agonistic behaviour in *Carcinus maenas* (Crustacea: Decapoda). J. Exp. Mar. Biol. Ecol. 375, 57–63.

Dixson, D.L., Munday, P.L., Jones, G.P., 2010. Ocean acidification disrupts the innate ability of fish to detect predator olfactory cues. Ecol. Lett. 13, 68–75.

Duffy, J.E., Thiel, M., 2007. Evolutionary Ecology of Social and Sexual Systems: Crustacean as Model Organisms. Oxford University Press, New York.

Dunlap, J.C., Loros, J.J., DeCoursey, P., 2004. Chronobiology: Biological Timekeeping. Sinauer Associates Incorporated Publishing, Sunderland, Massachusetts.

Dusenbery, D.B., 1992. Sensory Ecology: How Organisms Acquire and Respond to Information. W.H. Freeman and Company, New York.

Farmer, A.S., 1973. Age and growth in *Nephrops norvegicus* (Decapoda: Nephropidae). Mar. Biol. 23, 315–325.

Farmer, A.S., 1974a. The functional morphology of the mouthparts and pereiopods of *Nephrops norvegicus* (L.) (Decapoda: Nephropidae). J. Nat. Hist. 8, 121–142.

Farmer, A.S.D., 1974b. Reproduction in *Nephrops norvegicus* (Decapoda: Nephropidae). J. Zool. 174, 161–183.

Farmer, A.S.D., 1975. Synopsis of data on the Norway lobster, *Nephrops norvegicus* (Linnaeus, 1758). FAO Fish. Synop. 112, 1–97.

Fernández de Miguel, F., Aréchiga, H., 1994. Circadian locomotor activity and its entrainment by food in the crayfish Procambarus clarkii. J. Exp. Biol. 190, 9–21.

Garm, A., Watling, L., 2013. The crustacean integument: setae, setules, and other ornamentation. In: Watling, L., Thiel, M. (Eds.), Functional Morphology and Diversity. Oxford University Press, Oxford, pp. 167–198.

Gherardi, F., Atema, J., 2005. Memory of social partners in hermit crab dominance. Ethology 111, 271–285.

Gherardi, F., Daniels, W.H., 2003. Dominance hierarchies and status recognition in the crayfish *Procambarus acutus acutus*. Can. J. Zool. 81, 1269–1281.

Gherardi, F., Tiedemann, J., 2004. Chemical cues and binary individual recognition in the hermit crab *Pagurus longicarpus*. J. Zool. 263, 23–29.

Gherardi, F., Tricarico, E., Atema, J., 2005. Unraveling the nature of individual recognition by odor in hermit crabs. J. Chem. Ecol. 31, 2877–2896.

Goodall, C.A., 1988. The sensory detection of water borne vibrational stimuli and their motor effects in the Norway lobster, *Nephrops norvegicus* (L.). PhD thesis, University of Glasgow, Faculty of Science, Glasgow.

Goodall, C., Chapman, C., Neil, D., 1990. The acoustic response threshold of the Norway lobster, Nephrops norvegicus (L.) in a free sound field. In: Wiese, K., Krenz, W.D., Tautz, J., Reichert, H., Mulloney, B. (Eds.), Frontiers in Crustacean Neurobiology. Advances in Life Sciences. Birkhäuser, Basel, pp. 106–113.

Hallberg, E., Skog, M., 2011. Chemosensory sensilla in crustaceans. In: Breithaupt, T., Thiel, M. (Eds.), Chemical Communication in Crustaceans. Springer, New York, pp. 103–121.

Hardege, J., Terschak, J.A., 2011. Identification of crustacean sex pheromones. In: Breithaupt, T., Thiel, M. (Eds.), Chemical Communication in Crustaceans. Springer, New York, pp. 373–392.

Hardege, J.D., Bartels-Hardege, H.D., Fletcher, N., Terschak, J.A., Harley, M., Smith, L., Davidson, L., Hayden, D., Müller, C.T., Lorch, M., Welham, K., Walther, T., Bublitz, R., 2011. Identification of a female sex pheromone in Carcinus maenas. Mar. Ecol. Prog. Ser. 436, 177–189.

Hardeland, R., Balzer, I., Poeggeler, B., Fuhrberg, B., Uria, H., Behrmann, G., Wolf, R., Meyer, T.J., Reiter, R.J., 1995. On the primary functions of melatonin in evolution: mediation of photoperiodic signals in a unicell, photooxidation, and scavenging of free radicals. J. Pineal Res. 18, 104–111.

Hazlett, B., 1969. "Individual" recognition and agonistic behavior in Pagurus bernhardus. Nature 222, 268–269.

Hazlett, B., 2011. Chemical cues and reducing the risk of predation. In: Breithaupt, T., Thiel, M. (Eds.), Chemical Communication in Crustaceans. Springer, New York, pp. 355–370.

Herring, P., 2002. The Biology of the Deep Ocean. Oxford University Press, Oxford.

Hochachka, P.W., Somero, G.N., 2002. Biochemical Adaptations: Mechanism and Process in Physiological Evolution. Oxford University Press, New York.

Hock, K., Huber, R., 2005. Modelling the acquisition of social rank in crayfish: winner and loser effects and self-structuring properties. Behaviour 143, 325–346.

Honculada-Primavera, J., Lebata, J., 1995. Diel activity patterns in Metapenaeus and Penaeus juveniles. Hydrobiologia 295, 295–302.

Howard, F.G., 1989. The Norway Lobster. Scottish Fisheries Information Pamphlet, 7, pp. 1–15.

Hsu, Y.Y., Wolf, L.L., 2001. The winner and loser effect: what fighting behaviours are influenced? Anim. Behav. 61, 777–786.

Huges, D.A., 1968. Factors controlling emergence of the pinks shrimp Penaeus duorarum, from the substrate. Biol. Bull. Mar. Biol. Lab., Woods Hole 134, 48–59.

Jerlov, N.G., 1968. Optical Oceanography. Elsevier, Amsterdam.

Jivoff, P., Hines, A.H., 1998. Female behaviour, sexual competition and mate guarding in the blue crab, Callinectes sapidus. Anim. Behav. 55, 589–603.

Johnson, M.E., Atema, J., 2005. The olfactory pathway for individual recognition in the American lobster Homarus americanus. J. Exp. Biol. 208, 2865–2872.

Johnson, M.L., Gaten, E., Shelton, P.M.J., 2002. Spectral sensitivities of five marine decapod crustaceans and a review of spectral sensitivity variation in relation to habitat. J. Mar. Biol. Assoc. 82, 835–842.

Johnson, N.S., Yun, S.S., Thompson, H.T., Brant, C.O., Li, W., 2009. A synthesized pheromone induces upstream movement in female sea lamprey and summons them into traps. Proc. Natl. Acad. Sci. U.S.A. 106, 1021–1026.

Jury, S.H., Chabot, C.C., Watson, W.H., 2005. Daily and circadian rhythms of locomotor activity in the American lobster (Homarus americanus). J. Exp. Mar. Biol. Ecol. 318, 61–70.

Kamio, M., Derby, C.D., 2011. Approaches to a molecular identification of sex pheromones in blue crabs. In: Breithaupt, T., Thiel, M. (Eds.), Chemical Communication in Crustaceans. Springer, New York, pp. 393–412.

Karavanich, C., Atema, J., 1998. Individual recognition and memory in lobster dominance. Anim. Behav. 56, 1553–1560.

Katoh, E., 2011. Sex, pheromone and aggression in Norway lobsters (Nephrops norvegicus): for a better future of Scampi. PhD thesis, University of Hull, Hull.

Katoh, E., Johnson, M., Breithaupt, T., 2008. Fighting behaviour and the role of urinary signals in the maintenance of dominance of Norway lobsters, Nephrops norvegicus. Behaviour 145, 1447–1464.

Kirk, J., 1996. Light and Photosynthesis in Aquatic Ecosystems. Cambridge University Press, Cambridge.

Krang, A.S., Rosenqvist, G., 2006. Effects of manganese on chemically induced food search behaviour of the Norway lobster, Nephrops norvegicus (L.). Aquat. Toxicol. 78, 284–291.

Kronfeld-Schor, N., Dayan, T., 2003. Partitioning of time as an ecological resource. Annu. Rev. Ecol. Syst. 34, 153–181.

Lee, C.L., Fielder, D.R., 1982. Agonistic behaviour and the development of dominance hierarchies in the freshwater prawn, *Macrobrachium australiense* Holthuis, 1950 (Crustacea: Palaemonidae). Behaviour 83, 1–17.

Lipcius, R.N., Edwards, M.L., Herrnkind, W.F., Waterman, S.A., 1983. In situ mating behaviour of the Spiny lobster, *Panulirus argus*. J. Crust. Biol. 3, 217–222.

Lorance, P., Trenkel, V.M., 2006. Variability in natural behaviour, and observed reactions to an ROV, by mid-slope fish species. J. Exp. Mar. Biol. Ecol. 332, 106–119.

Main, J., Sangster, G.I., 1985. The Behaviour of the Norway Lobster *Nephrops norvegicus* (L.), During Trawling. Scottish Fisheries Research Reports, Vol. 34. pp. 1–23.

McGregor, P.K., Peake, T., 2000. Communication networks: social environments for receiving and signalling behavior. Acta Ethol. 2, 71–81.

McMahon, A., Patullo, B.W., Macmillan, D.L., 2005. Exploration in a T-maze by the crayfish *Cherax destructor* suggests bilateral comparison of antennal tactile information. Biol. Bull. 208, 183–188.

McVean, A.R., 1982. Autotomy. In: Bliss, D.E. (Ed.), The Biology of the Crustacea, vol. 4. Academic Press, New York, pp. 107–132.

Menesatti, P., Aguzzi, J., Costa, C., García, J.A., Sardà, F., 2009. Video-image analysis for microcosm experiments on activity rhythms with multiple individuals of Norway lobster, *Nephrops norvegicus* (L.). J. Neurosci. Methods 184, 161–168.

Mercier, A., Sun, Z., Sandrine, B., Hamel, J.F., 2011. Lunar rhythms in the deep sea: evidence from the reproductive periodicity of several marine invertebrates. J. Biol. Rhythms 26, 82–86.

Mesterton-Gibbons, M., 1999. On the evolution of pure winner and loser effects: a game-theoretic model. Bull. Math. Biol. 61, 1151–1186.

Moller, T.H., Jones, D.A., 1975. Locomotor rhythms and burrowing habits of *Penaeus semisulcatus* (de Haan) and *P. monodon* (Fabricius) (Crustacea: Penaeidae). J. Exp. Mar. Biol. Ecol. 18, 61–77.

Moore, P.A., 2007. Agonistic behaviour in freshwater crayfish: the influence of intrinsic and extrinsic factors on aggressive encounters and dominance. In: Duffy, J.E., Thiel, M. (Eds.), Evolutionary Ecology of Social and Sexual Systems: Crustacean as Model Organisms. Oxford University Press, New York, pp. 90–114.

Morello, E.B., Froglia, C., Atkinson, R.J.A., 2007. Underwater television as a fishery-independent method for stock assessment of Norway lobster (*Nephrops norvegicus*) in the central Adriatic Sea (Italy). ICES J. Mar. Sci. 64, 1116–1123.

Mrosovsky, N., Hattar, S., 2005. Diurnal mice (Mus musculus) and other examples of temporal niche switching. J. Comp. Physiol. A Neuroethol. Sens. Neural Behav. Physiol. 191, 1011–1024.

Munday, P.L., Dixson, D.L., Donelson, J.M., Jones, G.P., Pratchett, M.S., Devitsina, G.V., Doving, K.B., 2009. Ocean acidification impairs olfactory discrimination and homing ability of a marine fish. Proc. Natl. Acad. Sci. U.S.A. 106, 1848–1852.

Naylor, E., 1985. Tidally rhythmic behavior of marine animals. Symp. Soc. Exp. Biol. 39, 63–93.

Naylor, E., 2005. Chronobiology: implications for marine resources exploitation and management. Sci. Mar. 69, 157–167.

Naylor, E., 2010. Chronobiology of Marine Organisms. Cambridge University Press, Cambridge.

Naylor, E., Atkinson, R.J., 1972. Pressure and the rhythmic behaviour of inshore marine animals. Symp. Soc. Exp. Biol. 26, 395–415.

Neil, D.M., Miyan, J.A., 1986. Phase-dependent modulation of auxiliary swimmeret muscle-activity in the equilibrium reactions of the Norway lobster, *Nephrops norvegicus* L. J. Exp. Biol. 126, 157–179.

Neil, D.M., Wotherspoon, R.M., 1982. Structural specializations, fluid-flow and angular sensitivity in the statocyst of the lobster *Nephrops norvegicus*. J. Physiol. (Lond.) 329, 26–27.

Newland, P.L., Chapman, C.J., 1989. The swimming and orientation behaviour of the Norway lobster, *Nephrops norvegicus* (L.), in relation to trawling. Fish. Res. 8, 63–80.

Newland, P.L., Neil, D.M., 1987. Statocyst control of uropod righting reactions in different planes of body tilt in the Norway lobster, *Nephrops norvegicus*. J. Exp. Biol. 131, 301–321.

Newland, P.L., Neil, D.M., 1990. The tail flip of the Norway lobster, *Nephrops norvegicus*. 2. Dynamic righting reactions induced by body tilt. J. Comp. Physiol. A 166, 529–536.

Newland, P.L., Neil, D.M., Chapman, C.J., 1992. Escape swimming in the Norway lobster. J. Crust. Biol. 12, 342–353.

Norman, C.P., Jones, M.B., 1991. Limb loss and its effect on handedness and growth in the velvet swimming crab *Necora puber* (Brachyura: Portunidae). J. Nat. Hist. 25, 639–645.

Oakley, S.G., 1979. Diurnal and seasonal changes in the timing of peak catches of Nephrops norvegicus reflecting changes in behaviour. In: Naylor, E., Hartnoll, R.G. (Eds.), Cyclical Phenomena in Marine Plants and Animals. Pergamon Press, Oxford, pp. 367–373.

Olsén, K.H., 2011. Effects of pollutants on olfactory mediated behaviours in fish and crustaceans. In: Breithaupt, T., Thiel, M. (Eds.), Chemical Communication in Crustaceans. Springer, New York, pp. 507–529.

Parker, G.A., 1974. Assessment strategy and the evolution of fighting behavior. J. Theor. Biol. 47, 223–243.

Peschel, N., Helfrich-Förster, C., 2011. Setting the clock—by nature: circadian rhythm in the fruitfly Drosophila melanogaster. FEBS Lett. 585, 1435–1442.

Popper, A.N., Salmon, M., Horch, K.W., 2001. Acoustic detection and communication by decapod crustaceans. J. Comp. Physiol. A: Sens. Neural Behav. Physiol. 187, 83–89.

Puig, P., Palanques, A., Guillén, J., García-Ladona, E., 2000. Deepslope currents and suspended particle fluxes in and around the Foix submarine canyon (NW Mediterranean). Deep-Sea Res. Pt. I 47, 343–366.

Raffaelli, D., Bell, E., Weithoff, G., Matsumoto, A., Cruz-Motta, J.J., Kershaw, P., Parker, R., Parry, D., Jones, M., 2003. The ups and downs of benthic ecology: considerations of scale, heterogeneity and surveillance for benthic-pelagic coupling. J. Exp. Mar. Biol. Ecol. 285–286, 191–203.

Reeder, P.B., Ache, B.W., 1980. Chemotaxis in the Florida spiny lobster *Panulirus argus*. Anim. Behav. 28, 831–839.

Refinetti, R., 2006. Circadian Physiology. Francis and Taylor, New York, 667 pp.

Rice, A.L., Chapman, C.J., 1971. Observations on the burrows and burrowing behaviour of two mud-dwelling decapod crustaceans, *Nephrops norvegicus* and Goneplax rhomboides. Mar. Biol. 10, 330–342.

Sancar, A., 2003. Structure and function of DNA photolyase and cryptochrome blue-light photoreceptors. Chem. Rev. 103, 2203–2237.

Sandeman, D.C., 1989. Physical properties, sensory receptors and tactile reflexes of the antenna of the Australian freshwater crayfish *Cherax destructor*. J. Exp. Biol. 141, 197–217.

Sardà, F., 1995. A review (1967–1990) of some aspects of the life history of *Nephrops norvegicus*. ICES Mar. Sci. Symp. 199, 78–88.

Sardà, F., Aguzzi, J., 2012. A review of burrow counting as an alternative to other typical methods of assessment of Norway lobster populations. Rev. Fish Biol. Fish. 22, 409–422.

Schmidt, M., Mellon, D., 2011. Neuronal processing of chemical information in crustaceans. In: Breithaupt, T., Thiel, M. (Eds.), Chemical Communication in Crustaceans. Springer, New York, pp. 123–147.

Schmitt, B.C., Ache, B.W., 1979. Olfaction: responses of a decapod crustacean are enhanced by flicking. Science 205, 204–206.

Shelton, P.M.J., Gaten, E., Chapman, C.J., 1985. Light and retinal damage in *Nephrops norvegigus* (L.). Proc. R. Soc. Lond. B 226, 217–236.

Sheppard, C., 2000. Seas at the millennium: an environmental evaluation. Global Issues and Processes, vol. 3. Pergamon Press, Amsterdam and Oxford.

Simon, T.W., Edwards, D.H., 1990. Light-evoked walking in crayfish: behavioural and neuronal responses triggered by the caudal photoreceptor. J. Comp. Physiol. A 166, 745–755.

Skog, M., 2009. Male but not female olfaction is crucial for intermoult mating in European lobsters (*Homarus gammarus* L.). Chem. Senses 34, 159–169.

Smith, K.N., Herrnkind, W.F., 1992. Predation on early juvenile spiny lobsters Panulirus argus (Latreille): influence of size and shelter. J. Exp. Mar. Biol. Ecol. 157, 3–18.

Smith, L.D., Hines, A.H., 1991. The effect of cheliped loss on blue crab *Callinectes sapidus* Rathbun foraging rate on soft shell clams *Mya arenaria*. J. Exp. Mar. Biol. Ecol. 151, 245–256.

Smith, K.L., Ruhl, H.A., Bett, B.J., Billett, D.S.M., Lampitt, R.S., Kaufmann, R.S., 2009. Climate, carbon cycling, and deep-ocean ecosystems. Proc. Natl. Acad. Sci. 106, 19211–19218.

Sneddon, L.U., Huntingford, F.A., Taylor, A.C., Clare, A.S., 2003. Females sex pheromone-mediated effects on behaviour and consequence of male competition in the shore crab (*Carcinus maenas*). J. Chem. Ecol. 29, 55–68.

Stephan, F.K., Zucher, I., 1972. Circadian rhythms in drinking behaviour and locomotor activity of rats are eliminated by hypothalamic lesions. Proc. Natl. Acad. Sci. U.S.A. 69, 1583–1586.

Strauss, J., Dircksen, H., 2010. Circadian clocks in crustaceans: identified neuronal and cellular systems. Front. Biosci. 15, 1040–1074.

Streiff, R., Mira, S., Castro, M., Cancela, M.L., 2004. Multiple paternity in Norway Lobster (*Nephrops norvegicus* L.) assessed with microsatellite markers. Marine Biotechnol. 6, 60–66.

Tautz, J., Plummer, M.R., 1994. Comparison of directional selectivity in identified spiking and nonspiking mechanosensory neurons in the crayfish *Orconectes limosus*. Proc. Natl. Acad. Sci. U.S.A. 91, 5853–5857.

Tautz, J., Masters, W.M., Aicher, B., Markl, H., 1981. A new type of water vibration receptor on the crayfish antenna 1. Sensory physiology. J. Comp. Physiol. 144, 533–541.

Taylor, R.C., 1967. The anatomy and adequate stimulation of a chordotonal organ in the antennae of a hermit crab. Comp. Biochem. Physiol. 20, 709–717.

Thiel, M., Breithaupt, T., 2011. Chemical communication in crustaceans: research challenges for the twenty-first century. In: Breithaupt, T., Thiel, M. (Eds.), Chemical Communication in Crustaceans. Springer, New York, pp. 3–22.

Thiel, M., Lovrich, G.A., 2011. Agonistic behaviour and reproductive biology of squat lobsters. In: Poore, G.C.B., Ahyong, S.T., Taylor, J. (Eds.), The Biology of Squat Lobsters. CSIRO Publishing/CRC Press, Melbourne/Boca Raton, pp. 223–247.

Tinbergen, N., 1963. On aims and methods of ethology. Z. Tierpsychol. 20, 410–433.

Tosini, G., Aguzzi, J., 2005. Effects of space flight on circadian rhythms. In: Sonnenfeld, G. (Ed.), Experimentation with the Animal Model in Space, vol. 10. Elsevier, Amsterdam, pp. 165–174.

Tuck, I.D., Atkinson, R.J.A., 1995. Unidentified burrow surface trace from the Clyde Sea area. Neth. J. Sea Res. 34, 331–335.

Van Son, T.C., Thiel, M., 2007. Anthropogenic stressors and their effects on the behavior of aquatic crustaceans. In: Duffy, J.E., Thiel, M. (Eds.), Evolutionary Ecology of Social and

Sexual Systems, Crustaceans as Model Organisms. Oxford University Press, Oxford, pp. 413–441.

Wagner, H.J., Kemp, K., Mattheus, U., Priede, I.G., 2007. Rhythms at the bottom of the deep-sea: cyclic current flow changes and melatonin patterns in two species o demersal fish. Deep-Sea Res. Pt. I 54, 1944–1956.

Wang, Z., Gerstein, M., Snyder, M., 2009. RNA-Seq: a revolutionary tool for transcriptomics. Nat. Rev. Genet. 10, 57–63.

Weissburg, M.J., 2011. Waterborne chemical communication: stimulus dispersal dynamics and orientation strategies in crustaceans. In: Breithaupt, T., Thiel, M. (Eds.), Chemical Communication in Crustaceans. Springer, New York, pp. 63–83.

Wiese, K., 1976. Mechanoreceptors for near-field water displacements in crayfish. J. Neurophysiol. 39, 816–833.

Wine, J.J., 1984. The structural basis of an innate behavioural pattern. J. Exp. Biol. 112, 283–319.

Winston, M.L., Jacobson, S., 1978. Dominance and effects of strange conspecifics on aggressive interactions in hermit crab Pagurus longicarpus (Say). Anim. Behav. 26, 184–191.

Yang, J., Dai, Z., Yang, F., Yang, W., 2006. Molecular cloning of Clock cDNA from the prawn Macrobrachium rosenbergii. Brain Res. 1067, 13–24.

CHAPTER FOUR

The Reniform Reflecting Superposition Compound Eyes of *Nephrops norvegicus*: Optics, Susceptibility to Light-Induced Damage, Electrophysiology and a Ray Tracing Model

Edward Gaten[*], **Steve Moss**[†], **Magnus L. Johnson**[‡,1]

[*]Biology Department, University of Leicester, Leicester, United Kingdom
[†]School of Biological, Biomedical and Environmental Sciences, University of Hull, Hull, United Kingdom
[‡]Centre for Environmental and Marine Sciences, University of Hull, Scarborough, United Kingdom
[1]Corresponding author: e-mail address: m.johnson@hull.ac.uk

Contents

Abstract

The large reniform eyes of the reptant, tube-dwelling decapod *Nephrops norvegicus* are described in detail. Optically these reflecting superposition compound eyes are a little unusual in that they are laterally flattened, a feature that may enhance their sensitivity in that region, albeit at the expense of resolution. Electrophysiological and anatomical investigations suggest that the eyes are tuned to appropriate spectral and temporal sensitivities in the long and short term through movement of proximal pigments and possibly rhabdom adaptation. Although exposure to ambient surface light intensities is shown to cause damage to the retinal layer, especially in deeper living animals, there is no evidence yet that demonstrates an impact of eye damage on their survival. It is suggested that experimentation on marine decapods, with sensitive eyes, requires that particular attention is paid to their light environment.

Advances in Marine Biology, Volume 64
ISSN 0065-2881
http://dx.doi.org/10.1016/B978-0-12-410466-2.00004-2

© 2013 Elsevier Ltd.
All rights reserved.

Keywords: Invertebrate vision, Dim environments, Visual ecology, Discard injury, Visual physiology

1. INTRODUCTION

All large crustaceans have compound eyes. The apposition type is the simplest form of such eyes and is common in all primitive Crustacea and some brachyuran crabs, amphipods and isopods. In this type of eye, each photoreceptive rhabdom has a single aperture through which light enters such that the scene that the animal observes is simply a matrix of 'pixels', one for each facet (Figure 4.1A). Resolution at the level of individual receptors in this case is determined by the angular separation of the ommatidia. Generally, apposition eyes are limited in low light conditions by the small aperture through which light passes to reach the rhabdom (Land and Nilsson, 2002). In compound eyes, different solutions have appeared which serve to increase the absolute sensitivity of the eye. Most crustaceans living in

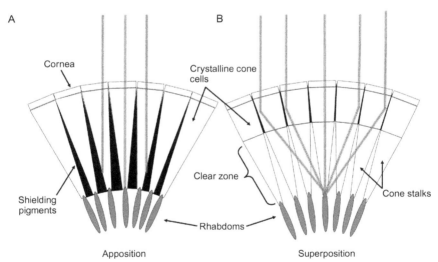

Figure 4.1 (A) Schematic diagram of a cross section through several ommatidia of an apposition compound eye as found in many arthropods. Light from a particular direction will only be detected by the photoreceptive rhabdom if it passes through the axial facet. (B) A schematic demonstration of the principle of a reflecting superposition compound eye as found in *Nephrops* and many other long bodied decapods. Parallel light is reflected by the multilayer reflectors along the edges of the crystalline cone cells towards the axial rhabdom, thus several facets redirect light to the appropriate photoreceptor.

low light conditions use superposition optics, in which light from one direction is focussed onto the target rhabdom via a large number of facets, resulting in an increase in sensitivity of up to three orders of magnitude (Figure 4.1B). Refracting superposition optics are used in various crustacean taxa while parabolic superposition optics have been described in some hermit crabs and swimming crabs (Nilsson, 1989). However, the majority of decapod crustaceans have reflecting superposition eyes which can easily be identified by the fact that they have square corneal facets arranged in a lattice and a distinct clear zone between the photoreceptive rhabdom layer and the dioptric layer (Gaten, 1998; Land, 1976; Vogt, 1975).

The large eyes of *Nephrops norvegicus* (hereafter referred to as *Nephrops*) are arguably the most recognisable feature of the animal, resulting in its generic name, derived from the Greek *nephros* (kidney) and *ops* (eye). The eyes of *Nephrops* have attracted the attention of researchers because of their size which facilitates anatomical and physiological investigations and, in Europe, their ease of availability as they are a commercially targeted species.

Light intensity affects the ecology and behaviour of this species and the ways in which the animals are commercially exploited. However, as these are covered fully in this volume elsewhere we will concentrate on physiological and anatomical aspects of *Nephrops* vision.

2. EYE STRUCTURE

The eye of *Nephrops* has been described in detail (Aréchiga and Atkinson, 1975; Loew, 1976); the following account is a brief description of this large and unusual example of a reflecting superposition eye. The eye is one of the largest of all crustacean eyes measuring up to 10 mm along the antero–posterior axis, and it is borne on a short moveable eyestalk. The eye is atypical among decapods with superposition optics in its lack of spherical symmetry, the distinctive kidney shaped eye being oval when viewed from the lateral aspect. More than 10,000 ommatidia are present in the eye with around 3000 of these displaying eyeshine in the dark-adapted eye (Figure 4.2A). The corneal facets are square except for a small region close to the dorsal margin of the eye where the facets are hexagonal. This was formerly described as a growth zone where newly formed hexagonally packed ommatidia develop and mature (Parker, 1890; Shelton et al., 1981). However, this region uses apposition optics and more likely has a special function in relation to viewing the brighter downwelling light (Gaten, 1994; Tokarski and Hafner, 1984).

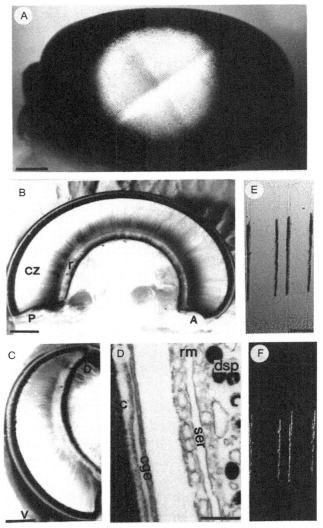

Figure 4.2 (A) Lateral view of a dark-adapted eye showing oval shape, square facets and brilliant eyeshine. (B) Longitudinal section through a fixed eye to show the increased length of ommatidia anteriorly (A) and posteriorly (P). The clear zone (cz) extends from the rhabdom layer (r) to the cone layer (marked by the dark distal shielding pigment). (C) In this transverse section of the eye, the gradual increase in ommatidial length from dorsal (D) to ventral (V) can be seen. (D) Electron micrograph of the reflecting multilayer (rm) in a transverse section of the mid-cone region. The reflecting layers are separated by smooth endoplasmic reticulum (ser) and cytoplasm. c, crystalline cone; cce, cor-neagenous cell extension; dsp, shielding pigment granules. (E) Phase contrast micro-graph showing shielding pigment in the distal pigment cells between the crystalline cones. (F) Polarising micrograph of the same section showing the birefringent reflecting multilayer in the distal pigment cells. Scale bars: A, B, C = 1.0 mm; D = 1.0; and E, F = 50 μm.

Sections through the eye reveal distal crystalline cone cell and proximal retinula cell layers, separated by a clear zone (Figure 4.2B). In longitudinal sections of the eye, the longest ommatidia are anterior and posterior, some 20% longer than those from the centre of the eye (Figure 4.2B). The lengths of the ommatidia increase from dorsal to ventral (Figure 4.2C).

The ommatidial structure is typical of a reflecting superposition eye (Land and Nilsson, 2002), consisting of a distal dioptric apparatus (for focusing the light) and a proximal retinula cell layer where the image is formed. Distally, square corneal facets, up to 60 µm across, are secreted by two corneagenous cells. Underlying the externally visible facet there are four cone cells that secrete a central crystalline cone. Proximally four cone cell processes cross the clear zone as a quadripartite cone stalk (or crystalline tract). The cones are separated by distal pigment cells which contain both reflecting pigment and dark shielding pigment (Figure 4.2D). In phase contrast micrographs of longitudinal sections the distal shielding pigment can be seen surrounding the middle 2/3 of the cone (Figure 4.2E). The reflecting pigment is birefringent as is revealed using crossed polaroids. It extends approximately 1/2 the length of the cones (Figure 4.2F). The reflecting pigment granules are arranged in the form of a reflecting multilayer (Land, 1972) surrounding the sides of each cone and can be seen in transverse sections of the cones (Figure 4.2D). The distal pigment cells do not extend beyond the proximal end of the cone, except in light-damaged eyes (Gaten, 1988), and neither these cells nor the pigment granules move in response to light (Shelton et al., 1986).

There are eight retinula cells in each ommatidium and each retinula cell contributes microvilli to a central, light-sensitive spindle-shaped rhabdom (Figure 4.3A). In cross section, the distal part of the rhabdom is surrounded by the four lobes of retinula cell 8, each contributing microvilli that make up the distal part of the rhabdom (Figure 4.3C). The lobes of R8 are in turn surrounded by the cell bodies of retinula cells 1–7 (Figure 4.3C). The cone stalk separates to form four separate cone cell processes (Figure 4.4B) each of which lies at one corner of the rhabdom (Figure 4.3C). The major part of the spindle-shaped rhabdom is formed by the remaining retinula cells (R1–R7). The proximal rhabdom appears banded due to the alternating layers of orthogonally orientated microvilli (Figure 4.3D). The retinula cell nuclei are located just distal to the rhabdoms. Below the rhabdom, the retinula cells form axons that extend proximally through the basement membrane to the first optic neuropil.

A tapetum is present behind the rhabdoms. It is formed by reflecting pigment cells, each of which extends over several ommatidia. The reflecting

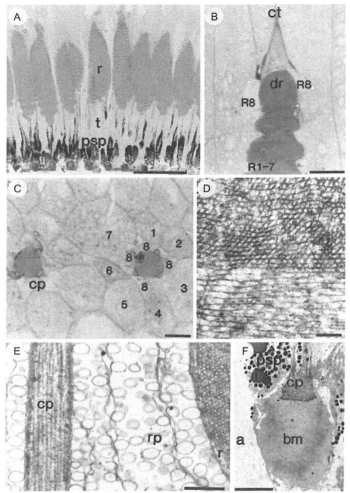

Figure 4.3 (A) Light micrograph of a dark-adapted eye showing fusiform rhabdoms (r), backed by tapetal cells (t). The proximal shielding pigment (psp) is withdrawn behind the tapetal cells, close to the basement membrane. (B) In this light micrograph, two of the lobes of retinula cell 8 (R8) can be seen either side of the distal rhabdom (dr) just below where the crystalline tract (ct) divides. The cytoplasm of R8 is noticeably free of cellular inclusions, compared to the surrounding retinula cells (R1–R7). (C) Light microscopy of this transverse section at the distal rhabdom level shows the rhabdoms surrounded by four lobes of cell R8 and by the retinula cells (R1–R7). The cone cell processes (cp) lie at the corners of the distal rhabdoms. (D) The banded appearance of the main rhabdom, seen in this electron micrograph of a longitudinal section, is due to orthogonally arranged layers of microvilli. (E) The tapetal cells are packed with reflecting pigment granules (rp) which surround the rhabdom (r) closely in places. A cone process (cp) can be seen in this longitudinal section passing proximally between the tapetal cells. (F) The retinula cell axons (a) pass through the basement membrane (bm). The cone cell processes (cp) rejoin and anchor to the basement membrane. In this longitudinal section of a dark-adapted ommatidium the proximal shielding pigment (psp) is close to the basement membrane. Scale bars: A = 100; B, C = 10; D = 0.2; E = 1.0; and F = 5 μm.

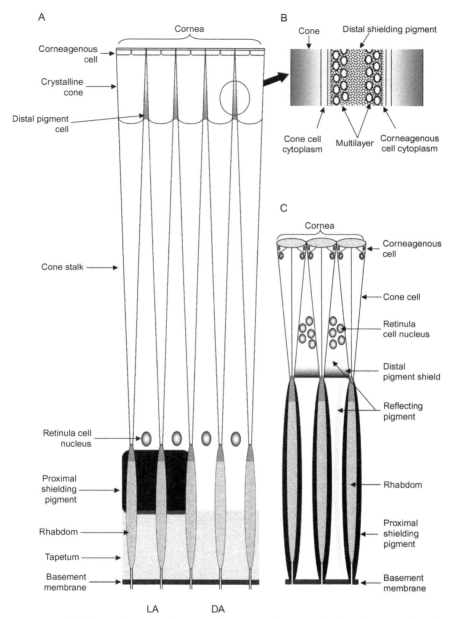

Figure 4.4 (A) Semi-schematic diagram of a group of ommatidia from the eye of *Nephrops*, shown in the light-adapted (LA) and dark-adapted (DA) states. The main difference between the two adaptional states is that the proximal shielding pigment migrates from around the basement membrane to a position higher up the rhabdom layer. (B) An enlarged view of part of a distal pigment cell between two crystalline cones. Distal shielding pigment granules and a reflecting multilayer are present within the distal pigment cells. (C) Semi-schematic diagram of a first zoeal eye demonstrating that at this stage it resembles an apposition type eye.

pigment granules are spherical, about 0.4 μm in diameter and densely packed within the tapetal cells (Figure 4.3E). The basement membrane is formed of a layered collagenous sheet with the retinula cell axons passing through regularly spaced holes (Figure 4.3F).

Light-induced migration of retinula cell screening pigments has been observed in most decapod species (Kleinholz, 1961; Land, 1981; Parker, 1932). Some migration of proximal shielding pigment occurs in response to light in *Nephrops*, although the distal shielding pigment and the reflecting pigment do not migrate during light adaptation (Shelton et al., 1986). Shielding pigment granules are present in the retinula cells, with the numbers of granules being related to the depth at which the animals are caught (Gaten et al., 1990). The pigment is present close to the basement membrane in dark-adapted eyes (Figure 4.4A) and it migrates distally to shield the rhabdom during light adaptation (Shelton et al., 1986). The extent and speed of the migration is independent of the light levels although the normal fully dark- and light-adapted positions also vary with the habitat depth (Gaten et al., 1990).

3. EYE DEVELOPMENT

Nephrops eggs are laid in the autumn and attached to the pleopods (Chapter 6) where they remain throughout development (Farmer, 1975). After the first few divisions, development may be arrested for up to 7 months. After recommencement of development in the spring, the embryos mature rapidly. There are usually two or three planktotrophic zoeal stages.

The eyes of the larval animals can generally be regarded as of the apposition type (Figure 4.4C). At any stage in the embryonic development of the eye of *Nephrops* a gradient of increasing ommatidial maturity can be seen from posterior to anterior (Figure 4.5A), so the temporal sequence of developmental stages can therefore be displayed spatially in a longitudinal section of the eye. The monolayer of epidermal cells merges into eye tissue at the posterior margin of the eye. As the depth of the retina increases, each cell becomes elongated and appears to extend from the cornea to the basement membrane (Figure 4.5B).

The next region, moving anteriorly, is where cell differentiation occurs. Distal invaginations are seen occasionally in the region where clusters of retinula cells sink away from the cornea (Figure 4.5C). The retinula cell clusters change shape during development, with cross sections of the early retinula cell clusters showing that initially they are arranged in a square pattern

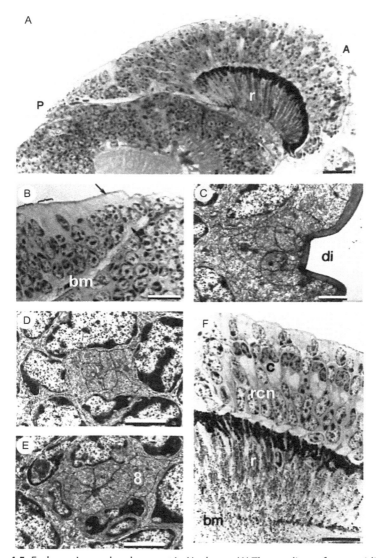

Figure 4.5 Embryonic eye development in *Nephrops*. (A) The gradient of ommatidial size from posterior (P) to anterior (A) can be seen in this light micrograph. (B) At the posterior eye margin, epidermal cells become elongate (arrowed) and appear to extend from the cornea to the basement membrane (bm). Anterior to this is the region where differentiation of the cells and rhabdom formation occurs (bracketed). (C) Electron micrograph of the region of differentiation. Distal invaginations (di) of the cornea occur where clusters of retinula cells sink away from the cornea. (D) The square clusters of retinula cells are found in the most posterior part of the region marked with a bracket in B. (E) The seven main retinula cells become wedge-shaped in transverse section, eventually interdigitating in the centre to form a presumptive rhabdom. The retinula cell R8 is positioned to one side of the main cluster. (F) The most mature (anterior) ommatidia found in the embryos are similar in structure to those seen in the larval eye. The distal cone cell layer is largely filled with cone cells (c) and retinula cell nuclei (rcn). The rhabdom (r) layer is heavily pigmented from the level of the retinula cell nuclei down to the basement membrane (bm). Scale bars: A = 200; B = 100; C, D, E = 2.0; and F = 100 μm.

(Figure 4.5D). The eighth retinula cell is always positioned posterior to the other seven cells. The latter are arranged in three rows and become progressively wedge-shaped, eventually meeting centrally (Figure 4.5E). In the most mature ommatidia, rudimentary microvilli can be seen interdigitating in the centre of the cluster to form a presumptive rhabdom.

As the retinula cells move proximally, clusters of cone cells and corneagenous cells are found below the cornea. By the time clear ommatidia are recognisable, the cone cells have also sunk away from the cornea (Figure 4.5F). However, the distal extension of the cone cells always retains contact with the cornea between the corneagenous cells.

In the most anterior ommatidia of the mature embryo the cone and corneagenous cells are arranged as in the first zoea. The unpigmented cone cell layer and the heavily pigmented rhabdom layer each occupy half of the length of the ommatidia. Some shielding pigment can be seen in the retinula cells and occasional strings of larger pigment granules are present between the retinula cells. Groups of eight axons can be seen passing through the basement membrane.

Embryonic development of the eyes of *Nephrops* closely follows the pattern seen in the lobster *Homarus americanus* (Hafner and Tokarski, 2001) and crayfish *Procambarus clarkia* (Hafner and Tokarski, 1998). Harzsch and Hafner (2006) concluded that a conserved ontogenetic process of retinal development in the Tetraconata was supported by the similarities observed in several crustacean species.

The eye of the first zoea (Figure 4.6A) is a typical larval apposition eye with the cone closely apposed to the thin rhabdom. It is similar to those found in the other larval crustaceans (Nilsson, 1983). The cornea consists of circular biconvex facet lenses, arranged hexagonally. In each ommatidium, two corneagenous cells and four cone cells are grouped together in an inverted cone shape beneath the facet (Figure 4.7A). There is no crystalline material contained within the cone cells. The proximal ends of the cone cells abut the distal end of the rhabdom.

The area between the cones contains the retinula cell nuclei, although the retinula cell cytoplasm is mostly within the rhabdom layer. Proximal shielding pigment is distributed throughout the retinula cells and is visible alongside the rhabdoms (Figure 4.6B). The top of the rhabdom layer is marked by an extensive pigment shield consisting of larger pigment grains than those found between the rhabdoms (Figure 4.5B). A layer of reflecting pigment cells is also found here, overlaying the shielding pigment. The processes of these reflecting pigment cells extend proximally between the rhabdoms.

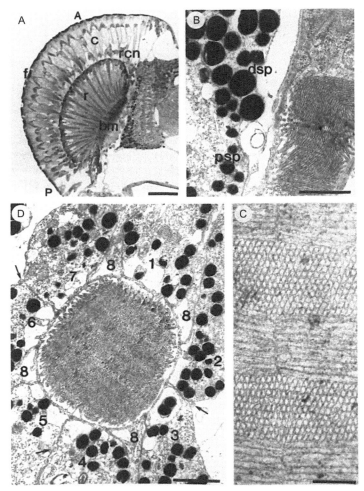

Figure 4.6 Eye structure in the first zoea of *Nephrops*. (A) Light micrograph of a longitudinal section through the eye showing the gradient of ommatidial size from posterior (P) to anterior (A). bm, basement membrane; c, cone cell; f, corneal facet; rcn, retinula cell nuclei; r, rhabdoms. (B) Electron micrograph of the region where the cone cells abut the rhabdom. The difference in size can be seen between the proximal shielding pigment (psp) that surrounds the rhabdoms, and the large shielding pigment grains (lsp) that form the pigment shield above the rhabdom layer. (C) Electron micrograph of a rhabdom in LS showing alternating layers of orthogonally orientated microvilli as seen over most of the length of the rhabdom. (D) Just proximal to the distal rhabdom, seven large retinula cells (1–7) and four lobes of R8 (8) are found around the rhabdom. Cone cell processes (arrowed) extend down to the basement membrane between the retinula cells. The rhabdom is surrounded by a pallisade of expanded cisternae of smooth endoplasmic reticulum. Scale bars: A = 100 μm; B = 1 μm; C = 10 μm; D = 1 μm.

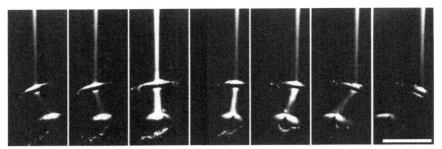

Figure 4.7 A narrow beam of light is redirected at the crystalline cone layer to a super-position focus. The eye has been rephotographed at intervals as the beam is traversed across the eye. The light is redirected to the same region of the retinula cell layer. The beam is made visible by the addition of fluorescein to the seawater. Scale bar = 2 mm.

The rhabdoms are thin, cylindrical and extend almost down to the basement membrane. They are formed by the interlocking rhabdomeres of seven retinula cells over most of their length (R1–R7). The rhabdoms have a banded appearance over most of their length due to the orthogonal layering of the microvilli (Figure 4.6C). At the distal end of the rhabdom, R8 contributes a small rhabdomere in which no layering is seen (Figure 4.6B). In cross section, four lobes of the R8 cell body can be seen between the cell bodies of the regular retinula cells (Figure 4.6D). The cytoplasm of R8 is free of shielding pigment and most other cellular inclusions (Figures 4.6D). The rhabdom is surrounded by cisternae of smooth endoplasmic reticulum, which form a pallisade (Figure 4.6D). The pallisade constitutes a low refractive index layer around the rhabdom, which may cause the rhabdom to act as a light-guide (Horridge and Barnard, 1965) or may function to keep scattering and absorbing particles away from the rhabdom (D.G. Stavenga, personal communication). Groups of eight retinula cell axons penetrate the basement membrane below the rhabdoms.

4. OPTICS

Since Exner's (1891) work on optical mechanisms in insects and crustaceans, compound eyes have been routinely divided into apposition and superposition eyes. The existence of superposition optics was brought into question when Kuiper (1962) found that crayfish cones were of a low, more or less constant refractive index. This rendered them incapable of functioning as lens cylinders and thus apparently incapable of contributing to superposition image formation. Several authors subsequently questioned whether the superposition mechanism existed (reviewed by Horridge, 1975). The discovery of reflecting superposition optics (Land, 1976; Vogt, 1975) resolved the theoretical problems of image formation by homogeneous cones of constant refractive index. Image formation in superposition eyes has now been demonstrated in several species.

A superposition image is clearly formed by the eyes of *Nephrops* as can be demonstrated using a modification of the technique of Land et al. (1979). A reducing telescope is used to direct a fine beam of light onto the cornea, and the path of this beam inside the eye observed through a small hole cut in the cornea. The redirection of a series of parallel rays within the cone cell layer of a dark-adapted *Nephrops* eye is shown in Figure 4.7. When these rays are superimposed onto a diagram of the eye the focus can be seen to occur in the rhabdom layer (Figure 4.8). Spherical aberration is seen in those rays furthest from the optical axis, resulting in these rays being focused more distally than those close to the optical axis.

Eyeshine is a characteristic seen in most superposition eyes. It occurs when light is reflected back out of the eye by a tapetum without being absorbed during its passage through the rhabdoms (Figure 4.2A). In *Nephrops*, the tapetum acts as a diffuse reflector, made up of reflecting pigment granules, and it enhances sensitivity by effectively increasing the length of the rhabdom. Eyeshine observations are of optical importance as the diameter of the eyeshine patch at the cornea represents the effective aperture of the superposition eye (Kunze, 1979).

In many crustacean superposition eyes, the distal pigment acts as an iris (Stavenga, 1979), reducing the diameter of the eyeshine patch during light adaptation through pigment migration. Upon illumination of a dark-adapted eye of *Nephrops*, the width of the eyeshine patch remains relatively constant, because of its non-migrating distal pigment, but the eyeshine brightness decreases within a few minutes, due to proximal pigment obscuring the tapetum (Figure 4.9).

Regional variations in the eyeshine of *Nephrops* are seen, due to two factors already mentioned: the kidney shape of the eye and the presence of shorter apposition ommatidia dorsally. When isolated dark-adapted eyes are photographed from various directions, variations in the exposed eye area and eyeshine area are seen in both vertical and horizontal planes (Figure 4.10). The area of the eye decreased anteriorly and posteriorly from a maximum when the eye was viewed laterally (Figure 4.11A). When observed along the dorsal to ventral axis, the apparent area of the eye fell symmetrically either side of the lateral view (Figure 4.11B). The manner in which the area of

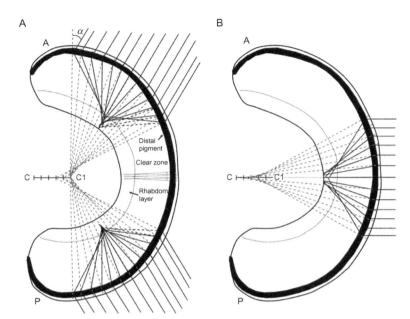

Figure 4.8 (A) Diagrammatic longitudinal section of the eye of *Nephrops*, showing parallel rays incident at 60° anterior (A) and posterior (P). The ommatidial axes ($\Delta\varphi = 1°$; every fifth axis shown) converge towards the local centre of curvature of the eye which lies between C and C1. Cones closer to the centre of the eye are centred more deeply. They thus focus rays more distally than if they were centred at C1, thereby improving the quality of the focus (thicker dotted lines show ray paths if centred at C1). (B) Diagrammatic longitudinal section of the eye of *Nephrops*, showing parallel rays incident on the central part of the eye. The ommatidial axes ($\Delta\varphi = 1°$; every fifth axis shown) converge towards the local centre of curvature of the eye which lies between C and C1. The rays are focused much deeper within the rhabdom layer than is the case with anterior and posterior rays.

eyeshine varied with viewing angle differed in many ways from that in which the eye area changed. The most noticeable difference in the horizontal plane is that the area of eyeshine remains more or less constant over the region from 90° anterior to 60° posterior. At the anterior and posterior margins eyeshine area falls rapidly. In the vertical plane, the eyeshine area is as large only in the ventro-lateral part of the eye. Dorsally, the eyeshine decreases rapidly and is more or less absent 30° from the horizontal (Figure 4.11B).

A dorso-ventral gradient of eyeshine brightness is seen in *Nephrops*, with the effective aperture of the eye varying from around 3000 facets in the ventral half of the eye down to a single facet dorsally where apposition optics is apparently in use. For a strictly benthic animal such as *Nephrops*, most visually mediated behaviour, such as territorial disputes, mate location and feeding,

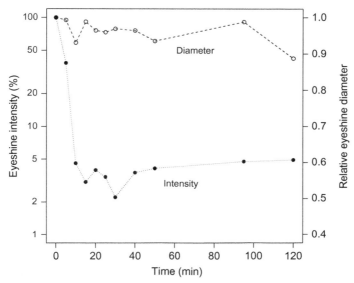

Figure 4.9 Variation in eyeshine patch diameter and intensity after exposure to daylight. Eyeshine intensity decreases rapidly but the diameter of the eyeshine patch remains fairly constant.

will occur at or below the horizontal. The distribution of bright eyeshine thus appears to coincide with the regions of both greatest interest and least light. The dorsal region of reduced eyeshine occurs in that part of the eye viewing objects in silhouette against the relatively bright downwelling light.

Determining the path of light within the eye requires knowledge of the refractive indices (RIs) of the active optical components of the eye. Interference microscopy has been used for many years in the determination of the RI of biological specimens (Hale, 1958). Several optical investigations have used the method to provide information on imaging systems in compound eyes. Apposition eyes have been studied in some depth, both in insects (e.g. honey bee: Valera and Wiitanen, 1970) and in crustaceans (*Artemia salina*: Nilsson and Odselius, 1981; *Cirolana borealis*: Nilsson and Nilsson, 1981), as have refracting superposition eyes (*Ephestia*: Cleary et al., 1977; euphausiids: Land and Burton, 1979; *Gennadus*, *Dardanus* and *Anaspides*: Nilsson, 1990) and reflecting superposition eyes (*Munida rugosa*: Gaten, 1994; crayfish: Vogt, 1980; *Cherax destructor*: Bryceson, 1981; *Macrobrachium rosenbergii*: Nilsson, 1983). However, interference microscopy has its limitations, especially in determining absolute, rather than differences in, refractive index (Kirschfeld and Snyder, 1975). The fact that due to their

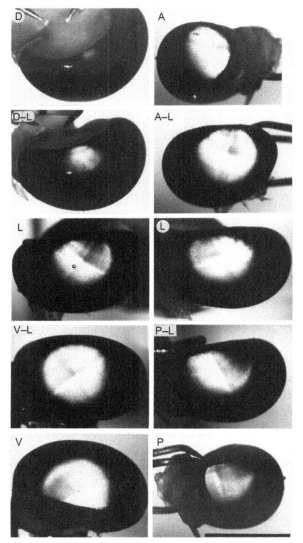

Figure 4.10 Photographs of eyeshine in an isolated, dark-adapted *Nephrops* eye, at 45°
intervals from dorsal to ventral (D to V) and from anterior to posterior (A to P). Although
there is a clear contrast between the dorsal and ventral eyeshine, around the horizontal
axis the eyeshine is relatively uniform. Scale bar = 5 mm.

fragility it is necessary to use fixed, rather than fresh, material means that the
precise RIs of most parts of the decapod eye remain unknown.

Rhabdoms and fragments of cornea were isolated from lightly fixed eyes
and the RI measured using the technique described in Gaten (1994)

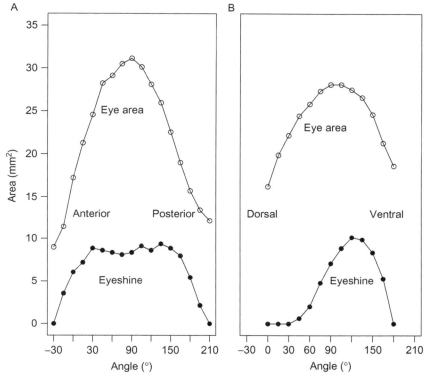

Figure 4.11 Variation in apparent eye area and eyeshine area along (A) the antero-posterior axis and (B) the dorso-ventral axis.

(Table 4.1). Cone stalks and attached crystalline cones were similarly isolated and the RI measured at intervals down the middle of the crystalline cones and the cone stalks. The RI of the crystalline cone is homogeneous over most of its length, decreasing sharply at the distal and proximal ends.

In the cone stalk, the RI decreases along a distal to proximal gradient (Figure 4.12). This RI gradient has been observed in other reflecting super-position eyes and approximately follows the function:

$$n_y = n_m \left[1 + \left(\frac{\sin \partial}{\gamma + \cos \partial - 1} \right)^2 \right]^{1/2} \qquad [4.1]$$

where n_y is the refractive index at y, the relative distance from the distal end of the cone, n_m is the refractive index of the surrounding medium and ∂ is the angle between the sides of the cone (Vogt, 1977).

Table 4.1 Refractive index (mean and s.d. of *n* measurements) of cornea (measured in the centre and at the edge of the facets), crystalline cone and rhabdoms

Location	RI	s.d.	n
Cornea: centre	1.425	0.004	35
Cornea: edge	1.420	0.003	35
Crystalline cone	1.412	0.007	12
Rhabdom	1.365	0.010	25

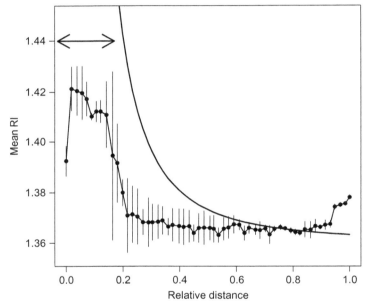

Figure 4.12 Variation in refractive index along the cone stalk from the region abutting the crystalline cone (0) to the most proximal part that connects to the rhabdom (1).

The measured RI gradient of the crystalline cone tract is consistently lower than predicted by the theoretical curve (Figure 4.12). This curve was calculated using a value for the external RI (nm) of 1.365, which is the value obtained for *Nephrops* cone cell cytoplasm in more distal regions of the cells (mean of 14 measurements, s.d. $= 0.007$). It was not possible to measure the RI in the immediate vicinity of the crystalline tracts. The only areas where the predicted curve differs markedly in shape from the experimental curve are in the crystalline cone and at the proximal end of the stalk.

The RI values obtained from the middle region of the cones all fell within the range 1.41–1.44. This is in line with previously published results from reflecting superposition eyes such as those of the freshwater shrimp, *M. rosenbergii* (Nilsson, 1983), and the crayfish *Astacus leptodactylus* (Vogt, 1980).

A high RI in the distal stalk is necessary to ensure that rays of light passing through the crystalline cone without striking the sides are reflected at a more proximal level by total internal reflection in the stalk. However, after a single reflection, either within the cone or the stalk, it is vital for the correct functioning of the eye that any rays leave the stalk when they next strike the wall lower down the stalk. To ensure that this happens, the RI gets progressively lower in more proximal regions of the stalk, leading to the characteristic RI gradient found in the decapod crystalline tract (Vogt, 1977).

Vogt's (1977) function relating the decrease in RI to distance down the cone and cone stalk agrees well with the experimental results, although the theoretical gradient predicts higher RI values than those measured here. The value used for the RI of the surrounding medium (1.365) was probably too high as this was calculated using lightly fixed cells from more distal regions of the cone cells. The values used by Vogt (1977) were estimates of the RI of unfixed cytoplasm (1.34 and 1.35) and these resulted in the theoretical curve being rather lower than his experimental results. In the region of the cone/stalk junction, the RI required by the theoretical curves exceeds reasonable physiological values. Distal to this the cone can only contribute to the superposition image using an additional reflection mechanism. This role is fulfilled by the reflecting multilayer (Land, 1972) that lines the sides of the cones (Vogt, 1977).

The proximal end of the stalk had a higher RI than the mid-section. This has also been recorded in *M. rosenbergii* (Nilsson, 1983). According to Nilsson (1983), this proximal increase in the RI serves to deflect stray light from the target rhabdom. This has the effect that light entering from another ommatidium is refracted at the tip of the stalk and forced to leave the opposite side of the rhabdom leading to a narrowing of the acceptance angle (Nilsson, 1983).

Spatial resolution in compound eyes is ultimately dependent on the interommatidial angle ($\Delta\varphi$), defined as the angular separation of the receptors. The method used was to photograph half-eyes, fixed and then cut along the midline. From these, the orientation of the cone axes can be determined accurately, in addition to the local radius of curvature of the eye. The best estimate of $\Delta\varphi$ obtained from photographs of fixed, bisected eyes was 0.94° (s.d. = 0.1), although it should be emphasised that such anatomical data can deviate from effective visual interommatidial angles (Stavenga, 1979).

Although all eyes with high spatial resolution have small interommatidial angles, the presence of the latter in reflecting superposition eyes does not necessarily mean that spatial resolution is high. Even though *Nephrops* has a very low value of $\Delta\varphi$ (0.94°), direct observation of the superposition light path suggests that each rhabdom should receive light over a wide angle. The spatial resolution is thus dependent on factors other than $\Delta\varphi$ and is best determined by electrophysiology. Angular sensitivity functions were determined by intracellular electrophysiology (Shelton and Gaten, 1996). The acceptance angles (half width of the angular sensitivity function) were 11.3° in the dark-adapted eye and 8.85° in the light-adapted eye. All of these measurements were made on central regions of the eye (due to practical constraints) and lower values would be expected in anterior and posterior regions of the eye.

The discrepancy between the interommatidial angle ($\Delta\varphi$) and the acceptance angle ($\Delta\rho$) in *Nephrops* must lead to oversampling of the environment with several ommatidia viewing a single point in space. Clearly, the ability of the eye to resolve detail depends not only on the anatomical distribution of the receptors, but also on the quality of the optics delivering the light to the receptors (Land, 1989). The spatial resolution determined electrophysiologically for the eyes of *Nephrops* suggests that resolving power is limited in these eyes by optical defects rather than anatomical constraints such as receptor density and orientation (Gaten and Shelton, 1993). Such optical defects include imperfect superposition of light rays (Nilsson, 1989) and cross over of rays within the rhabdom layer (Warrant and McIntyre, 1991).

Spherically symmetrical superposition eyes have the cone axes centred on the centre of curvature of the eye (Land et al., 1979). In *Nephrops*, the radius of curvature of the cornea varies across the eye due to the flattening of the central part of the eye. The paths of ray bundles across the eye, based on the optical data referred to above, can be visualised on a tracing of a longitudinal section of the eye (Figure 4.9). Every fifth cone axis (obtained from photographs of half-eyes) is drawn and the rays are plotted assuming that the angle of rays crossing the clear zone is equal to the angle between the incident ray and the ommatidial axis plus the cone taper angle. For both anterior and posterior regions of the eye a reasonably sharp focus is obtained in the distal part of the rhabdom layer (Figure 4.9A). Had all of the cone axes been centred on the same point (labelled C1 in figure), those rays closest to the centre of the eye would have been reflected to a point deeper in the rhabdom layer. However, these cones are centred somewhat deeper, leading to smaller anterior and posterior blur circles. The blur circles in these regions are estimated to have diameters of about 7–10 rhabdoms ($\sim$7° to 10°).

When the superposition of rays in the lateral part of the eye is plotted (Figure 4.8B) the predicted image is much worse. The centre of curvature of the crystalline cone layer is now much deeper within the eye (towards point C in Figure 4.8), resulting in incoming light being focussed well below the surface of the rhabdom layer. As a result, the blur circle on the surface of the rhabdom layer has a diameter of around 20 rhabdoms, covering an angle of 20°. This large blur circle is formed because the eye retains approximately the same interommatidial angle and radius of curvature of the rhabdom layer while decreasing the width of the clear zone. However, although there is a considerable reduction in spatial resolution as a result of the lateral flattening of the eye, its sensitivity in this region should not be affected as the amount of light entering the eye and the number of rhabdoms receiving the light remains unchanged.

In *Nephrops*, the disadvantage of loss of resolution resulting from the lateral flattening of the eye is presumably of little importance in an animal active at low light levels.

The spatial resolving power of a compound eye depends on the interommatidial angle, but is further related to three main factors: the photic range over which the animal is normally active, its mean velocity and the tasks for which the eyes are normally used (Snyder et al., 1977). *Nephrops* is active at low light levels, generally moves slowly and does not use its eyes for fine discrimination so we would expect its eyes to be adapted for high sensitivity even though its ommatidial array suggests that potentially it is capable of high spatial resolution (Warrant, 2006).

A low interommatidial angle/acceptance angle ratio (such as described here for *Nephrops*) and neural pooling are typically seen in animals active at low light levels and both have been implicated in the evolution of neural superposition eyes (Nilsson and Ro, 1994). For any given interommatidial angle, a large acceptance angle will result in more photons being collected with a concomitant improvement in the signal to noise ratio; a smaller acceptance angle would reduce the visual overlap between adjacent ommatidia, improving the spatial resolution (Nilsson and Ro, 1994). It is apparent from this low interommatidial angle/acceptance angle ratio that the eye of *Nephrops* is adapted for high sensitivity.

Reflecting superposition eyes are usually spherically symmetrical in the horizontal plane, with the rhabdom layer, crystalline cone layer and the cornea sharing a common centre of curvature (e.g. *Munida*; Gaten, 1994). However, in *Nephrops* the eye is flattened laterally, resulting in the centre of curvature of the cornea and cones (but not the rhabdom layer) being displaced proximally and the lengths of the ommatidia in the middle of the eye being shortened; both of these combine to produce underfocussing, with the distance between the retina and the crystalline cones less than the focal length of the dioptric apparatus. This results in an obvious loss of resolution in spite of the small inter-ommatidial angle. Underfocussing is seen in insect ocelli (Berry et al., 2007; Schuppe and Hengstenberg, 1993; Wilson, 1978) and in box jellyfish eyes (Nilsson et al., 2005) where poor quality images are produced in situations where advanced spatial information processing is not required.

5. ELECTROPHYSIOLOGY

Eyes function as matched filters in that they are adapted to extract the available, or important, spatial and spectral information out of the environment and to react as appropriate to temporally, and sometimes spectrally, varying signals. Evolution viewed through the lens of symmorphosis suggests that an understanding of the limitations of the eye in the spectral and temporal domains should provide some clues as to what the animal is looking for (Weibel et al., 1998). Compound eyes are generally accepted as being energetically expensive (Laughlin et al., 1998). The large size of the eyes in absolute and relative terms therefore indicates their importance to *Nephrops*. As a major component of the sensory machinery it could be expected that they would be well adapted to meet the needs of the animal.

Ambient light availability in the marine environment is significantly impacted by a variety of factors other than the position of the sun and cloud cover. The initial interaction of light with water involves a transition between media of different RIs. Light striking a flat water surface is refracted so that it becomes more vertical, if there are waves at the air:water interface then the interaction becomes more complex and a temporal element becomes important as the waves move across the surface. The absorption of light as it passes through the water varies depending on the solutes and colloids suspended in it (Kirk, 1996). Generally coastal waters have plenty of detritus and gelbstoff (organic breakdown products) in them so that light is absorbed fairly quickly with depth. The pigments in the water also modify the spectrum of light that does penetrate to any depth towards longer wavelengths. Reviewing the literature relating to spectral sensitivities of decapods, Johnson et al. (2002) found that the general variations in spectral sensitivity by habitat match those found in fish (Cronin, 1986; Partridge et al., 1992) with coastal species being more sensitive to longer wavelength light than deep-sea or pelagic inhabitants.

Electrophysiological recordings from *Nephrops* appear to confirm evidence from histological examinations that the rhabdom consists of two sections with different spectral sensitivities (Figure 4.13). From other electrophysiological investigations, 16 species have been found to have both

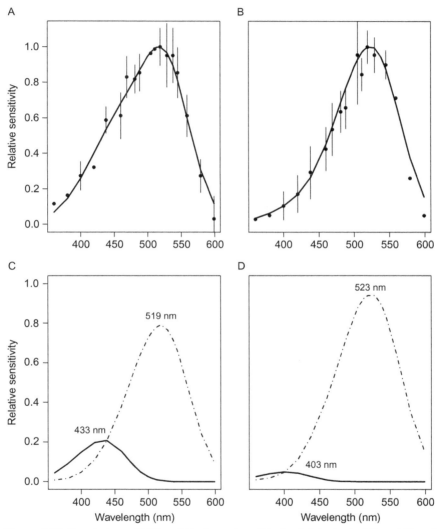

Figure 4.13 Spectral sensitivities of *Nephrops* (n = 4) in the light-adapted (A and C) and dark-adapted (B and D) conditions. (A) and (B) depict the overall sensitivity curve based on contributions from two pigments. (C) and (D) show the separate absorption templates (Stavenga et al., 1993) that contribute to the combined curve. *Redrawn from Johnson et al. (2002).*

long and shortwave pigments (Johnson et al., 2002) and there are numerous other studies that indicate that most long bodied decapods probably have two photopigments (Cummins and Goldsmith, 1981; Eguchi et al., 1973; Frank and Widder, 1994a,b, 1996; Gaten, 1992; Johnson, 1998). The spectral sensitivities of *Nephrops* correspond very well with the spectra for coastal decapods generally. The presence of a shortwave pigment is particularly evident in light-adapted *Nephrops*, where the description of spectral sensitivity with one rhodopsin template (Stavenga et al., 1993) is particularly inadequate. In light-adapted animals, where the reflective tapetum is obscured by proximal pigments that coat the rhabdom except for the very distal tip (Figure 4.13A and C), light passes through the rhabdom only once and the distal R8 rhabdomere thus contributes more to overall sensitivity. In dark-adapted animals light passes twice through the proximal, longer wavelength sensitive region before reaching the distal shortwave region again. This is emphasised by the fact that in light-adapted *Nephrops* the distal region appears to contribute 21% of the overall sensitivity curve while in dark-adapted animals it is a mere 5% (Figure 4.13B and D). The importance of the proximal region of the rhabdom layer in dark-adapted animals may be enhanced by the afocal nature of the flattened lateral area of the eye as discussed above which may mean that light is focussed more proximally (Shelton et al., 1986).

Johnson (1998) suggests that the possession of two pigments allows decapods to be both sensitive to the most common wavelengths in their habitats (as per the sensitivity hypothesis of Munz, 1958) and to be able to pick out less dominant but arguably more important frequencies with their short wavelength pigments (Lythgoe, 1968). Gaten et al. (2004) noted that in some decapod species the relative length of the distal shortwave receptor increased from dorsal to ventral. Looking upwards, marine organisms would be best to have enhanced sensitivity to silhouettes while laterally and ventrally shortwave bioluminescence would dominate the visual scene. The benthic infaunal crepuscular habit of *Nephrops* and the fact that it has a very broad depth range (10–800 m) may make the possession of two pigments useful in a range of situations. Short wavelength sensitivity may be useful in deeper water where there has been found to be an increase in bioluminescent activity near the sea bed (Craig et al., 2011), or generally as a mechanism for picking up on increased bacterial activity on detritus (Barak and Ulitzer, 1980).

The adaptability of the eyes of *Nephrops* may also be an important factor with regard to their responses to temporally varying stimuli. Slow moving, deep-sea animals that inhabit a permanently dark environment where downwelling light is of little importance have little need to adapt to varying background light intensities. Whatever they see is likely to be dim and important, so having their eyes set to maximum sensitivity all the time is the best strategy. In shallower water there will be more natural variation in ambient light intensity over the day and much of what is visible is likely to be unimportant. Animals in this situation need to have eyes that can be optimised for a range of conditions. For longer term variations in ambient light intensity the slow gain control provided by pigment movements will serve to ensure that the amount of light reaching the retinula cells maximises the information available and may protect the eye from damage through over-exposure (Gaten, 1990). Rapidly responding eyes are likely to be pointless in scotopic conditions where any dim object moving quickly is likely to be invisible no matter how effective the eyes are. It could actually be disadvantageous for an eye to adapt at all in low light conditions if gradual and small changes in light levels or slowly moving objects are important aspects of the visual scene. However, in mesopic or photopic conditions where more light is available, shorter term background variations produced, for example, by surface waves focussing light at depth (Čepič, 2008), more rapid gain control will be necessary. Without this ability it will not be possible to distinguish between ambient changes in light intensity and those caused by objects moving across or appearing in the visual field. The compromise between the need to ignore low frequency changes in intensity and systemic limits on the ability to respond to high frequency modulations leads to band-pass characteristics (McFarland and Loew, 1983). In comparative studies of insects and decapods it has been found that there is generally a good match between the frequency that an eye responds to optimally and the ambient levels they are adapted for (Howard, 1981; Moeller and Case, 1995; O'Carroll et al., 1996; Pinter, 1972). Animals from scotopic habitats tend to have 'low-pass' eyes that respond to low intensity signals very strongly but slowly. Animals from photopic habitats require greater changes in light intensity to demonstrate a response but respond more quickly. *Nephrops* is able to shift between scotopic and mesopic modes (Johnson et al., 2000). It exhibits sluggish sensitivity to all frequencies when dark-adapted but optimal responses to 1–5 Hz when light-adapted (Figure 4.14).

It is thought that this variation in response characteristic is mediated by various populations of potassium ion channels in the rhabdom membranes that are responsible for repolarisation (Weckström and Laughlin, 1995). In dark-adapted animals these channels open briefly in response to even very small depolarisations of the rhabdom. This leads to a slow repolarisation of the rhabdom. In the light-adapted state a different population of potassium channels respond to depolarisations of 20–40 mV by remaining open

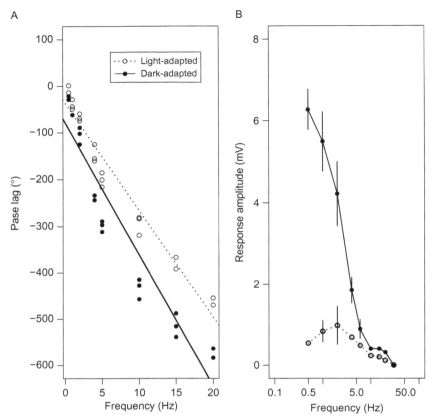

Figure 4.14 Temporal sensitivities of *Nephrops* ($n = 3$) in the light-adapted and dark-adapted conditions stimulated with a light varying in intensity sinusoidally. (A) Eyes have reduced lag in the light-adapted state. (B) Dark-adapted eyes respond most to low frequency signals while light-adapted eyes demonstrate band-pass characteristics. *Redrawn from Johnson et al. (2000).*

until the rhabdom repolarises. It seems possible that part of the explanation for the insignificant input of the distal blue-sensitive rhabdomere to the dark–adapted spectral sensitivity curve of *Nephrops* may be due to it being dominated by a light–adapted type population of potassium receptors.

6. LIGHT-INDUCED DAMAGE

The loss of visual input is for most animals potentially disastrous. Acute light-induced damage can result through photothermal, photomechanical and photochemical mechanisms (Youssef et al., 2011) and, although this can affect any animals exposed to excessive amounts of light, it is particularly dangerous for species whose eyes are adapted for high sensitivity. Deep-sea crustaceans generally have eyes adapted for maximum sensitivity to light through the possession of eyes with large apertures and broad rhabdoms, and in the case of decapods through the use of superposition optics. Many such animals have been shown to exhibit photodamage when exposed to light levels beyond their ambient range, including the isopod *C. borealis* (Nilsson and Lindstrom, 1983), krill *Meganyctiphanes norvegica* (Gaten et al., 2010) and vent shrimps *Rimicaris exoculata* (Herring et al., 1999). Photodamage in *Nephrops* was first described by Loew (1976).

Nephrops is particularly vulnerable to light-induced damage as, in addition to having very photosensitive eyes, it is the object of a significant fishery throughout Northern Europe (Chapter 7). Following capture, undersized *Nephrops* are returned to the sea often after having been exposed to daylight. Short exposure to ambient daylight (as little as 9 min at 2.29×10^{20} photons m^{-2} s^{-1}) is sufficient to cause significant damage to the retinula cell layer (Shelton et al., 1985). In dark-adapted eyes more than 75% of the retinal cell layer is damaged by 15 s exposure to daylight. Histological evidence of damage is visible as breakdown of the cell membranes and disruption of the rhabdom microvilli within 15 min of exposure; after 6 h, the retinula cell body layer is absent (Shelton et al., 1985). Animals preserved after being exposed to daylight for 2 h show complete loss of rhabdom structure and incursion of the shielding pigment (Figure 4.15B); 1 month later, the area is dominated by haemocytes with small amounts of disorganised membrane (Figure 4.15C).

Exposure of restrained animals to an artificial light source of known intensity reveals the time course of the retinal damage (Shelton et al., 1985). At high intensity (6×10^{20} photons m^{-2} s^{-1}), 1 min exposure was sufficient to cause total retinal damage in dark-adapted eyes, whereas in light-adapted eyes around 60–70% of the eye was damaged (Figure 4.16A). In these experiments, exposure to a lower intensity (1.5×10^{20} photons m^{-2} s^{-1}) left a small proportion of undamaged retina in both light- and dark-adapted states (Figure 4.16A). The threshold of photodamage was investigated by exposing restrained animals to lower light levels (Shelton et al., 1985). In dark-adapted animals, damage was detected at the lowest intensity used, 6×10^{18} photons m^{-2} s^{-1}, whereas in light-adapted animals the threshold was 3×10^{19} photons m^{-2} s^{-1} (Figure 4.16B). The proportion of the retina damaged by these 10 s exposures was directly proportional to the light intensity used (Shelton et al., 1985).

Although all *Nephrops* are susceptible to damage at relatively low light levels, the extent of the damage depends on the depth from which they were

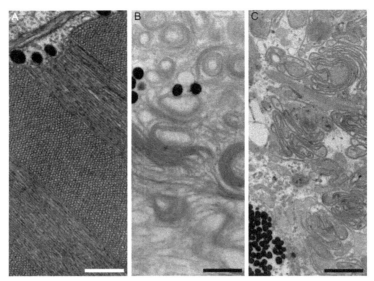

Figure 4.15 (A) Electron micrograph of an undamaged rhabdom showing alternating layers of microvilli cut in transverse and longitudinal section. Shielding pigment granules can be seen along the edge of the rhabdom. Scale bar = 1 μm. (B) After 2 h in daylight, the regular array of microvilli has broken down and shielding pigment is found within the rhabdom. Scale bar = 1 μm. (C) In eyes fixed 1 month after exposure, only a few membrane whorls remain where the rhabdom had been, together with haemocytes containing shielding pigment granules. Scale bar = 2 μm.

taken (Gaten et al., 1990), with animals caught at 18 m suffering 40% retinal damage compared to 80% in those from 135 m following the same light exposure (Figure 4.17). Animals from 18 m had around half of the concentration of proximal shielding pigment compared to those from 135 m, and these shielding pigments only migrated half as far up the rhabdoms during light adaptation in shallow-water animals compared to those from deeper water (Gaten et al., 1990). The shielding pigments protect the rhabdoms (Shelton et al., 1986) by screening the rhabdoms from excess light, by raising the refractive index adjacent to the rhabdom (thus reducing light retention by total internal reflection) and by isolating the tapetum (reducing the reflection of light back through the rhabdoms). As a result, the increase in damage seen in deep-water animals is due to the reduction in pigment concentration and the exposure of the distal half of the rhabdom. The differences between animals from different depths are unlikely to be due to genetic isolation as there is extensive mixing during the planktonic larval stages (Farmer, 1975).

However, even when the differences in screening pigment concentration and position are taken into account by exposing dark-adapted animals to light

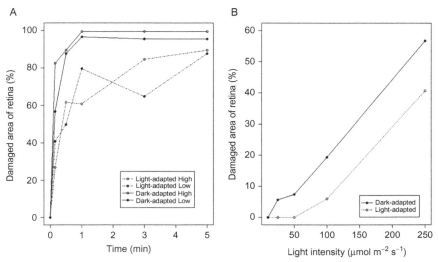

Figure 4.16 (A) Time course of damage to the retinula cell layer caused by exposure to tungsten light at 1000 (○) and 250 (filled circles) μmol m^{-2} s^{-1} (1 μmol m^{-2} s^{-1} = 6.023 × 10^{17} photons m^{-2} s^{-1}) in both light-adapted (dotted lines) and dark-adapted (solid lines) states. Each point represents one or two eyes. (B) Threshold for damage to the retinula cell layer; exposure of dark-adapted (filled circles) and light-adapted (○) eyes for 10 s to tungsten light of various intensities. Each point represents one eye. *Redrawn from Shelton et al. (1985).*

(with the screening pigment retracted below the reflective tapetum), there is still an increase in susceptibility to damage with depth (Gaten et al., 1990). Dark-adapted animals from all depths showed a positive correlation between light exposure and retinula cell damage. A possible explanation may be that there is a different visual pigment content in animals from different depths. The visual pigment (rhodopsin) has been shown in mammals to be responsible for acute photodamage to the retina, probably due to the photosensitising role of the all-trans-retinal chromophore found in rhodopsin (Rózanowska and Sarna, 2005). Sensitivity has been shown to depend on the rhodopsin content of a fly rhabdom (Razmjoo and Hamdorf, 1976), so a higher concentration of rhodopsin in the eyes of deep-water *Nephrops* would result in both increased sensitivity and greater susceptibility to light-induced damage.

It has also been suggested (Loew, 1976) that the slow rate of visual pigment regeneration in this species might result in a loss of structural integrity in the photoreceptor membrane. Maintaining *Nephrops* in constant darkness for 26 h before exposing them to damaging levels of light resulted in a significant decrease in the amount of damage, even though the shielding pigments

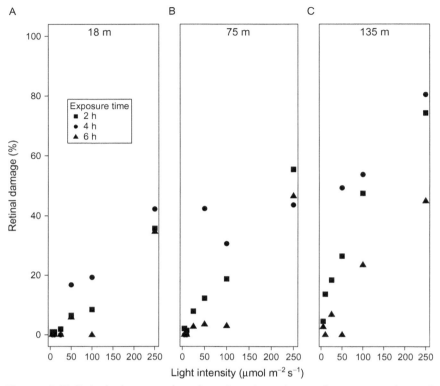

Figure 4.17 Retinal damage plotted against log photon fluence rate (1 μmol $m^{-2} s^{-1} = 6.023 \times 10^{17}$ photons $m^{-2} s^{-1}$) for animals taken from three different depths. At each depth, animals were exposed 2 h (triangles), 4 h (circles) and 26 h (squares) after capture. Each point is the mean of 2–5 animals. *Redrawn from Gaten et al. (1990).*

remained in the dark-adapted state (Gaten et al., 1990). This may have been as a result of the interruption to the normal turnover of photoreceptor membrane that is seen in all invertebrate rhabdoms; this process normally prevents the accumulation of photodamaged membrane within the eye (Fleissner and Fleissner, 2006). Alternatively, the protection of the rhabdoms may have occurred as a result of the release of heat shock proteins. These are produced in response to thermal stress or other metabolic shock in many taxa including crustaceans (Clark and Peck, 2009). They appear to provide protection for retinal cells against light-induced damage (Barbe et al., 1988).

Although photodamage is initially detectable only at the electron microscope level, within 15 min serious disruption of the rhabdoms and surrounding cells can be seen and after 6 h the retinula cell layer is completely disorganised (Shelton et al., 1985). This is followed by damage to the

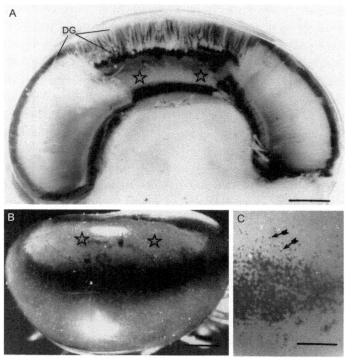

Figure 4.18 (A) An unfixed eye bisected 2 months after light damage, showing the absence of retinula cells centrally (∗∗), retraction of the crystalline tracts and redistribution of the distal shielding pigment (DG). (B) Whole *Nephrops* eye from an animal recaptured 1 year after exposure to sunlight. The regular array of undamaged dorsal facets (∗∗) can be contrasted with the darker, indistinct appearance of damaged areas. (C) Individual damaged ommatidia (arrowed) are identifiable by their darker appearance. All scale bars = 1 mm. *From Gaten (1988).*

crystalline tracts that cross the eye (Figure 4.18A) with the result that they no longer fill the proximal clear zone (Gaten, 1988). The crystalline cones become variable in appearance and no longer display the precise alignment seen in undamaged eyes. Around 2 months after exposure, the damage to the dioptric apparatus can be visualised externally in live *Nephrops*, appearing as a pale area when under infra-red illumination (Gaten, 1988). The damage is still apparent externally in animals recaptured 1 year (Figure 4.18B and C) after exposure to light and sectioning of the eyes confirms that there is no recovery from retinal damage with time.

The effects of blinding on the ecology of *Nephrops* has been investigated by observation of individuals that were caught and tagged during daylight

(Chapman et al., 2000) and by tagging and releasing animals following controlled exposure to artificial light (Shelton et al., 1985). Examination of tagged animals exposed to daylight (for around 5 min) and then recaptured up to 7 years later revealed a large range of eye damage, with 0–63.4% of the retina being destroyed (Chapman et al., 2000). No light measurements were taken at the times of capture or release so the variation in damage presumably relates to the ambient light at those times, the depth from which the animals were taken and the degree of exposure to direct sunlight. Animals released and recaptured following controlled exposure to known levels of natural light (Shelton et al., 1985) showed a similar variation in retinal damage.

When exposed to intense artificial illumination, retinal damage was complete in the case of dark-adapted eyes and ranged from 5% to 100% in light-adapted eyes (median 56.4%) illustrating the protective value of the retinal shielding pigments (Chapman et al., 2000). *Nephrops* (both with intact vision and 100% retinal damage) were recaptured after 1–3 years and analysed for survival rate, growth rate and reproductive potential. Over the 8 years of the experiment, it was found that the survival of the animals was independent of the extent of eye damage and that there was no evidence that eye damage reduced growth rates. Although females showed a slightly higher survival rate than males, there was no significant change in the number of females carrying eggs (Chapman et al., 2000). It would appear that, in spite of investing in such a large eye, blind *Nephrops* survive, grow and reproduce as well as fully sighted individuals. However, the results may have been skewed due to the possibility that blinded individuals may have foraged for longer periods, leading to an increased likelihood of being recaptured.

The results of these studies clearly show that *Nephrops* are very susceptible to light-induced damage, particularly in the dark-adapted state, and that the damage is permanent. However, there is no evidence to suggest that this would have any effect on the fishery. It is important, though, to bear in mind the potential for eye damage when collecting or maintaining specimens for experimentation or for population studies.

7. MODELLING THE PERFORMANCE OF THE EYE OF *NEPHROPS*

Anatomical and electrophysiological investigations suggest that the eyes of *Nephrops* are well adapted to a crepuscular lifestyle in generally low light conditions. However, there is much to be learned from models based on measurements and theory. Simple models that estimate the performance of the eye as a single unit have been developed and modified over the years. The accepted equation that describes the ratio of the number of photons absorbed per receptor to the number emitted per steradian from an extended source that are absorbed is:

$$S = (\pi/4)^2 D^2 (d/f)^2 \left[\frac{kl}{2.3 + kl} \right] \qquad [4.2]$$

where D is the diameter of the aperture, l is the length of the photoreceptor, k is the absorption coefficient of the rhabdom, d is the rhabdom diameter and f the focal length (Kirshfeld, 1974; Land, 1981; Warrant and Nilsson, 1998). However, while Equation (4.3) is adequate as a description of sensitivity, it does not capture the complexities of sensitivity under a range of internal pigment distributions and says nothing about resolution (or its relationship to sensitivity). Warrant and McIntyre (1991) pointed out that even if a super-position eye was perfectly focussed, because of the large aperture and correspondingly low F number, light from the periphery of the aperture would enter the target rhabdom at such an angle that light would bleed into adjacent rhabdoms thus degrading spatial resolution. Various solutions have evolved to deal with this problem including partial and complete sheaths of tapetal or shielding pigment and rhabdom morphology (Figure 4.19). *Nephrops* has rhabdoms with pointed distal tips that have been estimated to supplement the angle that light can be accepted over by about 12.5° (Gaten, 1992).

Using basic trigonometry it is possible to estimate the extent of the blur circle based on the refraction of light by the dioptric layer and the relationship between the radius of curvature of the rhabdom layer and the focal length (Figure 4.20).

If we assume a spherical eye, its diameter and the aperture diameter can be used to estimate the angle of incidence of light at the periphery of the aperture (θ). To calculate the point where light from the periphery of the aperture is incident on the distal region of the rhabdom layer (r) the angle at which light is redirected by the dioptric layer (β) must be known. This can be calculated by applying Snell's law and assuming the RIs of sea water (1.35; Kirk, 1996) and crystalline cones (1.412; Gaten, 1992) and accounting for the taper due to the interommatidial angle ($\Delta\varphi$):

$$\beta = \sin e^{-1}(0.95 \sin \theta) + \Delta\phi \qquad [4.3]$$

Since we know r, p and β we can apply the Sine rule for triangles:

$$\delta = 180 - \sin e^{-1}\left(r \sin \beta p^{-1}\right) \qquad [4.4]$$

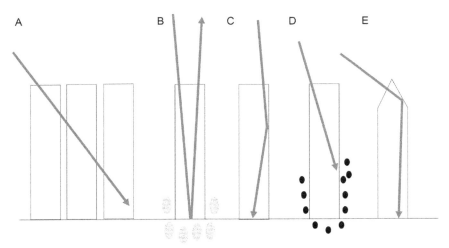

Figure 4.19 The fate of light rays when they encounter a rhabdom. (A) Passage through the target rhabdom to those adjacent. (B) Reflection by the tapetum. (C) Reflection by total internal reflection at the edge of the rhabdom due to differences in refractive index between the rhabdom and surrounding cytoplasm. (D) Absorption by proximal shielding pigment (enhanced by equal refractive indices). (E) Total internal reflection of rays coming from peripheral facets may be enhanced by rhabdom morphology such as pointed distal tips.

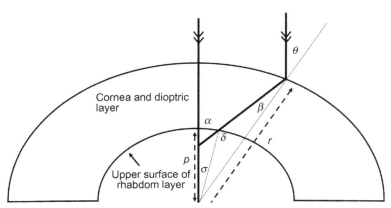

Figure 4.20 Vectors and angles involved in calculating the extent of the blur circle on the retina of a superposition compound eye. θ, angle of incidence with the cornea, β, angle of exit from the crystalline cone layer, r, radius of curvature of the cornea, p, radius of curvature of the rhabdom layer, σ, angular extent of the blur circle. See text for further details.

And the angle (σ) that defines the edge of the blur circle can be calculated as

$$\sigma = \delta + \alpha - 180 \qquad [4.5]$$

Based on these assumptions and from measurements of *Nephrops* eyes discussed above it is possible to estimate the extent of the blur circle at the anterior and lateral regions of the eye. The calculations based on anatomical measurements show good agreement with the diagrammatic ray-tracing approach, suggesting a larger blur circle laterally.

To explore the implications of the variation in eye parameters on sensitivity and resolution of superposition compound eyes a ray tracing model has been developed (Figure 4.21) using the programming language Python (Van Rossum, 2011; see also supplementary material in the web version of this chapter) and we apply it here to *Nephrops* (Johnson, 1998). The model initially assumes that superimposed light coming from a single facet will land on a single rhabdom and uses a combination of anatomical measurements and estimated parameters (Table 4.2) to calculate the distribution of light to and within the rhabdom layer. The model takes account of the blur circle extent, absorbance of light by the photopigments and the interaction of light with proximal shielding and reflecting pigments. It also accounts for the decreased per facet contribution of light towards the periphery of the aperture and behaviour of light within rhabdoms depending on reflecting/tapetal pigment extent and whether it is assumed that they are flat or tapered distally. Lateral and anterior regions of the *Nephrops* eye were modelled using the parameters in Table 4.2. Rhabdom dimensions of 180 μm length and 25 μm width and a facet diameter of 50 μm were assumed to be constant (Johnson, 1998). Although the model calculates specific correspondence between facets and rhabdoms it was assumed that light spreads evenly over the rhabdom layer from each facet to the extent of the blur circle.

The model suggests that the presence of any sort of tapetum in *Nephrops* improves sensitivity by about 10% but that increasing extent of reflective pigment along the rhabdom may actually reduce sensitivity (Figure 4.22A–C). This is likely to be a result of light passing between adjacent rhabdoms when the tapetum is basal so that light has a longer pathlength through photopigment. However, increasing tapetal extent does improve resolution. There is a clear impact of shielding pigment on sensitivity in both lateral (Figure 4.22D–E) and anterior regions of the eye (Figure 4.22F). Despite the fact that the aperture of the lateral region is only slightly larger than anteriorly, the model suggests that the combination of a larger blur circle

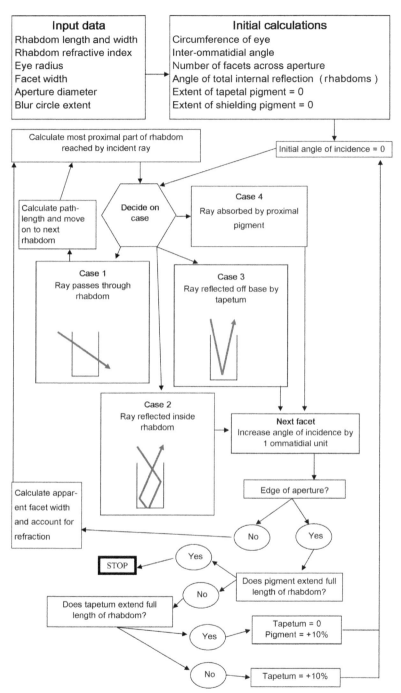

Figure 4.21 A flow chart depicting the ray tracing model of the *Nephrops* superposition compound eye. The model assumes parallel pencils of light enter the eye, starting with the axial facet and moving sequentially along the radius to the periphery of the aperture. The angle of incidence changes at each iteration because of the radius of curvature of the eye.

Table 4.2 Parameters for anterior and lateral regions of the eye of *Nephrops norvegicus* used in the ray tracing model

Parameter	Lateral	Anterior
Interommatidial angle ($\Delta\varphi$)	0.73°	0.84°
Radius of curvature of the eye (r)	3.9 mm	3.4 mm
Radius of curvature of retina (p)	2.8 mm	2.0 mm
Aperture radius	1.60 mm	1.53 mm
Blur circle extent (σ)	12.8°	8.34°
Blur circle extent (ommatidia)	18	10

and smaller interommatidial angle results in about 10% more sensitivity. It is notable, comparing figures for lateral and anterior regions of the eye that the relative contribution of the axial rhabdom to overall sensitivity is much less laterally compared with the anterior region of the eye.

It is likely that this difference can be linked to the smaller blur circle and much better resolution in the anterior region of the eye. The predicted resolution in this region approaches 5.5° for a fully sheathed rhabdom compared with a best of around 10° laterally. This is due to the more appropriate relationship between the radii of cornea and rhabdom layers. The model suggests that while the possession of pointed rhabdoms results in a slight improvement in sensitivity in the poorly focussed lateral region of the eye, it may actually be detrimental to resolution. This is a consequence of the rhabdoms within the larger blur circle individually more effectively capturing incident light. This makes the profile of absorbance across the retina broader and flatter than is the case with flat-ended rhabdoms. In the anterior region of the eye the possession of pointed rhabdoms appeared to have minimal effect. This is because very little light arrives at the distal tips of rhabdoms at an angle greater than 12.5°. In summary, the ray tracing model appears to confirm the suggestion based on anatomical and electrophysiological investigations that the anterior and lateral areas of the eye have different capabilities and are likely to serve slightly different functions. The anterior region of the eye with its better resolution could be considered analogous to the acute zones found in some apposition eyes (Land and Nilsson, 2002) or the temporal fovea of some hole-dwelling reef fish (Collin and Pettigrew, 1989). It seems also that the lateral flattening of the cornea of *Nephrops* could serve some advantage rather than simply represent a relaxation of evolutionary pressure in the absence of specific function.

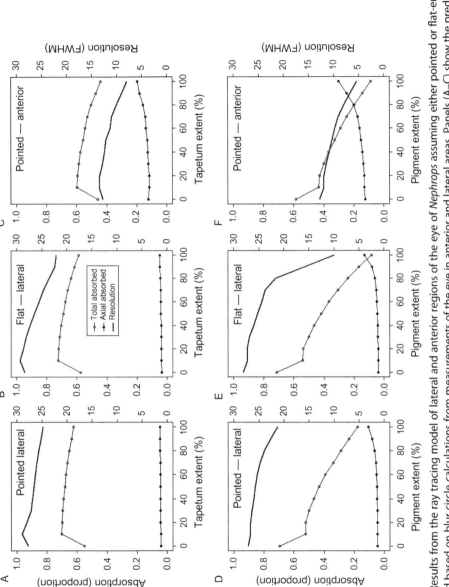

Figure 4.22 Results from the ray tracing model of lateral and anterior regions of the eye of *Nephrops* assuming either pointed or flat-ended rhabdoms and based on blur circle calculations from measurements of the eye in anterior and lateral areas. Panels (A–C) show the predicted effect of varying the extent of the tapetal sheath along the rhabdom. Panels (D–F) assume a tapetal sheath of 1/3 and show the predicted effect of varying the extent of a shielding pigment sheath along the rhabdom.

ACKNOWLEDGEMENTS

The authors gratefully acknowledge comments from Doekele Stavenga that significantly improved the work, the original inspiration from Pete Shelton and Colin Chapman to work on eyes and *Nephrops*, respectively, and the assistance of Nicola Dobson with Figure 4.4.

REFERENCES

Aréchiga, H., Atkinson, R.J.A., 1975. The eye and some effects of light on locomotor activity in *Nephrops norvegicus*. Mar. Biol. 32 (1), 63–76.

Barak, M., Ulitzer, S., 1980. Bacterial bioluminescence as an early indicator of marine fish spoilage. Eur. J. Appl. Microbiol. 10, 155–165.

Barbe, M.F., Tyrell, M., Gower, D.J., Welch, W.J., 1988. Hyperthermia protects against light damage in the rat retina. Science 241, 1817–1820.

Berry, R.P., Warrant, E.J., Stange, G., 2007. Form vision in the insect dorsal ocelli: an anatomical and optical analysis of the locust ocelli. Vision Res. 47, 1382–1393.

Bryceson, K.P., 1981. Focusing of light by corneal lenses in a reflecting superposition eye. J. Exp. Biol. 90, 347–350.

Čepič, M., 2008. Underwater rays. Eur. J. Phys. 29, 845–855.

Chapman, C.J., Shelton, P.M.J., Shanks, A.M., Gaten, E., 2000. Survival and growth of the Norway lobster, *Nephrops norvegicus* (L.), in relation to light-induced eye damage. Mar. Biol. 136, 233–241.

Clark, M.S., Peck, L.S., 2009. HSP70 heat shock proteins and environmental stress in Antarctic marine organisms: a mini-review. Mar. Genomics 2, 11–18.

Cleary, P., Deichsel, G., Kunze, P., 1977. The superposition image in the eye of *Ephestia kühniella*. J. Comp. Physiol. A 119 (1), 73–84.

Collin, S., Pettigrew, J., 1989. Quantitative comparison of the limits on visual spatial resolution set by the ganglion cell layer in twelve species of reef teleosts. Brain Behav. Evol. 34 (3), 184–192.

Craig, J., Jaimeson, A.J., Bagley, P.M., Priede, I.G., 2011. Naturally occurring bioluminescence on the deep-sea floor. J. Mar. Syst. 88 (4), 563–567.

Cronin, T., 1986. Optical design and evolutionary adaptation in crustacean compound eyes. J. Crust. Biol. 6 (1), 1–23.

Cummins, D., Goldsmith, T.H., 1981. Cellular identification of the violet receptor in the crayfish eye. J. Comp. Physiol. 142, 199–202.

Eguchi, E., Waterman, T.H., Akiyama, J., 1973. Localization of the violet and yellow receptor cells in the crayfish retina. J. Gen. Physiol. 62, 355–374.

Exner, S., 1891. The Physiology of the Compound Eyes of Insects and Crustaceans (R.C. Hardie, Translator). Springer-Verlag, Berlin.

Farmer, A.S.D., 1975. Synopsis of data on the Norway lobster, *Nephrops norvegicus*. FAO Fisheries Synopsis No. 112, pp. 1–97.

Fleissner, G., Fleissner, G., 2006. Endogenous control of visual adaptation in invertebrates. In: Warrant, E.J., Nilsson, D.-E. (Eds.), Invertebrate Vision, University Press, Cambridge.

Frank, T.M., Widder, E.A., 1994a. Comparative study of behavioural sensitivity thresholds to near UV and blue-green light in deep-sea crustaceans. Mar. Biol. 121, 229–235.

Frank, T.M., Widder, E.A., 1994b. Evidence for behavioural sensitivity to near-UV light in the deep-sea crustacean *Systellaspis debilis*. Mar. Biol. 118, 279–284.

Frank, T.M., Widder, E.A., 1996. UV light in the deep-sea: in situ measurements of downwelling light in relation to the visual threshold sensitivity of UV-sensitive crustaceans. Mar. Freshwater Behav. Physiol. 27, 189–197.

Gaten, E., 1988. Light-induced damage to the dioptric apparatus in *Nephrops norvegicus* and the quantitative assessment of the damage. Mar. Behav. Physiol. 13, 169–183.

Gaten, E., 1990. The ultrastructure of the compound eye of *Munida rugosa* (Crustacea: Anomura) and pigment migration during light and dark adaptation. J. Morphol. 205, 243–253.

Gaten, E., 1992. The anatomy and physiology of selected reflecting superposition eyes. Ph.D. thesis. University of Leicester, 203 pp.

Gaten, E., 1994. Geometrical optics of a galatheid compound eye. J. Comp. Physiol. A 175, 749–759.

Gaten, E., 1998. Optics and phylogeny: is there an insight? The evolution of superposition eyes in the Decapoda (Crustacea). Contrib. Zool. 67, 223–235.

Gaten, E., Shelton, P.M.J., 1993. Spatial resolution in benthic decapods, determined by electrophysiological measurement of acceptance angle. J. Physiol. 467, 371.

Gaten, E., Shelton, P.M.J., Chapman, C.J., Shanks, A.M., 1990. Depth-related variation in structure and functioning of the compound eyes of the Norway lobster *Nephrops norvegicus*. J. Mar. Biol. Assoc. 70, 343–355.

Gaten, E., Shelton, P.M.J., Nowel, M.S., 2004. Contrast enhancement through structural variations in the rhabdoms of oplophorid shrimps. Mar. Biol. 145, 499–504.

Gaten, E., Wiese, K., Johnson, M.L., 2010. Laboratory based observations of behaviour in northern krill (*Meganyctiphanes norvegica* Sars). Adv. Mar. Biol. 57, 233–255.

Hafner, G.S., Tokarski, T.R., 1998. Morphogenesis and pattern formation in the retina of the crayfish *Procambarus clarkii*. Cell Tissue Res. 293, 535–550.

Hafner, G.S., Tokarski, T.R., 2001. Retinal development in the lobster *Homarus americanus*: comparison with compound eyes of insects and other crustaceans. Cell Tissue Res. 305, 147–158.

Hale, A.J., 1958. The Interference Microscope in Biological Research, E.S. Livingstone, Edinburgh and London, 114 pp.

Harzsch, S., Hafner, G.S., 2006. Evolution of eye development in arthropods: phylogenetic aspects. Arthropod Struct. Dev. 35, 319–340.

Herring, P.J., Gaten, E., Shelton, P.M.J., 1999. Are vent shrimps blinded by science? Nature 398, 116.

Horridge, G.A., 1975. Optical mechanisms of clear-zone eyes. In: Horridge, G.A. (Ed.), The Compound Eye and Vision of Insects, Clarendon, Oxford, pp. 255–298.

Horridge, G.A., Barnard, P.B.T., 1965. Movement of palisade in locust retinal cells when illuminated. Q. J. Microsc. Sci. 106, 131–135.

Howard, J., 1981. Temporal resolving power of the photoreceptors of *Locusta migratoria*. J. Comp. Physiol. 144, 61–66.

Johnson, M.L., 1998. Aspects of vision in mesopelagic decapods. Ph.D. thesis. University of Leicester.

Johnson, M.L., Shelton, P.M.J., Gaten, E., 2000. Temporal resolution in marine decapods from coastal and deep-sea habitats. Mar. Biol. 136, 243–248.

Johnson, M.L., Gaten, E., Shelton, P.M.J., 2002. Spectral sensitivities of five marine decapods crustaceans and a review of spectral sensitivity variation in relation to habitat. J. Mar. Biol. Assoc. 82, 835–842.

Kirk, J.T.O., 1996. Light and Photosynthesis in Aquatic Ecosystems, Cambridge University Press, Cambridge.

Kirschfeld, K., Snyder, A.W., 1975. Waveguide mode effects, birefringence and dichroism in fly photoreceptors. In: Snyder, A.W., Menzel, R. (Eds.), Photoreceptor Optics. Springer-Verlag, Berlin.

Kirshfeld, K., 1974. The absolute sensitivity of lens and compound eyes. Z. Natureforsch. 29c, 592–596.

Kleinholz, L.H., 1961. Pigmentory effectors. In: Waterman, T.H. (Ed.), The Physiology of Crustacea, vol. 2. Academic Press, New York, pp. 133–169.

Kuiper, J.W., 1962. The optics of the compound eye. Symp. Soc. Exp. Biol. 16, 58–71.

Kunze, P., 1979. Apposition and superposition eyes. In: Autrum, H. (Ed.), Handbook of Sensory Physiology, vol. VII/6A. Springer-Verlag, Berlin, pp. 441–502.

Land, M.F., 1972. The physics and biology of animal reflectors. Prog. Biophys. Mol. Biol. 24, 75–106.

Land, M.F., 1976. Superposition images are formed by reflection in the eyes of some oceanic decapod Crustacea. Nature 263, 764–765.

Land, M.F., 1981. Optics and vision in invertebrates. In: Autrum, H. (Ed.), Handbook of Sensory Physiology, vol. VII/6B. Springer Verlag, Berlin, pp. 471–492.

Land, M.F., 1989. Variations in the structure and design of compound eyes. In: Stavenga, D. G., Hardie, R.C. (Eds.), Facets of Vision. Springer-Verlag, Berlin, pp. 90–111.

Land, M.F., Burton, F.A., 1979. The refractive index gradient in the crystalline cones of the eyes of a euphausiid crustacean. J. Exp. Biol. 82, 395–398.

Land, M.F., Nilsson, D.-E., 2002. Animal Eyes, Oxford University Press, Oxford.

Land, M.F., Burton, F.A., Meyer-Rochow, V.B., 1979. The optical geometry of euphausiid eyes. J. Comp. Physiol. 130, 49–62.

Laughlin, S.B., de Ruyter van Stevenwick, R.R., Anderson, J.C., 1998. The metabolic cost of information. Nat. Neurosci. 1 (1), 36–41.

Loew, E.R., 1976. Light, and photoreceptor degeneration in the Norway lobster *Nephrops norvegicus* (L.). Proc. R. Soc. Lond. B 193, 31–44.

Lythgoe, J.N., 1968. Visual pigments and visual range under water. Vision Res. 8, 977–1012.

McFarland, W.N., Loew, E.R., 1983. Wave induced changes in underwater light and their relation to vision. Environ. Biol. Fish. 8, 173–184.

Moeller, J.F., Case, J.F., 1995. Temporal adaptations of visual systems in deep-sea crustaceans. Mar. Biol. 124, 47–54.

Munz, F.W., 1958. Photosensitivie pigments from the retinae of certain deep-sea fishes. J. Physiol. 140, 220–235.

Nilsson, D.-E., 1983. Evolutionary links between apposition and superposition optics in crustacean eyes. Nature 302, 818–821.

Nilsson, D.-E., 1989. Optics and the evolution of the compound eye. In: Stavenga, D.G., Hardie, R.C. (Eds.), Facets of Vision, Springer-Verlag, Berlin, pp. 30–73.

Nilsson, D.-E., 1990. Three unexpected cases of refracting superposition eyes in crustaceans. J. Comp. Physiol. A 167, 71–78.

Nilsson, H.L., Lindstrom, M., 1983. Retinal damage and sensitivity loss of a light-sensitive crustacaean compound eye (*Cirolana borealis*). J. Exp. Biol. 107, 277–292.

Nilsson, D.-E., Nilsson, H.L., 1981. A crustacean compound eye adapted for low light intensities (Isopoda). J. Comp. Physiol. 143, 503–510.

Nilsson, D.-E., Odselius, R., 1981. A new mechanism for light–dark adaptation in the *Artemia* compound eye. J. Comp. Physiol. 143, 389–399.

Nilsson, D.-E., Ro, A.-I., 1994. Did neural pooling for night vision lead to the evolution of neural superposition eyes? J. Comp. Physiol. A 175, 289–302.

Nilsson, D.-E., Gilsén, L., Coates, M., Skogh, C., Garm, A., 2005. Advanced optics in a jellyfish eye. Nature 435, 201–205.

O'Carroll, D.C., Bidwell, N.J., Laughlin, S.B., Warrant, E.J., 1996. Insect motion detectors matched to visual ecology. Nature 382, 63–66.

Parker, G.H., 1890. The histology and development of the eye of the lobster. Bull. Mus. Comp. Zool. Harvard Univ. 20, 1–60.

Parker, G.H., 1932. The movements of retinal pigment. Ergebn. Biol. 9, 239–291.

Partridge, J.C., Archer, S.N., van Oostrum, J., 1992. Single and multiple visual pigments in deep sea fishes. J. Mar. Biol. Assoc. U.K. 72, 113–130.

Pinter, R.B., 1972. Frequency and time domain properties of retinular cells of the desert locust (*Schistocerca gregaria*) and the house cricket (*Acheta domesticus*). J. Comp. Physiol. 77, 383–397.

Razmjoo, S., Hamdorf, K., 1976. Visual sensitivity and the variation of total photopigment content in the blowfly photoreceptor membrane. J. Comp. Physiol. A 105, 279–286.

Rózanowska, M., Sarna, T., 2005. Light-induced damage to the retina: role of rhodopsin chromophore revisited. Photochem. Photobiol. 81, 1305–1350.

Schuppe, H., Hengstenberg, R., 1993. Optical properties of the ocelli of *Calliphora erythrocephala* and their role in the dorsal light response. J. Comp. Physiol. A 173, 143–149.

Shelton, P.M.J., Gaten, E., 1996. Spatial resolution determined by electrophysiological measurement of acceptance angle in two species of benthic decapod crustacean. J. Mar. Biol. Assoc. U.K. 76, 391–401.

Shelton, P.M.J., Shelton, R.G.J., Richards, P.R., 1981. Eye development in relation to moult stage in the European lobster, *Homarus gammarus* (L.). J. Conseil 39, 239–243.

Shelton, P.M.J., Gaten, E., Chapman, C.J., 1985. Light and retinal damage in *Nephrops norvegicus* (L.). Proc. R. Soc. Lond. B 226, 217–236.

Shelton, P.M.J., Gaten, E., Chapman, C.J., 1986. Accessory pigment distribution and migration in the compound eye of *Nephrops norvegicus* (L.) (Crustacea: Decapoda). J. Exp. Mar. Biol. Ecol. 98, 185–198.

Snyder, A.W., Stavenga, D.G., Laughlin, S.B., 1977. Spatial information capacity of compound eyes. J. Comp. Physiol. A 116, 183–207.

Stavenga, D.G., 1979. Pseudopupils of compound eyes. In: Autrum, H. (Ed.), Handbook of Sensory Physiology: Vision in Invertebrates, vol. VII/6A. Springer-Verlag, Berlin, pp. 357–502.

Stavenga, D.G., Smits, R.P., Hoenders, B.J., 1993. Simple exponential functions describing absorbance bands of visual pigment spectra. Vision Res. 33 (8), 1011–1017.

Tokarski, T.R., Hafner, G.S., 1984. Regional morphological variations within the crayfish eye. Cell Tissue Res. 235, 387–392.

Valera, F.G., Wiitanen, W., 1970. The optics of the compound eye of the honeybee (*Apis mellifera*). J. Gen. Physiol. 55, 336–358.

Van Rossum, G., 2011. The Python Reference Manual, Network Theory Limited, Bristol, UK, 150 pp.

Vogt, K., 1975. Optik des flubkrebsauges. Z. Naturforsch. 30c, 691.

Vogt, K., 1977. Ray path and reflection mechanisms in crayfish eyes. Z. Naturforsch. 32, 466–468.

Vogt, K., 1980. Die Spiegeloptik des Flussjrebsauges. J. Comp. Physiol. 135, 1–19.

Warrant, E.J., 2006. Invertebrate vision in dim light. In: Warrant, E.J., Nilsson, D.-E. (Eds.), Invertebrate Vision, Cambridge University Press, Cambridge.

Warrant, E.J., McIntyre, P.D., 1991. Strategies for retinal design in arthropod eyes of low F-number. J. Comp. Physiol. A 168, 499–512.

Warrant, E.L., Nilsson, D.-E., 1998. Absorption of white light in photoreceptors. Vision Res. 38 (2), 195–207.

Weckström, M., Laughlin, S.B., 1995. Visual ecology and voltage gated ion channels in insect photoreceptors. Trends Neurosci. 18, 17–21.

Weibel, E.R.C., Taylor, C.R., Bolis, L., 1998. Principles of Animal Design: The Optimization and Symmorphosis Debate. Cambridge University Press, Cambridge.

Wilson, M., 1978. The functional organisation of locust ocelli. J. Comp. Physiol. A 124, 297–316.

Youssef, P.N., Sheibani, N., Albert, D.M., 2011. Retinal light toxicity. Eye (Lond.) 25, 1–14.

CHAPTER FIVE

Stress Biology and Immunology in *Nephrops norvegicus*

Susanne P. Eriksson[*,1], **Bodil Hernroth**[†,‡], **Susanne P. Baden**[*]

[*]Department of Biological and Environmental Sciences-Kristineberg, University of Gothenburg, SE-45178 Fiskebäckskil, Sweden
[†]The Royal Swedish Academy of Sciences-Kristineberg, SE-45178 Fiskebäckskil, Sweden
[‡]Department of Natural Science, Kristianstad University, Kristianstad, Sweden
[1]Corresponding author: e-mail address: Susanne.eriksson@bioenv.gu.se

Contents

Abstract

The Norway lobster *Nephrops norvegicus* lives at low-light depths, in muddy substrata of high organic content where water salinities are high and fluctuations in temperature are moderate. In this environment, the lobsters are naturally exposed to a number of potential stressors, many of them as a result of the surficial breakdown of organic material in the sediment. This process (early diagenesis) creates a heterogeneous environment with temporal and spatial fluctuations in a number of compounds such as oxygen, ammonia, metals, and hydrogen sulphide. In addition to this, there are anthropogenically generated stressors, such as human-induced climate change (resulting in elevated temperature and ocean acidification), pollution and fishing. The lobsters are thus exposed to several stressors, which are strongly linked to the habitat in which the animals live. Here, the capacity of *Nephrops* to deal with these stressors is summarised. Eutrophication-induced hypoxia and subsequent metal remobilisation from the sediment is a well-documented effect found in some wild *Nephrops* populations. Compared to many other crustacean species, *Nephrops* is well adapted to tolerate periods of hypoxia, but prolonged or severe hypoxia, beyond their tolerance level, is common

Advances in Marine Biology, Volume 64
ISSN 0065-2881
http://dx.doi.org/10.1016/B978-0-12-410466-2.00005-4

© 2013 Elsevier Ltd.
All rights reserved.

in some areas. When the oxygen concentration in the environment decreases, the bio-availability of redox-sensitive metals such as manganese increases. Manganese is an essential metal, which, taken up in excess, has a toxic effect on several internal systems such as chemosensitivity, nerve transmission and immune defence. Since sediment contains high concentrations of metals in comparison to sea water, lobsters may accumulate both essential and non-essential metals. Different metals have different target tissues, though the hepatopancreas, in general, accumulates high concentrations of most metals. The future scenario of increasing anthropogenic influences on *Nephrops* habitats may have adverse effects on the fitness of the animals.

Keywords: Lobster, *Nephrops norvegicus*, Stress, Homeostasis, Climate change, Acidification, Pollutant, Disease, Immunity

1. INTRODUCTION

External factors that disrupt the internal environment of an organism are called stress factors (Cannon, 1926). The ability to control the internal environment is called homeostasis, and most biochemical processes strive to maintain some sort of internal balance, consuming energy and resources in the process. In a constantly changing internal and external environment, achieving a steady state is an almost impossible task. Normally, organisms are well adapted to live in their natural environment but, at the limit of their tolerance range for the intensity of any particular stressor, additional intensity (in magnitude, frequency or number) will have a strong impact on the survival of the population/species. Climate change has led to a significant change in the intensities of a number of abiotic factors, such as, salinity, temperature, oxygen availability and ocean acidification. The ecological consequences of such sudden environmental change (on a geological timescale) and increased stress will depend on the animals' ability to adapt to external and internal hypoxia, which sets their homeostatic capacities/limits. Adaptation occurs at all organisational levels, from gene to ecosystem. Within a species, the genetic diversity, physiological plasticity and behavioural responses will set the limits to how successfully the animal will cope with stress factors. From an evolutionary point of view, a stress response is an important aspect of the behaviour of all higher animal species, including crustaceans. The response to stress may be very individual in both space and time, however, and tolerance to single or multiple stressors can vary between and within individuals, sexes, ages and populations.

The Norway lobster *Nephrops norvegicus* lives and burrows in soft, muddy, marine sediment, typically of high organic content. The animals live in direct contact with the sediment, and any environmental changes are expected to be reflected in changes in the animals' physiology. In this environment, there are normally moderate seasonal fluctuations in salinity and temperature. The light level is typically low and the compound eye of *Nephrops* is well adapted to these low levels of light (Aréchiga and Atkinson, 1975). The eye may, however, be damaged if the animal is brought up to the daylight at the surface (i.e. during fishing). Eye damage is discussed in full in Chapter 4 and is therefore not included in this chapter. The availability of oxygen to a *Nephrops* habitat (at depths >20 m) depends on gas exchange with the atmosphere at the water surface. Elevated primary production locally increases the flux of organic carbon to the bottom sediments (Wassmann, 1990) and increases oxygen demand through bacterial and other metazoan respiratory processes (Graf et al., 1982). Accumulation of organic carbon is found mainly in enclosed sea areas or where water column stratification causes stagnant conditions (Diaz and Rosenberg, 2008). Increased water temperatures decrease the solubility of oxygen and, in addition, enhance the oxygen consumption of both organisms and substrate. Hypoxia is therefore often associated with the seasonal peak in the annual temperature cycle, which occurs around August in the northern hemisphere (Diaz and Rosenberg, 2008). Stratification is also strengthened by other differences in water density caused by, for example, salinity (Diaz and Rosenberg, 2008). Oxygen exchange between the seawater masses is limited, and in a stratified water column, a gradient of declining oxygen concentrations will result from a combination of water column and benthic oxygen demand setting up hypoxic conditions. Due to the high oxygen demand, oxygen passively diffuses only a few millimetres into compact, muddy and silty sediments (Revsbech et al., 1980). Aquatic sediments are, however, normally inhabited by a large diversity of bottom-dwelling animals, and through animal activity (i.e. bioturbation), oxygen is normally distributed much deeper into the sediment (Boudreau and Jørgensen, 2001). By introducing oxygen into anoxic sediment zones, the total area of oxic–anoxic boundaries as well as the surface available for diffusive exchange increases. Animal activity therefore strongly influences both the structure and chemical composition of the sediment as well as fluxes across the sediment–water interface (Boudreau and Jørgensen, 2001). The boundary zone separating surficial aerobic areas from subsurface anaerobic areas is called the redox potential discontinuity (Boudreau and Jørgensen, 2001; Diaz and Rosenberg, 2008). Its position and thickness is, apart from the bioturbation, also affected by the oxygen content of the bottom water, sedimentation of organic matter, sediment grain size, and temperature. Redox-sensitive metals such as manganese (Mn) and iron (Fe) are enriched in this zone, and often also along the oxic–anoxic boundary on the inner side of a well-irrigated burrow wall. Carbon reduces a number of compounds naturally occurring in the sediment that are potentially hazardous to organisms, such as Mn, Fe and sulphide. These compounds thus increase in bioavailability as they flux out of the sediment and into the water of animal burrows (Boudreau and Jørgensen, 2001; Eriksson and Baden, 1998).

Diagenesis is any physical and/or chemical process that changes sediment into sedimentary rock. This change is controlled by factors such as pressure, temperature, particle size, porosity/permeability, composition and water flow (e.g. Aller, 2004; Boudreau and Jørgensen, 2001; Revsbech et al., 1980). Early diagenesis (eodiagenesis) is the initial stage of diagenesis and takes place in the top sediment layer where deposition occurs (Aller, 2004). The main changes in this process are bioturbation, mineralogical changes, and compaction. Particles in the aquatic environment are buried and transformed in the surficial sediments and the upper few centimetres of the sediment and the sediment–water interface are the most physically, chemically and biologically active parts of deposits (e.g. Aller, 2004; Boudreau and Jørgensen, 2001; Revsbech et al., 1980). During eodiagenesis, the organic material imported from the water column is transformed and mineralized in surficial sediments through a series of biogeochemical reactions involving a number of compounds potentially hazardous to organisms, such as sulphide, Mn and Fe (see Table 5.1). It is

Table 5.1 Biogeochemical processes (early diagenesis) and products (in bold) in the sediment affecting the ecophysiology of *Nephrops norvegicus*

O_2 reduction $CH_2O + O_2 \rightarrow \mathbf{CO_2} + H_2O$
NO_3^- reduction $2CH_2O^- + NO_3^- + 2H^- \rightarrow 2CO_2 + \mathbf{NH_4^+} + H_2O$
$Mn(IV)$ reduction $CH_2O + 2MnO_2(s) + 4H^+ \rightarrow CO_2 + 2\mathbf{Mn^{2+}}(aq) + 3H_2O$
$Fe(III)$ reduction $CH_2O + 4FeOOH(s) + 8H^+ \rightarrow CO_2 + 4\mathbf{Fe^{2+}}(aq) + 7H_2O$
SO_4^{2-} reduction $CH_2O + 0.5SO_4^{2-} + H^+ \rightarrow CO_2 + 0.5\mathbf{H_2S}(g) + H_2O$

a well-documented fact that benthic fauna have a large impact on both sediment transport and chemically available surface areas, and thus, also on diagenetic reactions (Aller, 2004; Boudreau and Jørgensen, 2001). Knowledge of how sediment biogeochemistry in turn affects fauna is, however, still lagging behind. Since there is seldom such a thing as a chemical steady state in nature, but rather a mixture of transitional states and pseudo-steady states, questions of animal tolerance and physiological plasticity need to be investigated to increase our understanding of animal–chemical interactions.

The boundary zone separating surficial aerobic sediment from subsurface anaerobic sediment is called the redox potential discontinuity or redoxcline. Due to the steep concentration gradients, diagenetically generated solutes in the pore water diffuse into the water of animal burrows (Boudreau and Jørgensen, 2001). This process is further enhanced by hypoxic events that may follow eutrophication caused by nutrient pollution (Diaz and Rosenberg, 2008) and trophic cascade effects from overfishing (Casini et al., 2008). The micro-environmental chemical properties associated with processes in and around burrows are thus likely critical determinants of uptake of these solutes in biota.

Although the availability (or lack of it) of compounds involved in eodiagensis may be detrimental to benthic-dwelling organisms, they are still naturally occurring processes. Apart from these naturally occurring potentially hazardous compounds, other ambient stressors have been linked directly to human activities, especially during the past industrial centuries. Such is the case with the currently occurring global climate change with the concomitant elevated sea temperatures, ocean acidification, and sea level rise (Intergovernmental Panel on Climate Change, IPCC, 2007). Increased temperature generally affects both biological and chemical processes, in that it speeds up the rate at which they are occurring within certain boundaries. Whether the end result is likely to be positive or negative for a species depends mostly on their physiological capacity and their temperature tolerance (both maximum and minimum temperature). The industrial revolution has also resulted in increased turnover rates of natural resources, such as minerals, metals and fossil fuels (Zalasiewicz et al., 2011). Sediment particles adsorb and incorporate metal ions and organic pollutants, and the concentration in sediment can be several orders of magnitude higher compared to seawater (Hart, 1982). Sediment-burrowing animals, such as *Nephrops*, are thus prone to be exposed to pollutants both through their burrowing as well as through food and sediment intake. An overview of the stressors discussed in this chapter can be seen in Figure 5.1 and Table 5.2.

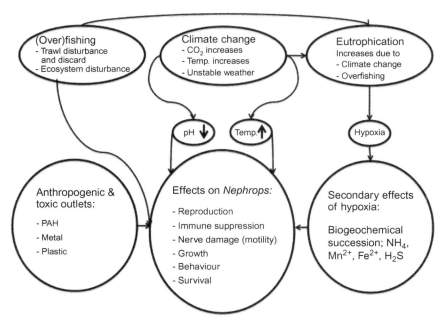

Figure 5.1 Stressors affecting *Nephrops norvegicus* in the habitat.

2. STRESS FACTORS

2.1. Temperature, salinity and dissolved gases

2.1.1 Temperature

Nephrops norvegicus is distributed from Iceland to the eastern Mediterranean at temperatures between 6 and 17 °C (Chapter 2) and each population has evolved to tolerate the spatial and temporal temperature variations experienced. Oxidative stress can be assessed by measuring advanced glycation end products (AGE) (Carney Almroth et al., 2012; Nyström, 2005). Experiments using this technique and exposing *Nephrops* from the Skagerrak to near-future (twenty-first century) temperature elevations of about 4 °C above normal in this area (from 14 to 18 °C) for 4 months did not suggest any oxidative stress. The elevated temperatures affected neither the immune defence nor the mortality rates of the lobster (Hernroth et al., 2012). When trawled, *Nephrops* may experience large temperature changes as they are hauled and then later discarded back into the sea. Survival is higher, stress lower and recovery faster in animals trawled during winter (when surface water is lower in temperature than the bottom water they inhabit) compared

Table 5.2 Stressors and their dose–response in *Nephrops norvegicus*

Stressor	Dose	Induced by	Nephrops stage	Sublethal effect	Reference
Temperature	18 °C (+trawl)	Fishing (trawl summer)	Adult	Increased crustacean hyperglycaemic hormone production, haemolymph acidosis	Lund et al. (2009) and Albalat et al. (2010)
Salinity	15 psu (<2 h)	Fishing (hauled through surface water)	Adult	Haemodilution, moribund (no tail flipping)	Harris and Ulmestrand (2004)
Oxygen (O₂)	50% saturation (sat)		Zoea	Induced change to adult haemocyanin with higher oxygen affinity	Spicer and Eriksson (2003)
	30% sat	Eutrophication and diagenesis	Juvenile	Raised body posture, escape swimming	Eriksson and Baden (1997)
	50% sat	Eutrophication and diagenesis	Adult	Increased irrigation and increased haemocyanin production	Hagerman and Uglow (1985), Hagerman and Baden (1988), Baden et al. (1990b), and Gerhardt and Baden (1998)
	30% sat	Eutrophication and diagenesis	Adult	Haemolymph alkalosis	Hagerman and Uglow (1985)
	20% sat	Eutrophication and diagenesis	Adult	Raised body posture	Baden et al. (1990b)

Continued

Table 5.2 Stressors and their dose–response in *Nephrops norvegicus*—cont'd

Stressor	Dose	Induced by	Nephrops stage	Sublethal effect	Reference
	15% sat	Eutrophication and diagenesis	Adult	Emerge from burrows, stop eating	Baden et al. (1990a,b)
	<10% sat	Eutrophication and diagenesis	Adult	Anaerobic metabolism (glycogen depletion, L-lactate and glucose build-up), haemocyanin depletion	Hagerman et al. (1990) and Baden et al. (1990b, 1994)
Ammonia (NH$_3$)	300 μM (+hypoxia)	Eutrophication and diagenesis	Adult	Haemolymph alkalosis, detoxification of high NH$_3$ in haemolymph	Hosie et al. (1991) and Bernasconi and Uglow (2011)
Hydrogen sulphide (H$_2$S)	50 μM	Eutrophication and diagenesis	Adult	Avoidance, detoxification of high H$_2$S in haemolymph	Visman (1991) and Butterworth et al. (2004)
	600 μM	Eutrophication and diagenesis	Adult	Bradycardia, slower scaphognathite beating	Taylor et al. (1999)
	800 μM	Eutrophication and diagenesis	Adult	Decreased respiration	Taylor et al. (1999)
Carbon dioxide (CO$_2$)	pH 7.7	Ocean acidification	Adult	Immunosuppression, increased AGE protein, haemolymph acidosis	Hernroth et al. (2012)
Manganese (Mn)	5 mg l^{-1}	Eutrophication, diagenesis, pollution	Adult	Decreased neuromuscular contractile force, induced hemocytopenia and apoptosis in haematopoietic cells (gradually increasing)	Holmes et al. (1999) and Oweson et al. (2006)

	Concentration	Driver	Life stage	Effect	Reference
	10 mg l^{-1}	Eutrophication, diagenesis, pollution	Adult	Decreased post-flip extension (during tail-flip swimming)	Baden and Neil (1998)
	15 mg l^{-1}	Eutrophication, diagenesis, pollution	Adult	Reduced total haemocyte counts decreased bactericidal capacity and increased susceptibility to infections	Oweson et al. (2006) and Oweson and Hernroth (2009)
	20 mg l^{-1}	Eutrophication, diagenesis, pollution	Adult	Immunosuppression (60% decrease in haemocytes), impaired haemocyanin synthesis	Baden et al. (2004) and Hernroth et al. (2004)
	5 mg l^{-1}	Eutrophication, diagenesis, pollution	Adult	Impaired ability to search for food	Kråg and Rosenqvist (2006)
Iron (Fe)	Not known	Diagenesis, pollution	Adult	Not known	
Lead (Pb)	Not known	Pollution	Adult	Not known	
Mercury (Hg)	Not known	Pollution	Adult	Not known	
Cadmium (Cd)	1 µg l^{-1}	Pollution	Adult	Metallothionein production	Canli et al. (1997)
Copper (Cu)	1 µg l^{-1}	Pollution	Adult	Metallothionein production	Canli et al. (1997)
Zinc (Zn)	8 µg l^{-1}	Pollution	Adult	Metallothionein production	Canli et al. (1997)

Continued

Table 5.2 Stressors and their dose–response in *Nephrops norvegicus*—cont'd

Stressor	Dose	Induced by	*Nephrops* stage	Sublethal effect	Reference
Cobalt (Co)	$6 \, \text{mg} \, \text{l}^{-1}$	Eutrophication, diagenesis, pollution	Adult	Decreased neuromuscular contractile force	Holmes et al. (1999)
Organic pollutants	Not known	Pollution	Adult	Not known	
Plastic	Not known	Fisheries, waste	Adult	Not known but of concern since majority of adults have plastic filaments in stomach	Murray and Cowie (2011)
Fishing gear	Trawl	Fishing	Adult	Increased egg loss, limb loss	Briggs et al. (2002), Eriksson (2006), and Ridgway et al. (2006)
	Trawl (+emersion, 2 h)	Fishing	Adult	Tail flipping, anaerobic metabolism (glycogen depletion, L-lactate build-up)	Gornik et al. (2010)
Disease	Hematodinium		Adult	Milky haemolymph, hyperpigmentation of the carapace and degraded muscle tissue	Field et al. (1992) and Stentiford and Shields (2005)

Doses are at levels shown to occur in the field, apart from pH and the metals cadmium, cobalt, copper and zinc, which are effects seen in the laboratory.

to animals caught in summer, when surface water temperature may increase to >20 °C (see below; Albalat et al., 2010; Lund et al., 2009). The stress hormone crustacean hyperglycaemic hormone (CHH) immediately after being trawled, was almost double in summer when the lobsters were caught at 18 °C compared to that at 5 °C during winter (295 and 170 pM, respectively) (Lund et al., 2009). In summer, lobsters also showed a combined metabolic and respiratory acidosis. Thus, the combination of stress from trawling and a temperature at the upper range of what *Nephrops* normally exhibits could create sublethal stress, which, when surface temperature reaches 21 °C, may even become lethal (Lund et al., 2009).

2.1.2 Salinity

Ambient salinity has an impact on the composition and osmolality of animals' internal fluids, and many marine species possess physiological abilities to regulate the osmotic pressure. Osmoregulation has been extensively studied in crustaceans (for a review, see Mantel and Farmer, 1983). *Nephrops* live in habitats of high salinity (32–39 psu; Chapter 2), and physiologically the species is described as being stenohaline, that is, intolerant of reduced salinities, with a suggested lower salinity limit of 29–30 psu restricting its distribution (Poulsen, 1946). Throughout its geographic range, salinity is generally high all the way to the surface. However, in the Kattegat/Skagerrak, *Nephrops* normally live in high salinities (33–34 psu), but when caught they are raised through a low-salinity surface layer and if discarded descend back through this gradient to the sea bed. Harris and Ulmestrand (2004) simulated such conditions by exposing *Nephrops* for short (<2 h) periods to low salinity water (15 psu) alone or in combination with emersion. The results show that discarded *Nephrops* experienced significant haemodilution and gained mass rapidly. When returned to high salinity water, the animals slowly recovered in terms of response to stimuli and stress behaviour such as tail flipping. Delayed effects included abdominal stiffness, swelling and mortalities (25–42% overall). The *Nephrops* showed an LC_{50} value of 25 psu and 100% mortality at 21 psu, which is significantly higher than past estimates in fishery due to the stress of this brief exposure to low-salinity surface water (Harris and Ulmestrand, 2004).

2.1.3 Oxygen

$$O_2 \text{ reduction}: \quad CH_2O + O_2 \rightarrow CO_2 + H_2O$$

Oxygen saturation (dissolved oxygen) in sea water is a relative figure that changes with salinity, temperature and air pressure. *Nephrops* generally live in areas of high salinity (32–39 psu) and low to medium temperatures (6–17 °C). In these waters, full oxygen saturation (100%) equals ≈8–10 mg oxygen l^{-1}. An oxygen saturation of 20% on a *Nephrops* ground thus equals approximately 1.6–2 mg oxygen l^{-1}. Normoxia is considered to be normal levels of oxygen with regard to the physiological responses of living organisms and hypoxia is when oxygen is deficient for aerobic organisms (Tyson and Pearson, 1991). *Nephrops* grounds normally exhibit oxygen concentrations of 6–9.4 mg l^{-1} (Chapter 2); thus, oxygen levels are high but the water is not fully saturated. Benthic habitats are naturally prone to occasional hypoxia, which is aggravated by eutrophication. Thus, living in benthic soft sediments requires a better natural tolerance to hypoxia than living in the pelagic. Of the marine invertebrate phyla, crustaceans are generally the most susceptible to hypoxia (e.g. Diaz and Rosenberg, 1995; Haselmair et al., 2010).

Hypoxic areas around the world including *Nephrops*-inhabited areas are unfortunately still increasing (Diaz and Rosenberg, 2008). However, reports on *Nephrops* affected by hypoxia are mainly from the SE Kattegat area between Denmark and Sweden. Incidents of hypoxia affecting *Nephrops* are also briefly reported from Lough Hyne Ireland (Kitching et al., 1976). The Kattegat is shallow (mean depth 23 m) and has a strong halocline at 15–20 m reinforced by a thermocline during summer and autumn, which makes the area susceptible to oxygen deficiency (Rosenberg, 1985; Svensson, 1975).

A negative correlation between oxygen concentration and catch efficiency of *Nephrops* was found in the Kattegat indicating that *Nephrops* emerge from their burrows when oxygen is declining. As a result, the catch efficiency doubled in the SE Kattegat in 1974–1976 at 50% oxygen saturation (when measured 1 m above the sediment) (Bagge and Munch-Petersen, 1979). Because females tend to stay in the burrows, the sex ratios of the catches are normally skewed. During normoxia, about 25% females were found in the catch, which increased to more than 50% during initial hypoxia (Baden et al., 1990b; Bagge and Munch-Petersen, 1979; Thomas and Figueiredo, 1965). Between 1984 and 1987, the female ratio varied between 7% and 38%, whereas in September of 1988, females represented 74–77%, possibly indicating that the males had already been caught during preceding years and the females had all been forced out of the burrows due to the severe hypoxia of down to 10% oxygen saturation (0.5 m above the bottom) hypoxia (Baden and Pihl, 1996; Baden et al., 1990a,b).

Eutrophication-induced hypoxia can result in lethal oxygen concentrations for *Nephrops*. In the autumn of 1985, fishermen reported dead and dying *Nephrops* (up to 50% of the total catch) for the first time in the SE Kattegat. This was repeated in 1986. In 1988, the oxygen deficiency was the worst since 1974 (10% O_2 saturation when measured 0.5 m above the sediment) and extended 4000 km^2 for 3 autumn months. No *Nephrops* were caught in research trawls from the autumn of 1988 to 1991 (Baden and Pihl, 1996; Baden et al., 1990a,b).

Hypoxic periods in the autumn often coincide with the settlement of juvenile *Nephrops*. In experiments, juvenile *Nephrops* follow the response pattern of the adults during progressive hypoxia (see later), but they are more sensitive and respond at higher oxygen concentrations (Eriksson and Baden, 1997). Repeating the research trawling, SE Kattegat in 1995 showed a slight recovery of mainly larger *Nephrops* males (potentially migrating into the areas) at the border of the affected area but no *Nephrops* in the previously most affected areas (S.P. Baden and L. Pihl, unpublished data). Whether the *Nephrops* population in SE Kattegat has recovered completely in this specific area (ICES square 4257) since 1995 is unknown as no research trawling has been carried out for historical comparison. In 2002, a serious hypoxic event occurred again in the SE Kattegat and may have significantly reduced a potential recovered population (Ärtebjerg et al., 2003). These hypoxic events below the halocline of course affected the whole benthic ecosystem, and caused migration and death of fish (Pihl, 1989) and emergence and death of all benthic invertebrate species (Baden et al., 1990a; Rosenberg, 1985; Rosenberg and Loo, 1988; Rosenberg et al., 1990). As a consequence, large-scale research projects were financed during the late 1980s to clarify the effects of eutrophication-induced hypoxia on the whole ecosystem and especially to find biomarkers of hypoxia and a dose–response relationship between eutrophication and degree of hypoxia (Rosenberg et al., 1990).

Nephrops emerge from their burrows at oxygen concentrations below 50% O_2 saturation (when measured 1 m above the sediment) (Bagge and Munch-Petersen, 1979) or below about 15% O_2 saturation when measured 0.5 m above the sediment (Baden et al., 1990a,b). In experiments, *Nephrops* exposed to 20% O_2 saturation and juvenile *Nephrops* exposed to 30% O_2 saturation stood high above the sediment with extended legs for better irrigation with their pleopods in order to extract more oxygen but were still eating when offered food. Juvenile *Nephrops* increased the flight activity at hypoxia of 30–25% O_2 saturation and the LT_{50} was 21.5 h. At below 15% O_2 saturation, adult *Nephrops* supported their flexed legs with the abdomen and the claws and stopped eating, and O_2 saturations <10% were lethal to adult lobsters within a couple of days (Baden et al., 1990b). During hypoxia, even the potential food items for *Nephrops* are affected. As crustaceans are more sensitive to hypoxia than other invertebrate phyla (Diaz and Rosenberg, 1995), the natural food selection of *Nephrops* from hypoxic areas mirrors this sensitivity. Comparing stomach analyses of *Nephrops* from hypoxic SE Kattegat (1986–1988) with reference areas in the NE Kattegat, crustaceans, normally being the most preferred food item, decreased with 30% in occurrence in *Nephrops* stomachs from hypoxic areas, and, most interestingly, the stomach weight/body weight ratio declined from 0.8–1.2 to 0.1 (Baden et al., 1990b).

The succession of responses to hypoxia in invertebrates is thoroughly reviewed by Grieshaber et al. (1994). *Nephrops* is no exception to this succession, and during the years following the events of serious hypoxia in the SE Kattegat, research to clarify the effects on *Nephrops* was extensive, which explains the bulk of literature on *Nephrops* from this period. In contrast to demersal fish, *Nephrops* is not reported to escape from large areas of hypoxia but rather to adapt (Hagerman and Baden, 1988; Pihl, 1989). The first reaction of adaptation in order to fulfil the oxygen demand in crustaceans is an increased rate of ventilation performed by the scaphognathites in the gill chambers (Grieshaber et al., 1994). Increased scaphognathite activity in *Nephrops* was initiated at 52% oxygen saturation and doubled at 26% oxygen saturation (Hagerman and Uglow, 1985). In burrow-living crustaceans, irrigation has an important burrow ventilatory function and, together with ventilation, renews the water to respiratory surfaces. *Nephrops* use their pleopods for irrigation. Irrigation can be quantified in artificial burrows in the laboratory (Figure 5.2) and when comparing the irrigation behaviour of *Nephrops*

Figure 5.2 *Nephrops norvegicus* in an artificial burrow in the laboratory. The animal's pleopod activity is quantified through the transparent glass under infrared light (photo: Susanne Eriksson). (For colour version of this figure, the reader is referred to the online version of this chapter.)

after 3 h exposure to normoxia and to 30% oxygen saturation Gerhardt and Baden (1998) found a 10-fold higher pleopod activity of male than female *Nephrops* in normoxia, whereas no difference was found between the sexes with regard to hypoxia. This unexpected sex-related difference was suggested by the authors to originate from a higher restlessness of the males observed in the artificial burrows but could be a result of better female adaptation to generally reduced oxygen in the burrows where they spend more time than males, as previously mentioned. Ovigerous females, however, showed highly increased pleopod irrigation during hypoxia (30% O_2 saturation). *Nephrops* carrying eggs at stage 4 (40% of embryonic time; Dunthorn, 1967) had a two orders of magnitude higher pleopod stroke rate than hypoxic non-ovigerous females, and a 20 times higher pleopod stroke rate than normoxic ovigerous females. *Nephrops* carrying eggs in stage 8 (88% of embryonic time) had a general increase in pleopod beat rate of about 5–10 times that when carrying eggs in stage 4. Again, females under hypoxia had a nine times higher pleopod stroke rate than non-ovigerous females and twice that of normoxic ovigerous females (Eriksson et al., 2006). After exposing males to 7 h of progressive hypoxia, the accumulated irrigation activity was more than four times the normoxic activity. The stroke frequency doubled between 50% and 60%, oxygen saturation reaching a maximum frequency at 20%, after which the activity reached a critical point below 10% and dropped to near-normoxic values at 5% saturation. Thus, increased ventilation and irrigation are initiated simultaneously at about 50–60% oxygen saturation and the maximum frequency is reached around 20–25% O_2 saturation (Gerhardt and Baden, 1998; Hagerman and Uglow, 1985). Initiation of higher ventilation and irrigation levels is comparable to that of other crustacean species living on the shallow sediment surface such as *Palaemon adspersus* (Hagerman and Uglow, 1981), whereas permanent burrow-living crustaceans such as Thalassinids are more tolerant (Astall et al., 1997). It is suggested by Gerhardt and Baden (1998) that the critical point of oxygen saturation (<10% oxygen saturation) is where the energy expenditure of high pleopod frequency exceeds beneficial gain, with the probable initiation of physiological adaptations as also suggested by Grieshaber et al. (1994).

During hypoxia when motility and feeding are reduced, internal energy resources, such as glycogen, are fuelling glycogenesis (Grieshaber et al., 1994). During starvation for 7 months in a continuous supply of non-filtrated water, the glycogen concentrations in muscle and hepatopancreas were reduced to 3% and 10%, respectively. When severe hypoxia (below 10% O_2 saturation) and hypoxia-induced starvation are superimposed, the decline in glycogen resources providing increased glucose and lactate concentrations in the haemolymph is fast and may indicate a shift to anaerobic metabolism (Baden et al., 1994; Hagerman et al., 1990). Exposure for 6 days to 10% oxygen saturation resulted in an 85% reduction in muscle glycogen compared to controls. On checking the muscle and hepatopancreas glycogen concentration of *Nephrops* exposed to hypoxia in the field, the concentration was found to be reduced to about 30% of *Nephrops* from normoxic areas (Baden et al., 1994). See Figure 5.3 for haemolymph sampling, typically collected by inserting a needle directly into the infrabranchial sinus at the base of the last walking leg.

Crustacean haemocyanin adapts to environmental hypoxia, for example by a negative Bohr effect (increase in haemolymph pH with decreasing O_2 saturation) in order to increase the efficiency of oxygen loading of haemocyanin at the gills and release at the tissue (Bridges, 2001). Four hours after exposure to 30% O_2 saturation, the haemolymph pH increases from about 7.8 to 8.0 due to the Bohr effect (Hagerman and Uglow, 1985). Hypoxia–related increase in haemocyanin in crustaceans was first observed by Senkbiel and Wriston (1981) in *Homarus americanus* and empirically confirmed in *Nephrops* by Hagerman and Uglow (1985). Further, they found a positive linearity between haemocyanin and oxygen-carrying capacity indicating the advantage of having increased haemocyanin during hypoxia.

Figure 5.3 Sampling of haemolymph in egg-carrying female *Nephrops norvegicus* (photo: Mikael Dahl). (For colour version of this figure, the reader is referred to the online version of this chapter.)

During the events of serious hypoxia (O_2 saturations <10%) in the SE Kattegat in 1986–1987, haemocyanin was sampled from *Nephrops* at five locations and compared with a normoxic reference location on the Scottish east coast (Baden et al., 1990b; Hagerman and Baden, 1988) to moderate hypoxic locations in the NE Skagerrak (Baden et al., 1990b). Surprisingly, extremely low mean haemocyanin levels of 0.04–0.16 mM l^{-1} were found in *Nephrops* in the Kattegat, whereas the Scottish reference station had a mean haemocyanin level of 0.66 mM l^{-1}. In the Skagerrak *Nephrops* exposed to moderate hypoxia (O_2 saturation between 20% and 40%), the mean haemocyanin was 1.2 mM l^{-1} in October with extreme levels of 1.4 mM l^{-1}.

The idea of using the haemolymph concentration of field *Nephrops* as a biomarker of exposure to hypoxia has been explored. To interpret the field concentrations correctly, a series of laboratory experiments with controlled exposures were performed. *Nephrops* were exposed to long-term (20–40 days or until death) hypoxia (50–10% O_2 saturation) and either fed (Hagerman et al., 1990) or starved (Baden et al., 1990b) during exposure. In normoxia, the haemocyanin of *Nephrops* remained stable when they were fed or decreased slowly when they were starved. Under moderate hypoxia (50–15% O_2 saturation), the haemocyanin increased as predicted and decreased after a while depending on the degree of hypoxia and access to food. In both experiments, *Nephrops* exposed to 10% O_2 saturations showed a rapid decrease in haemocyanin and died after about 3–4 days. Oxidative reduction of surplus organic matter in the sediment gives rise to hypercapnia as well as hypoxia. It has been speculated that hypercapnia causes haemocyanin dissociation and brings about a reduction in functional haemocyanin under intense hypoxia. After exposure to <10% O_2 saturation, haemolymph was oversaturated with CO_2 and effervescent when sampled (S.P. Baden, unpublished data). This supersaturation of the haemolymph may be exacerbated by the fact that CO_2 increases externally as well during hypoxia, making the CO_2 gradient less steep.

In conclusion, haemocyanin concentrations indicate whether *Nephrops* have been exposed to moderate (elevated haemocyanin) or serious hypoxia or moderate hypoxia for a long period (low haemocyanin).

Although mean haemocyanin levels showed statistically significant changes in response to hypoxic exposure, the inter-individual variation was large. In normoxic conditions, the haemocyanin concentration may indicate the prevailing access to food of high quality. In a virgin unexploited population of *Nephrops* (Loch Torridon), a sex difference existed with males that were more active on the bottom, and had a significantly higher haemocyanin level than females in burrows (Baden et al., 1990b). This difference the sexes was also found in trawled and creeled areas along the Swedish west coast (Eriksson, 2006) and may be interpreted as a better access to food for males, who stored the surplus in the haemocyanin protein. Alternatively, the females may have a higher rate of haemocyanin catabolism when food is limited.

The trade-off between storing energy in haemocyanin and using it during hypoxia was examined by Spicer and Baden (2001). *Nephrops* were exposed to 24 h of moderate and serious hypoxia of about 30% and 20% O_2 saturation. The results indicated that the synthesis of haemocyanin was much faster than previously found when measured first after weeks of hypoxia. Depending on the initial individual haemocyanin value, levels could more than double after only 24 h of hypoxic exposure. However, an unexpected decrease in haemocyanin was observed when the initial concentration was high. Thus, an optimal individual haemocyanin (trade-off) value depending on the hypoxic situation and initial storage of haemocyanin seems to exist.

As the oxygen saturation of water decreases with increasing temperature, the oxygen requirement (Q10) of the organism increases and hypoxic stress is exacerbated. With an expected temperature increase of 4 °C during the twenty-first century due to global change, areas with hypoxia are expected to increase (Intergovernmental Panel on Climate Change, IPCC, 2007). A meta-analysis of 363 experiments representing benthic animals of 108 species and 10 taxonomic groups showed that the median lethal oxygen concentration increases with temperature, and especially temperatures greater than 20 °C (Vaquer-Sunuyer and Duarte, 2011).

2.1.4 Ammonia

$$\text{Nitrate reduction}: \quad 2CH_2O^- + NO_3^- + 2H^- \rightarrow 2CO_2 + NH_4^+ + H_2O$$

Besides hypoxia, burrow-living organisms have to cope with ammonia deriving from microbial activity in the burrow as well as from their own excretory products.

Ammonia concentrations in seawater are low ($\approx 10\ \mu M$), but 200–300 μM has been recorded from porewater under normoxic conditions (Hopkinson, 1987; Hosie et al., 1991). Under hypoxia, ammonia efflux from the sediment may reach 100 $\mu M\ m^2 h^{-1}$ (Kristensen, 1984). As a result of both internal and external exposure to toxic ammonia, the normoxic haemolymph ammonia concentration in *Nephrops* may vary between 130 and 170 μM. Under hypoxia (≈15% in 6 h), the internal ammonia concentration decreases with about 30%, helped by a simultaneous increase in turnover time of haemolymph ammonia (Hosie et al., 1991). In the same experiment, *Nephrops* were exposed to hypoxic water enriched with 300 μM ammonia resulting in an internal ammonia concentration increase, but it is not known if this derived from the protein metabolism of the animal, from an influx from the enriched surrounding water or both in unknown proportions. The combination of enrichment and hypoxia resulted in less ammonia being accumulated than with enrichment alone. Combinations of enrichment and hypoxia resulted in haemolymph alkalosis (0.2–0.6 units) and this most prominently with hypoxic treatments (Hosie et al., 1991).

Accumulation of internal ammonia is of commercial concern particularly in supply chains of live crustaceans. *Nephrops* may survive transport of more than 72 h if kept immersed, moist, and cold even though they experience a hypoxia-induced ammonia concentration of about 800 μM accumulated during longer journeys. Bernasconi and Uglow (2011) found that the concentration would have been even higher if *Nephrops* were not able to activate purineolytic detoxification of ammonia by converting some of it to less toxic or non-toxic urea and uric acid, which the animals store.

2.1.5 Hydrogen sulphide

$$SO_4^{2-} \text{ reduction}: \quad CH_2O + 0.5SO_4^{2-} + H^+ \rightarrow CO_2 + 0.5H_2S(g) + H_2O$$

One further step into the diagenetic succession (Table 5.1) is the reduction of sulphate to hydrogen sulphide. Compared to the effects of hypoxia alone, the combination of hypoxia and hydrogen sulphide may shorten the LT_{50} by at least 50% in crustaceans (Diaz and Rosenberg, 1995; Hagerman, 1998; Vaquer-Sunuyer and Duarte, 2010). Hydrogen sulphide is toxic to most animals as it blocks oxygen uptake in the blood and inhibits the respiratory cytochrome c oxidase in the mitochondria (Nicholls, 1975). However, some animals experience hydrogen sulphide exposure naturally and have evolved strategies to avoid poisoning (Childress, 1987). For the burrow-dwelling *Nephrops*, irrigation not only gets rid of hypoxia, ammonia and hypercapnia, but also oxidises toxic sulphide to non-toxic thiosulphate, sulphite or sulphur (Visman, 1991). However, when digging burrows in the substrate, contact with hydrogen sulphide cannot be avoided, which is why a certain sulphide tolerance is necessary. In the body, sulphide must be oxidised in the hepatopancreas to thiosulphate, this being an energetically more efficient oxidation than to sulphite and sulphate (Johns et al., 1997). See Figure 5.4 for an internal view of a hepatopancreas in the cephalothorax of a *Nephrops*. The behaviour and physiology of *Nephrops* exposed to sulphide were investigated by Butterworth et al. (2004) and the haemocyanin properties by Taylor et al. (1999). Sulphide concentrations recorded in *Nephrops* burrows range from 0 to 122 µM. The concentrations in burrows of *Nephrops* and Thalassinid crustaceans may, however, show local variations and even reach concentrations of up to 1591 µM (Johns et al., 1997; Taylor et al., 1999). *Nephrops* retreated slightly from contact with water containing 50 µM sulphide and tail-flipped away from water containing 500 µM. In general, *Nephrops* tolerated water of 150 µM sulphide well but in 600 µM and above, the heart and scaphognathite rate decreased. The oxygen consumption increased with exposure to 200 µM probably due to oxidation

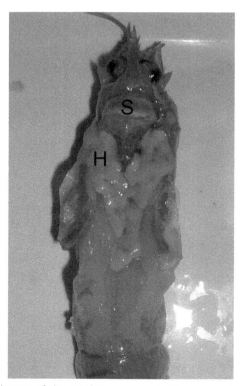

Figure 5.4 Internal view of the *Nephrops* stomach (S), surrounded by the H-shaped hepatopancreas (H) (photo: Susanne Eriksson). (For colour version of this figure, the reader is referred to the online version of this chapter.)

of hydrogen sulphide to thiosulphate, after which it decreased when exposed to 800 μM sulphide and above. LT_{50} is 22.5 h at 500 μM.

2.1.6 Carbon dioxide

About half of the escalating level of CO_2 in the atmosphere, produced by anthropogenic emissions, has been absorbed by the oceans, where it causes a decline in seawater pH (Sabine et al., 2004; Zeebe et al., 2008). During the last 200 years, the average pH in the oceans has decreased by approximately 0.1 units (Royal Society, 2005), and it is predicted to decrease by approximately 0.4 units by the end of the twenty-first century (Caldeira and Wickett, 2003). This so-called ocean acidification alters the seawater chemistry, shifting the equilibrium towards more dissolved CO_2, hydrogen ions (H^+), carbonic acid (H_2CO_3) and bicarbonate (HCO_3^-), and decreasing the amount of carbonates due to buffering (Doney et al., 2009; Feely et al., 2004). At present, the pH of ocean water is the lowest in at least 650,000 years (Feely et al., 2008) and it is still in decline. Assuming a 'business as usual' regime, the oceans are calculated to reach a pH of 7.7 (Intergovernmental Panel on Climate Change, IPCC, 2007) by year 2100, which is close to three times more acidic than the current pH of 8.1. Crustaceans have been considered tolerant to ocean acidification because of their retained capacity to calcify during subnormal pH. However, Hernroth et al. (2012) recently reported that 4 months' exposure to a decrease of 0.4 pH units significantly suppressed the immune system in *Nephrops* (see later). The acidified water also increased AGE products (advanced glycation end products), indicating stress-induced protein alterations. The extracellular pH of the *Nephrops* haemolymph was reduced by approximately 0.2 units; thus, the species has either limited pH compensation or limited buffering capacity. The effect of carbon dioxide on early larval stages of *Nephrops* is yet unknown, but Arnold et al. (2009) exposed early larval stages of cultured *Homarus gammarus* to levels of carbon dioxide concentrations predicted to occur in the year 2100 ($\sim$1200 ppm CO_2). The authors found a reduction in carapace mass during the final stage of larval development in CO_2-acidified seawater with a reduction in the amount of both calcium and magnesium. Acidosis or hypercapnia appeared to interfere with normal homeostatic function, but no increased mortality was seen. These two studies are the first of many on this and similar topics currently being carried out around Europe, and we are likely to have a clearer picture shortly of how the increase in ambient carbon dioxide levels, as well as lower ocean pH, will affect *Nephrops* and other nephropids.

2.2. Metals and organic pollutants

2.2.1 Metals in general

Marine invertebrates are frequently used as bioindicators of metal pollution in the aquatic environment because of their ability to concentrate metals from the surrounding medium and to integrate temporal variations in metal concentrations (Rainbow, 1996). Marine organisms accumulate trace metals from two different sources: the sea water and ingested material. For many organisms, the key determinant that influences metal accumulation is the specificity of the metal. Metals in solution are usually considered more bio-available than those associated with solids, although the higher concentrations in food, sediment, and suspended particulates often render the diet a potentially important source of metals to aquatic organisms. The availability of metals from food may depend, in turn, on the chemical form and binding strength of the metal to the foodstuff, food quality, particle size, particle type, and position of the species in the food chain. These, together with indirect effects on bioaccumulation (related to food availability and feeding strategy), need to be considered when deriving models of bioaccumulation. A metal that is accumulated mainly from ingested food is expected to become concentrated in an animal's internal soft tissues rather than on exoskeletons or exposed surface tissues, such as gills (Tessier and Turner, 1995). Benthic fauna live near or in direct contact with the sediment. Sedimentary metals available to benthos in excess of physiological requirements must be detoxified, and numerous studies have been conducted to investigate metal tolerance in aquatic organisms (i.e. Bryan, 1976). Quantification of metal accumulation and processing in benthos is also important because of the potential impact on human food quality and health.

Accumulation of dissolved metals is achieved firstly through uptake across the cell membrane, typically at a permeable respiratory surface. Metals dissolved in the sea are partitioned in equilibria between complexing ligands and the free metal ion released from these would be the form available for binding to cell membrane transport ligands and therefore represents the bio-available form of the metal (Rainbow, 1996; Tessier and Turner, 1995). Membrane metal transport systems are frequently regulated by the cell. Cellular regulatory systems, based upon detection of a change in metal ion activity, induce changes that modify the flux of metal ions across the cell (Tessier and Turner, 1995). Borderline metal ions such as Mn, Fe, zinc (Zn) and copper (Cu) have an extensive affinity for metal-binding donor atoms and ligands promoting cell uptake. The affinities of borderline metals for atoms (often sulphur or nitrogen) in functional groups of proteins and other molecules enable them to interfere in a wide spectrum of biological processes as active centres of metalloenzymes and integral parts of respiratory pigments, or as more loosely attached activators. Available metals that are surplus to requirements must be detoxified temporarily or permanently. A high metal concentration in itself does not necessarily imply a risk of toxicity; it is the much lower concentration of metabolically available metal that has not been detoxified that is potentially dangerous (Rainbow, 1996). Since the site of final detoxification is usually an organ other than the site of uptake (often in a derivate of the alimentary tract such as the hepatopancreas), internal transport is typically via the blood system. Products involved in the detoxification of metals in invertebrates include metal-rich inorganic deposits or granules and soluble metal-binding proteins (such as metallothioneins).

In general, metals that have an essential biochemical role, such as the metals mentioned previously, are regulated, while for non-essential metals such as mercury (Hg), cadmium (Cd) and silver (Ag), there is little evidence for control of accumulation (Rainbow, 1996). Under constant ambient conditions, the underlying control on tissue burdens is provided by the net balance between influx and efflux of metals and, in general, metals that exchange rapidly tend to be accumulated less efficiently than metals that exchange slowly. Accumulation may give rise to body concentrations in excess of four orders of magnitude above background in non-regulating organisms (Rainbow, 1996).

2.2.2 Manganese

$$Mn(IV)\,reduction: \quad CH_2O + 2MnO_2(s) + 4H^+$$
$$\rightarrow CO_2 + 2Mn^{2+}(aq) + 3H_2O$$

Mn is naturally found in high concentrations in the soft sediment typical of *Nephrops* grounds where it is one of the important oxidants involved in early diagenesis (Table 5.1). Sediment concentrations may reach concentrations of up to 50 mg g^{-1} dry weight (for review, see Baden and Eriksson, 2006). During oxic conditions in the bottom water, the sediment porewater contains 0.2–24 mg l^{-1}, whereas bottom water concentrations are between 0.2 and 16 μg l^{-1}. During hypoxic and anoxic conditions in the bottom water, the Mn concentration has been shown to increase by several orders of magnitude up to 22 mg l^{-1} or more locally, for example, close to industries (Magnusson et al., 1996; Weinstein et al., 1992). Mn is an essential metal to all organisms and is particularly involved as a cofactor in numerous enzyme activities (Simkiss and Taylor, 1989). Two of the most well-known Mn-containing polypeptides are arginase and Mn-containing superoxide dismutase (Mn-SOD). Mn-SOD is present in eukaryotic mitochondria as well as in most bacteria where it plays an important role in handling the toxic effects of superoxide (Gutteridge and Halliwell, 2000). The catholic affinity for metal-binding donor atoms and ligands also means that the borderline metal ions (such as Mn) can have toxic effects by replacing each other in metabolically important molecules or by binding onto biomolecules. In this way, they may directly block the function or interact indirectly by altering the specific conformation of the biomolecules. The competitive binding of metals by organic ligands has direct ecological relevance to crustaceans. Cu binds more strongly and will replace Mn in a bond with an organic ligand, which in turn will replace calcium (Ca). The affinity of these metal ions for organic ligands offers them a route across the cell membrane by binding with transport ligands and promotes a concentration of free metal ions across the membrane, against the concentration gradient (Rainbow, 1997). Entry of Ca ions into the cells is through specific Ca channels powered by an active transport pump. Mn in low concentrations has been shown to pass through the Ca channels (Tessier and Turner, 1995). At higher concentrations, however, Mn acts as an inhibitor (Fatt and Ginsborg, 1958; Hagiwara and Takahashi, 1967; Mounier and Vassort, 1975), most likely due to the high binding affinity, but low mobility, of Mn^{2+} (Hagiwara and Takahashi, 1967).

Since Mn is an essential metal, it is required in at least a minimum concentration for an animal to be able to fulfil its metabolic functions. When discussing the basic body requirements of Mn, it is also important, to differentiate between metabolically active soft tissues and relatively inert tissues such as the exoskeleton. The theoretical requirement of Mn for crustaceans has been calculated to be 3.9 µg Mn g^{-1} dw (White and Rainbow, 1987). This calculation is based on whole animals, including the exoskeleton where most of the Mn occurs, and this number is thus likely to be an overestimation. Lowest Mn concentrations in *N. norvegicus* have been found in animals from the pristine Faeroe Islands, with a total of 8 µg Mn g^{-1} dw (Eriksson and Baden, 1998), but the concentration in the soft tissues of the same animals was only 2.5 µg Mn g^{-1} dw (Baden and Eriksson, 2006). As Mn bioavailability increases, it accumulates in biota. Human concern about metals focuses mainly on highly toxic, rare and unessential metals, such as Pb, Hg and Cd (see below). Although Mn is an essential metal, it is also an unforeseen toxic metal in the aquatic environment. The role, routes and effects of Mn in crustaceans was reviewed by Baden and Eriksson (2006). Although the uptake and elimination of Mn in internal tissues is rapid (typically a couple of days in most tissues), it may reach toxic levels that decrease the fitness of organisms. As Mn bioavailability increases, its uptake is predominately through the water. It is mainly accumulated in the nerve tissue (including brain), hepatopancreas, haemocyanin and the male reproductive system.

Uptake of Mn from the water occurs mainly over the gill membrane and during hypoxia, the mean gill concentration in *Nephrops* may increase by 30 times to 1.6 mg Mn g^{-1} dw (Eriksson and Baden, 1998). A black layer of precipitated Mn was found on *Nephrops* gills from hypoxic areas in SE Kattegat (Baden et al., 1990b; Figure 5.5). It was suggested that this precipitation may physically hinder oxygen uptake over the gills. After gill uptake, Mn is transported in the haemolymph to destination tissues, either dissolved in the plasma or bound to the haemolymph proteins, mainly the respiratory protein haemocyanin (Baden and Neil, 1998). At ambient concentrations of ≤ 10 mg Mn l^{-1}, the *Nephrops* haemolymph plasma holds the same

Figure 5.5 Side view of *Nephrops norvegicus* cephalothorax with the exoskeletal part covering the gills removed. *Nephrops* origin from normoxic (top) and hypoxic (bottom) areas. Precipitated manganese is seen as a brown layer on the gills of the animal from the hypoxic site (photo Susanne Baden). (For colour version of this figure, the reader is referred to the online version of this chapter.)

concentration as the ambient water though the Mn concentrations of the protein and cell fraction keep increasing (Baden and Neil, 1998). The biological half-life for Mn in *Nephrops* haemolymph is short, ≈ 24 h (Baden et al., 1999). Exposure to 20 mg l^{-1} Mn, however, decreases the number of haemocytes by 60% (Hernroth et al., 2004) impairing their bactericidal capacity (see later and Oweson and Hernroth, 2009).

The hepatopancreas is the endpoint of Mn accumulation; this tissue does not reach a steady state but keeps accumulating Mn at a relatively slow rate (Baden et al., 1999). Simkiss (1981) found that Mn in invertebrates can be detoxified by incorporation into Ca-rich phosphate granules in the gut, which has not yet been found in *Nephrops*. Some metals induce metallothionein in *Nephrops*, others do not (see later and Canli et al., 1997). No one has yet looked for metallothionein in Mn-exposed *Nephrops*, but it has been shown that Mn induces metallothioneins in the closely related *H. gammarus (vulgaris)* (Bryan and Ward, 1965). The synthesis of haemocyanin is primarily recognised to take place in the hepatopancreas (for review, see Taylor and Antiss, 1999). Exposure to Mn in combination with hypoxia shows that Mn inhibits haemocyanin synthesis, which is needed to cope with hypoxia (Baden et al., 2004). The daily recommended Mn intake for humans is 2.5–5 mg Mn day^{-1}, and the mean muscle concentration in *Nephrops* is 3 μg Mn g^{-1} dw (Eriksson and Baden, 1998; Lourenco et al., 2009); Mn consumption is thus not likely to pose a threat to humans. When *Nephrops* are exposed to 10 mg Mn l^{-1} for 3 weeks, their muscle extension and, consequently, their swimming capacity is affected negatively (Baden and Neil, 1998).

Due to its chemical properties, Mn is found in the highest concentrations in the calcified parts of crustaceans in general, mainly in the exoskeleton, gills and the gastric mill of the stomach (for review, see Baden and Eriksson, 2006). Over 98% of the total Mn content of nephropids, *N. norvegicus* (Baden et al., 1995) and *H. gammarus* (Bryan and Ward, 1965) is found in the exoskeleton. The Mn incorporated in the matrix of the exoskeleton is believed to have little effect on the animals. The amount of Mn found in the exoskeleton of intermoult individuals primarily depends on the Mn concentration to which the animals are exposed during the calcification process at post-moult, rather than the current ambient Mn concentrations (Baden and Eriksson, 2006; Eriksson, 2000a; Eriksson and Baden, 1998). See Figure 5.6 for Ca storage in gastroliths just prior to moulting. Moulting has been suggested as one possible way for decapods to dispose of excess unwanted metals (Bryan and Ward, 1965). Although crustaceans on

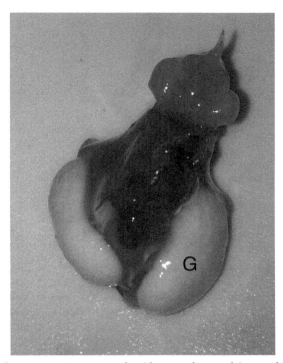

Figure 5.6 *Nephrops norvegicus* stomach with amorphous calcium carbonate stored in the gastroliths (G) prior to moult (photo: Susanne Eriksson). (For colour version of this figure, the reader is referred to the online version of this chapter.)

occasion eat part or all of their cast exuviae, preliminary data on the Mn uptake of *Nephrops* from food suggests that Mn incorporated in exoskeletal parts is not easily accessible when ingested (Baden and Eriksson, 2006; Figure 5.7). Moulting might thus serve as an important regulator of the Mn content provided there are low Mn^{2+} concentrations in the water at the time of moult. As the Mn bound to the exoskeleton is not released during recovery in non-dosed seawater, the concentration can work as a biomarker for a previous exposure to hypoxia. Baden and Neil (2003) found this to be especially true for *Nephrops* antennae waving in the Mn-rich water and thus getting a homogenous layer of precipitated Mn.

The female reproductive system of *Nephrops* is quite robust against ambient concentrations of 10–20 mg Mn l^{-1}. The Mn concentration of the oocytes, during maturation and throughout most of the embryogenesis, remains very stable regardless of ambient Mn concentrations (Eriksson, 2000b). The Mn concentration of fertilised eggs is stable at around

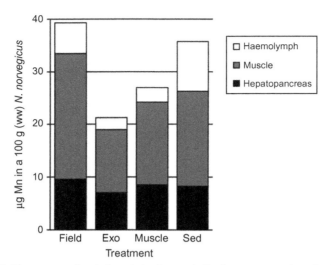

Figure 5.7 Manganese load found in internal *Nephrops norvegicus* body tissues (haemolymph, muscle and hepatopancreas) after being fed for 2 weeks from different food sources with natural manganese content. The manganese concentration in the food was Exo = exoskeleton (5.4 μg Mn g^{-1} dw), muscle (1.4 μg Mn g^{-1} dw) and Sed = sediment (145 μg Mn g^{-1} dw). Results are compared to concentrations found in animals from the field. *Modified from Norstedt (2004) and Baden and Eriksson (2006).*

5.5 μg g^{-1} dw egg^{-1} during the first five embryonic stages ($\approx$ 6 months) (Dunthorn, 1967). At the end of the embryonic development, the membrane around the egg gives no protection against external Mn, and dissolved Mn^{2+} passes through and is taken up by the embryo (Eriksson, 2000b). No mortality has been seen in late *Nephrops* embryos when exposed to 10 mg l^{-1}, and they are thus a lot more tolerant than, for instance, the intertidal crab *Metacarcinus* (*Cancer*) *anthonyi* in which severe mortality in embryos occurs even at 10 μg Mn l^{-1} (Macdonald et al., 1988). The Mn concentration in the *Nephrops* male reproductive organs is normally relatively high (33.2 μg g^{-1} dw) (Eriksson, 2000a), probably explained by the large amount of acidic mucopolysaccharide that protects the sperm when delivered to the female spermatheca. The negative charge of this substance attracts the positively charged metals such as Mn. *In vitro* exposure of male *Nephrops* reproductive organs to 20 mg Mn l^{-1} shows that the sperm mass concentration increases from 80 to 140 μg g^{-1} dw (Baden and Eriksson, 2006).

A primary target tissue for Mn is the central nervous system. Effects and toxicity of Mn in vertebrate nerve systems is well documented but is still quite unknown in invertebrates (Baden and Eriksson, 2006). The *Nephrops* brain may reach concentrations of $250\ \mu g\ Mn\ g^{-1}$ dw when exposed to $10\ mg\ Mn\ l^{-1}$ (Baden and Neil, 1998), which is in the same range as brain concentrations found in field animals during hypoxic events (Eriksson and Baden, 1998). Normal muscle extension is affected by commonly occurring field concentrations of Mn, which can also have an impact on neuromuscular performance. Exposure of muscle preparations to concentrations $>5\ mg\ l^{-1}$ Mn decreases the contractile force successively until total abolition at concentrations $>320\ mg\ l^{-1}$ (Holmes et al., 1999). Exposure of *Nephrops* to $10\ mg\ l^{-1}$ Mn affects the free tail-flip swimming by reducing the post-flip extension by about 40%, whereas the flip flexion is unaffected. The extension involves a chemical neuromuscular synapse affected by Mn, whereas the flexion is elicited primarily by an electrical synapse and is thus not affected by Mn (Baden and Neil, 1998).

The perception of food via chemosensory organs is also affected by Mn. Concentrations of up to $20\ mg\ Mn\ l^{-1}$ significantly increase the mean flick frequency by about 15–25%, whereas the frequency decreases significantly in exposures of more than $20\ mg\ Mn\ l^{-1}$ both in animals exposed to Mn and in Mn in combination with hypoxia. It thus seems that Mn affects the perception of odour at the aesthetascs possibly either by physical precipitation of Mn on the aesthetascs and/or by chemical action as a neurotoxin in such a way that increasing the flick frequency may compensate for a reduced perception of the stimulus (Baden and Eriksson, 2006). Mn significantly increases the reaction time to food odour stimuli and the ability of animals to perceive a source (Krång and Rosenqvist, 2006). The authors conclude that the ability to detect and find food can be reduced in areas where Mn concentrations are high, with possible consequences at individual and population levels.

The ingestion of Mn could potentially be quite significant as Mn can occur in sediment at concentrations of up to $50\ mg\ g^{-1}$ and large amounts of sediment are frequently found in the stomachs of *Nephrops* (Baden and Eriksson, 2006). However, uptake from water via the gills seems to be the most important path of Mn into aquatic crustaceans during hypoxic situations when bioavailable dissolved Mn is at high concentrations (Baden and Eriksson, 2006). Experiments with [54]Mn show that excretion in nephropids *N. norvegicus* and *H. gammarus* occurs only partly through the antennary glands and the faeces (Baden et al., 1995; Bryan and Ward, 1965). The major portion of loss has been suggested to take place via the body surface (Baden and Eriksson, 2006).

In conclusion, *Nephrops* are exposed to high concentrations of Mn in their habitat, and bioavailability increases to toxic levels during hypoxia.

Mn is found mainly in the calcified parts of the animals where it is incorporated into the Ca matrix. Mn in excess mainly affects the haemolymph and the immune defence as well as the nervous system and the nerve–muscle interaction. In addition, the oxides and hydroxides of Mn act as efficient scavengers of other metals (Glasby, 1984; Petersen et al., 1995). When Mn remobilisation occurs in the sediment, these metals are simultaneously released into the water column with Mn^{2+}. Elevated Mn leakage from the sediment, for whatever reason, is thereby also an indicator of other metals becoming more bioavailable to benthic living organisms.

2.2.3 Iron

$$Fe(III)\,reduction: \quad CH_2O + 4FeOOH(s) + 8H^+$$
$$\rightarrow CO_2 + 4Fe^{2+}(aq) + 7H_2O$$

Fe behaves in the same way as Mn in that it is a redox-sensitive metal involved in early diagenesis (Table 5.1), and it is found in high concentrations in most marine sediments. It is also enriched along the oxic–anoxic boundary on the inner side of animal burrow walls though it takes lower oxygen concentrations to release it from the sediment compared to Mn (Sundby et al., 1986). It is, like Mn, often overlooked when measuring metals in biota with the comment of 'unlikely to be toxic'. As with Mn, Fe is an essential trace element for living organisms, required by most organisms for growth and development. Fe toxicity has been investigated in a number of invertebrates, though not in *Nephrops*. In crustaceans, Fe was found to suppress embryos of the yellow crab *Metacarcinus (Cancer) anthonyi* from hatching at concentrations of 1–10 mg Fe l^{-1} (MacDonald et al., 1988). Embryonic EC_{50} values were 100–1000 mg Fe l^{-1}.

Fe appears to be mainly taken up over the gills in crustaceans (like Mn) and then transported via the haemolymph (i.e. Canli and Furness, 1993a; Depledge, 1989). Ferric iron precipitated on the gills of the American lobster *H. americanus* have been shown to hinder oxygen uptake, thus affecting respiration (Jansen and Groman, 1993). Earlier findings have shown that half the Fe load in crustaceans is found in the exoskeleton (i.e. Depledge, 1989; Jansen and Groman, 1993), and crustaceans have been found to have Fe storage cells in the hepatopancreas (Bryan, 1979).

Measurements of Fe concentrations in *Nephrops* are scarce. Canli and Furness (1993a) measured a suite of metals in *Nephrops* from the Clyde, Scotland, and found the highest Fe concentrations of 0.9–1.2 mg g^{-1} dw in the gill tissue. Lower concentrations were found in exoskeleton (156–214 $\mu g\ g^{-1}$ dw), hepatopancreas (102–138 $\mu g\ g^{-1}$ dw) and muscle tissue (21–37 $\mu g\ g^{-1}$ dw). Lourenco et al. (2009) measured a suite of metals in the edible abdominal muscle in *Nephrops* from the Portuguese coast and found a wet weight concentration of 20 μg Fe g^{-1} tissue (equalling roughly around 90 μg Fe g^{-1} dw), almost three times higher than in the animals caught in Scotland. Eriksson and Baden collected *Nephrops* for Mn measurements at three sites along the Swedish west coast during autumn 1995; one in the Skagerrak (Gråskär) and two in the Kattegat (Klåback and Laholm Bay) (see map in Eriksson and Baden, 1998). The material was also measured for Fe although till date this data has not been published. The two sites in Kattegat are known to frequently experience autumnal hypoxia. Concentrations of Fe and Mn were measured in sediment, exoskeleton, haemolymph, and brain and also in the hepatopancreas and muscle tissue ($n = 12$) for Fe. The sediment concentration of Fe_{tot} (49–65 mg g^{-1} dw) exceeded by about 100 times the Mn_{tot} concentration (0.36–0.63 mg g^{-1} dw). Despite this, the concentrations of Fe and Mn in the animals were in the same order of magnitude in both the haemolypmh (ranging 2–5 and 1–7 $\mu g\ g^{-1}$ wet weight for Fe and Mn, respectively) and the exoskeleton (ranging between 82–779 and 182–436 $\mu g\ g^{-1}$ dw for Fe and Mn, respectively). The lowest Fe concentration in the exoskeleton (82 $\mu g\ g^{-1}$ dw) was found in the animals from Laholm Bay in Kattegat (the most hypoxic site). The brain contained about 10 times more Fe than Mn (ranging between 238–3220 and 24–193 $\mu g\ g^{-1}$ dw for Fe and Mn, respectively) with the highest concentration (3220 $\mu g\ g^{-1}$ dw) found in the animals from Laholm Bay. Whether Fe concentrations at that level have any sublethal effects on the *Nephrops* is not known. The average Fe concentration in the hepatopancreas of the Kattegat animals was also three times higher (331 μg Fe g^{-1} dw) than that in both the lobsters caught in the Skagerrak (96–140 μg Fe g^{-1} dw) as well as in Scotland (see above). The average concentration in muscle tissue of *Nephrops* from the Swedish sites was 19–35 μg Fe g^{-1} dw, thus about the same as in the lobsters from Scotland and less than half of that in animals from Portugal (Canli and Furness, 1993a; Lourenco et al., 2009).

2.2.4 Other metals

Nephrops is a commercially exploited decapod of which humans consume mainly the abdominal muscle (and in some areas also the hepatopancreas, sometimes called 'butter' because of its high-fat content). Metals may thus be important not only to monitor toxicity to the lobsters themselves but also to measure concentrations for human health safety reasons. Human concern in this regard focuses mainly on highly toxic, rare and unessential metals such as lead (Pb), Hg (both organic, MeHg, and inorganic, Hg) and Cd. The influence of these metals on synaptic transmission is reviewed for crustaceans by Fingerman et al. (1996). Cu and Zn, although essential to *Nephrops,* have also been studied. Cu plays an important role in the oxygen-carrying blood protein haemocyanin and both metals are important in enzyme processes. They are naturally found in higher concentrations in metabolically active tissues such as the hepatopancreas (Canli and Furness, 1993a). The toxic effects of the non-essential metals are considered to be caused by their tendency to substitute for Cu and Zn and compete effectively for binding to biological ligands (for review, see Viarengo and Nott, 1993). Metal ions are ranked according to their affinity for SH groups in the following order: $Hg^{2+} > Cu^+ > Cd^{2+} > Cu^{2+} > Zn^{2+}$. Substitution of essential elements may alter the structure and function of the molecules and affect cell physiology.

Metal uptake and concentrations have been studied extensively by Canli and coauthors (Canli and Furness, 1993a,b, 1995; Canli and Stagg, 1996). They showed that exposure to MeHg, Hg, Cu, Cd, Pb and, to some extent, Zn results in accumulation in *Nephrops* tissues. Exposure to 10 Cu l^{-1} increases Cu concentrations in carapace, gills, tail muscle and ovary, but not in external eggs or the hepatopancreas (which is naturally high). Exposure to 100 µg Zn l^{-1} increases Zn concentration in the hepatopancreas and gills in particular, and also in the carapace and ovary, but not in tail muscle or external eggs. Higher metal concentrations in the carapace appeared to be mainly due to adsorption. In general, as seen for Mn and Fe earlier, metals taken up from seawater result in high concentrations in the gills (Cd, MeHg, Hg, Pb). The highest load of Pb has been found in the carapace and hepatopancreas. Accumulation of Hg, MeHg and Cd in the gills is greater in smaller animals. After having traversed the gill membranes, the metals have different target tissues. For example, most of the Cd accumulated ends up in the hepatopancreas, regardless of whether the intake was via food or water, and larger animals accumulate higher concentrations than smaller ones. MeHg differentially accumulates in abdominal muscle, increases in concentration with animal size and varies in its distribution with sex (Canli and Furness, 1993b). Accumulation of Hg in the hepatopancreas was greater in males than in females (Canli and Furness, 1993b) but the reverse (lower in males than in females) was found in muscle tissue (Barghigiani et al., 2000).

Nephrops measured for Hg in the Clyde, Scotland, had an average muscle tissue concentration of 0.5 μg Hg g^{-1} dw (Canli and Furness, 1993a). *Nephrops* with a carapace length of >15 mm caught in the Thyrrenian Sea, Italy, were (from a human perspective) considered contaminated with Hg with concentrations of >0.5 μg g^{-1} ww (~2 μg g^{-1} dw) (Barghigiani et al., 2000). The Hg in that area originates from mining, and is transported by runoff from land. The concentrations found in the field animals are slightly higher than those found in laboratory animals exposed to 10 μg Hg l^{-1} for 30 days (1.7 μg g^{-1} dw muscle tissue) (Canli and Furness, 1993b). No mortality was seen in *Nephrops* exposed to 10 μg Hg l^{-1} ;thus, concentrations are at a sub-lethal level, at least in the short term, but mortality was 50% within 2 days at 100 μg Hg l^{-1}.

Metallothionein has been measured as a stress indicator in *Nephrops* exposed to metals (Canli et al., 1997). Metallothioneins are a group of soluble, heat-stable, cysteine-rich and low-molecular-weight proteins with a strong affinity for metals. Exposure to several different concentrations of metals, the lowest being 1 μg l^{-1} for Cd and Cu and 8 μg l^{-1} for Zn, induced metallothionein production in animals, though differences were found between the sexes (Canli et al., 1997). There was a significant correlation between Cd concentrations and metallothionein in both the gill and hepatopancreas of males and females. Cu in the hepatopancreas also showed positive relationships with metallothionein concentrations in males, but not in females. The authors suggested that metallothionein in the gill and hepatopancreas of *Nephrops* could be used as a sensitive tool to detect Cd contamination (but not Cu or Zn contamination) in the lobsters. '*In vivo*' experiments carried out with the same concentrations on the activities of gill ATPases (Canli and Stagg, 1996) show a significant difference between sex and size response and the ATPases are thus not suitable as general indicators of metals.

Apart from the metals mentioned above, Holmes et al. (1999) have also investigated the effect of the naturally occurring metal cobalt (Co) released together with Mn during hypoxia, but to a much smaller extent. The study focused on the effects of Co (and Mn) on force production by the abdominal superficial flexor muscle of *Nephrops*. Co ions have a similar mode of action and at the same concentrations as Mn mentioned above, but as this metal is present in much smaller concentrations than Mn, this is not considered a problem to the animal.

Arsenic is a toxic metalloid, also of concern to humans. Concentrations of arsenic in the muscle tissue of wild-caught *Nephrops* outside of Portugal show arsenic concentrations of 30 μg g^{-1} ww ($\sim$120 μg g^{-1} dw). Though *Nephrops* contained higher concentrations than other edible animals sampled, the levels pose no threat, at least to humans (Anacleto et al., 2009).

2.2.5 Organic pollutants

As with metals, organic pollutants are of interest to monitor because they can reach humans through shellfish consumption, but data are scarce in the literature. Polycyclic aromatic hydrocarbons (PAHs) were analysed in *Nephrops* from the Central Adriatic Sea (Perugini et al., 2007). Acenaphthene, benzo(a)pyrene, chrysene, dibenz(a,h)anthracene, benzo(ghi)perylene and indeno(1,2,3-cd)pyrene were present at levels below the instrumental detection limit, but fluorene, phenanthrene, anthracene, fluoranthene, benz(a)anthracene, benzo(b)fluoranthene and benzo(k)fluoranthene were detected at different concentrations. The composition of the PAH pattern was dominated by the presence of PAHs with 3-rings (62%) followed by those with 4-rings (37%) and 5-rings (1%). *Nephrops* showed a three times lower contamination than the fish from the same study. No significant correlation was observed between weight or length and total PAHs. In 1993, an oil tanker spilled 85,000 tonnes of crude oil outside Shetland, United Kingdom, which led to a 7-year restriction on landing *Nephrops* due to contamination (Davies and Topping, 1997). The reference background value of total PAHs was $\approx$ 35 ng g^{-1} ww (range 8–140 ng g^{-1} ww). PAHs measured in *Nephrops* muscle tissue in the vicinity of an aluminium smelter in Loch Leven during 1999–2000 showed PAH concentrations of 387 ng PAH g^{-1} ww on average, with the highest concentration recorded being 573 ng PAH g^{-1} ww (McIntosh et al., 2001). Half of the PAHs originated from benzoperylene, but the three acute toxic substances benz(a)anthracene, benzo(a)pyrene and dibenz(a,h)anthracene were in low concentrations and added up to <9 ng g^{-1} ww. The *Nephrops* were thus interpreted as suitable for human consumption (McIntosh et al., 2001). In 2002, the oil tanker Prestige spilled 50,000 tonnes of heavy oil on the shelf outside northern Spain (Sánchez et al., 2006). During sampling, there was an indication of decreased densities of *Nephrops* in the areas most heavily affected, where 300 kg km^{-2} of tar aggregates were found. However, no traces of oil were found in any of the *Nephrops* stomachs investigated.

One study from the Clyde Sea has addressed how much plastic *Nephrops* consume (Murray and Cowie, 2011). They found that plastic contamination was high with 83% of the animals sampled containing plastics, mainly filaments, in their stomachs. Tightly tangled balls of plastic strands were found in 62% of the animals studied but were least prevalent in animals that had recently moulted. No significant difference in plastic load was observed between males and females. Raman spectroscopy indicated that some of the microfilaments identified from gut contents could be sourced to fishing waste. *Nephrops* fed with fish seeded with strands of polypropylene rope were found to ingest but not to excrete the strands. This may thus pose a potential threat to the *Nephrops* in the future. Recently, nanoparticulate plastics in sea-water have been examined because they tend to increase as a consequence of decomposition of plastic debris and through sewage systems discharging such particles released from cosmetics, clothes, etc. Virtually nothing is known about their effects on sea life, but nanoparticulates have been reported to neg-atively affect food intake and growth of blue mussels (Zuykov et al., 2011), and should be considered in future studies on *Nephrops*.

2.3. Fishery-induced stress

Nephrops experience physical as well as physiological stress during commer-cial fishing operations, that is, increased injuries and exercise while in the trawl and exposure to low salinity as well as aerial and light exposure during sorting of the catch. Stress caused by trawling affecting metabolism and res-piration as well as generating injury and limb loss in *Nephrops* is well docu-mented (Albalat et al., 2010; Bergmann and Moore, 2001; Gornik et al., 2010; Harris and Andrews, 2005; Lund et al., 2009; Schmitt and Uglow, 1997; Spicer et al., 1990). Females carrying eggs are also known to lose eggs from the abrasion during trawling (i.e. Briggs et al., 2002 and for review Chapter 6). Creeled animals are generally in better condition than those from trawled areas and their condition improves between spring and autumn (Bernasconi and Uglow, 2008; Eriksson, 2006). In addition, traditional trawl fisheries for *Nephrops* are known to generate considerable amounts of by-catch. Most organisms are dead or moribund when discarded, and when reaching the sea floor, they are utilised within days by scavengers including *Nephrops* (Bergmann and Moore, 2001; Bergmann et al., 2001, 2002). Hence, fishery discard may be viewed as subsidies to the benthic community.

Limb loss is common in decapod crustaceans, and the animals can autotomise a limb if they are physically provoked (for review, see Juanes and Smith, 1995). It is considered to be an adaptation to avoid predators and limit injury. But although autotomy may lead to survival for the moment, the loss of a limb can result in long-term functional and energetic costs, such as decreased growth, foraging efficiency and mating success. The animal also becomes more vulnerable to aggressive behaviour from others. *Nephrops* can autotomise their claws (chelipeds) when physically stressed, for example, during the trawling procedure. Females are more susceptible to damage compared with the males (Ridgway et al., 2006). Claw symmetry (paired cutters) has been suggested as an indicator of limb loss, with the highest occurrence found in females from trawled sites (Eriksson, 2006). Although the absence of a crusher cheliped is not linked to the percentage muscle dry weight of individuals, the mean was lower in trawled females than in other groups.

Creeled animals were generally less stressed with significantly lower L-lactate and CHH concentrations in the haemolymph than trawl-caught animals (Ridgway et al., 2006). When animals are trawled, they try to escape through swimming by tail flipping (Newland et al, 1992). Laboratory simulated trawl-catch of *Nephrops* with intense exercise and a 2 h emersion show premature onset of anaerobic glycolysis, with glycogen depletion and build-up of L-lactate concentrations (Gornik et al., 2010). The elevated stress level in trawled animals has been found to be the same regardless of trawl duration (1 or 5 h) although longer trawls increase injury levels. The stress level is affected by the catching time of the day with morning-caught animals being more stressed than those caught in the evening.

The recovery of animals captured by trawls is seasonal, and is faster in animals caught in the winter (4 h) compared to in the summer (24 h) (Albalat et al., 2010; Lund et al., 2009). This might be explained by a more stable temperature when brought from bottom to the surface during winter (Lund et al., 2009). In summer, when brought from 12 to 21 °C, few animals were alive at the surface and approximately 95% had died after 24 h. Stresses of capture elicit high haemolymph lactate contents and glucose concentrations. In winter, a potential metabolic lactic acidosis may be compensated for by a marked respiratory alkalosis, with significantly increased haemolymph pH and decreased CO_2 total content and partial pressure. *Nephrops* captured during the summer demonstrate an initially combined metabolic and respiratory acidosis. The capture stresses elicited very high haemolymph CHH titres, significantly higher in summer than in winter.

3. IMMUNOLOGY

3.1. Immunity of crustaceans

Physico-chemical barriers, such as shell and mucous layer, constitute a first line of defence against potential pathogens of crustaceans. Within the body, haemocytes are the key players in the immune defence. They are developed from pluripotent stem cells in the hematopoietic tissue, which in decapods is localised in the dorsal membrane covering the stomach cavity (Chaga et al., 1995). Here, the progenitor cells undergo proliferation, differentiation and maturation into morphologically and functionally distinct cell populations, the so-called hyaline, semigranular and granular cells, before being released to the haemolymph (Johansson et al., 2000; Söderhäll et al., 2003). The number of free haemocytes can vary and can, for instance, decrease dramatically during an infection (Lorenzon et al., 1999; Oweson and Hernroth, 2009; Smith and Söderhäll, 1983).

As in most invertebrates the immune defence is complex, efficient and highly developed although not including the adaptive components of the vertebrate immune systems. One of the main mechanisms of the cellular defence of decapods is phagocytosis by the hyaline cells, and in the case of larger particles or large numbers of pathogens, this mechanism is complemented by encapsulation. The granular haemocytes are, through degranulation, able to release a pro-phenol oxidase activating system (pro-PO-AS) (Cerenius and Söderhäll, 2004; Johansson and Söderhäll, 1985; Söderhäll and Cerenius, 1998), analogous to the vertebrate complement system. This is a zymogen package, which is activated through contact with bacterial components such as lipopolysaccharides and peptidoglycans. The foreign molecules are recognised by pathogen-associated molecular patterns, the so-called PAMPs (Cerenius and Söderhäll, 2004; Lee and Söderhäll, 2001; Lee et al., 2000), which through serine cleavage of prophenoloxidase activate a cascade of reactions creating the bactericidal enzyme phenoloxidase. The enzyme catalyses the oxidation of phenols to highly reactive and toxic quinone intermediates. Quinones are further polymerised non-enzymatically to melanin, which encapsulates the intruders and any damaged tissue (Söderhäll and Cerenius, 1998). The cell adhesive and opsonic protein, peroxinectin, is associated to the pro-PO system and can stimulate phagocytosis by hyaline cells (Jiravanichpaisal et al., 2006) or encapsulation by stimulating semi-granular cells (Johansson et al., 2000). These defence mechanisms are highly efficient and, therefore, inhibiting substances are necessary to prevent self-damage of the tissue (Aspán et al., 1990; Hergenhahn et al., 1987; Liang et al., 1997).

A large and increasing number of constitutive and inducible antimicrobial peptides (AMPs) have been registered as an important part of the immune defence in most phyla. Bindings of pattern recognition receptors to PAMPs initiate production of AMPs. In crayfish, haemocyanin has been demonstrated to take part in this defence by releasing the antimicrobial peptide astacidin 1 in response to bacterial compounds (Lee et al., 2003).

3.2. Immune suppression

Increasingly, the relationship between diseases and environmental stress has been drawing attention, not the least because of issues around the aquaculture of decapods. However, infectious diseases are found in both cultured and wild crustaceans. As burrowing organisms of soft bottom sediments, *Nephrops* have to face quite rapid changes in oxygen content, which when reduced will increase the amount of bioavailable metal ions, such as Mn. Although essential for many enzymatic reactions, it has been found that they act as immune suppressants when in surplus. Exposing *Nephrops* to realistic Mn levels (20 mg l^{-1}, 10 days) in bottom waters of Skagerrak during events of hypoxia (Magnusson et al., 1996) resulted in the total number of haemocytes becoming reduced by 60%. It is suggested that this is mostly because Mn inhibits proliferation and maturation of the haematopoietic precursor cells (Hernroth et al., 2004) and induces apoptosis of both these cells and circulating haemocytes. Apoptosis and hemocytopenia were already induced at a concentration of 5 mg Mn l^{-1} and responded in a dose-dependent manner with increasing rates at 10 and 20 mg Mn l^{-1} (Oweson et al., 2006). Moreover, Hernroth et al. (2004) demonstrated that Mn inhibits the degranulation process by 80%, and thereby the release of pro-PO-AS, and also reduces the conversion of the zymogen into phenoloxidase (Hernroth et al., 2004).

Studies on yeast cells, *Saccharomyces cerevisiae*, reveal that Mn utilises the same pathways as for cell trafficking of Ca (Cizewski Culotta et al., 2005) including widely conserved transport proteins. Briefly, the handling of an excess of Mn in yeast is through entering the Golgi apparatus via Ca and Mn ion transporting ATPase and then leaving the cell by secretory vesicles derived from Golgi. Mn could also be stored in vacuoles to avoid intracellular interactions and the uptake of the metal could be regulated by degradation of transport proteins. However, it has been shown that Mn accumulates in the haemolymph of *Nephrops* and can reach a threefold concentration in the surrounding water (Oweson and Hernroth, 2009). This incompetence in clearing excess Mn was at an exposure concentration of 15 mg l^{-1}, shown to correlate with impaired bactericidal capacity when encountering sublethal doses of the pathogen *Vibrio parahaemolyticus*. Severe infections of the parasitic dinoflagellate *Hematodinium* sp. have been found in

Nephrops from the west coast of Scotland (Field et al., 1992) where high tissue concentrations of Mn have also been recorded (Baden and Neil, 1998). One might speculate that Mn affects the outcome of host–parasite interactions as shell disease in blue crab (*Callinectes sapidus*) has shown to be correlated to elevated Mn concentrations as the only element out of 13 investigated in an estuary outside North Carolina, US (Weinstein et al., 1992).

Immune suppression of *Nephrops* has also been demonstrated when lobsters were exposed to seawater conditions mimicking the climate scenario predicted to occur in the ocean at the end of this century (Royal Society, 2005). The lobsters were unable to compensate for the lowered pH of the haemolymph, resulting in effects on the haemocytes (Hernroth et al., 2012). After 4 months of exposure, haemocyte numbers were reduced by 50% and their phagocytic capacity reduced by 60%. Furthermore, a 50% increase in protein alteration (measured as AGE levels in the tissue of hepatopancreas) was registered, indicating imbalance in the antioxidant defence. The negative effects on immune response and tissue homeostasis in *Nephrops* exposed to acidified water were more pronounced at higher temperatures (12–18 vs. 5 °C), which may reflect a thermal adaptation to keep basal metabolism energy efficient (Clarke and Fraser, 2004).

Further climate change will generate stress situations with bioavailable Mn in combination with ocean acidification, which most probably will reinforce the immune suppression of *Nephrops* and thereby their susceptibility to infectious diseases.

3.3. Diseases

Nephrops norvegicus is host to a number of pathogens and symbionts. Some are sessile and attach to the exoskeleton, but a few of these epibionts are infective. The diseases so far registered in *Nephrops* have been thoroughly reviewed by Stentiford and Neil (2011). In many populations of *Nephrops*, in both the Mediterranean and the north Atlantic, the commensal *Symbion pandora* has been found attached to the mouth parts of the *Nephrops* (Funch and Kristensen, 1995; Morris, 1995) seemingly without doing harm to the host. However, a number of important parasites, some of which have been shown to have a significant detrimental effect on host populations, have been reported. The most significant of these parasites is a dinoflagellate of the genus *Hematodinium* (Field et al., 1992) which has caused epidemic infections among several crustacean species worldwide (Kuris and Lafferty,

1992; Maclean and Ruddell, 1978; Small, 2012). Molecular sequence data indicate that there are at least two different species of the parasite, *H. perezi* and another, currently unnamed, the latter of which has been associated to infections in the Northern Hemisphere (Small, 2012). Infected hosts become indolent and symptoms include milky haemolymph, hyper-pigmentation of the carapace and degraded texture of the muscle tissue (Field et al., 1992; Stentiford and Shields, 2005). The infection can be fatal and has had a significant impact on wild populations and associated fisheries of *Nephrops* (Stentiford, 2011). Adult lobsters could also suffer from the infective trematode *Stichocotyle nephrosis*. Their larvae are found in cysts on the gut wall and after excystment, the worm can reach a length of 8 mm (MacKenzie, 1963).

Other parasites have a great impact on the early reproductive stages such as infestations of the ciliate *Zoothamnium* of *Nephrops* larvae. The parasitic polychete worm *Histriobdella homari* is believed to infect the branchial chamber and egg mass, but some authors suggest that the relationship is symbiotic (Lerch and Uglem, 1996; Shields et al., 2006).

Shell disease syndrome, which affects a wide range of crustaceans is caused by chitinivorous bacteria. Degradation of the exoskeleton leads to melanised lesions (Smolowitz, et al., 1992) and can even be lethal if the underlying tissues become infected. However, shell diseases of *Nephrops* are rarely reported compared to those of other decapod crustaceans (Bell et al., 2006; Ziino et al., 2002). Likewise, reports of viral infections of the *Nephrops* are lacking. However, this might reflect a lack of studies rather than a true host resistance. Potentially, undiscovered pathogens may have enormous impacts on larvae and juveniles and thereby affect the populations.

Environmental changes causing stressful conditions that interfere with cellular energy metabolism and the production of ATP will subsequently suppress the energy required for immune defence (Sheldon and Verhulst, 1996) and may give opportunistic pathogens an advantage in the arms race of hosts and parasites. White Spot Syndrome is an emergent viral infection (WSSV) that has caused high mortality rates in cultured shrimp (Sánchez-Paz, 2010). When experimentally challenged with this virus, *Nephrops* became infected but without developing the disease. The susceptibility indicates, however, a potential risk that should be considered, especially following the recent introduction of WSSV to areas such as European waters (Bateman et al., 2012). The future scenario with increased levels of such environmental stressors that have been shown to be immunosuppressive to *Nephrops* will further enhance the hazard of infections.

4. CONCLUSIONS

Nephrops norvegicus inhabits an environment where the need for oxidants is high as bacteria degrade organic material in the sediment. Free oxygen penetrates only a few millimetres into the sediment before it is used as the primary oxidant and bound in other chemical substances. Bioturbation by *Nephrops* and other burrowing organisms promotes oxidation–reduction reactions in sediments by increasing the total area of oxic–anoxic boundaries and, thereby, the surface available for diffusive exchange. The ecophysiological effects of hypoxia on *Nephrops* (as well as other benthic fauna) appear to be a function of the amplitude and absolute value as well as the temporal variation in oxygen concentration. Elevated organic input (such as in the case of eutrophication) may seasonally or locally result in hypoxia on *Nephrops* grounds, with secondary effects leading to increased exposure to a number of other potentially hazardous substances, such as ammonia, metals and hydrogen sulphide. *Nephrops* is adapted to deal with these substances at concentrations that are found during normoxic conditions in the bottom water; however, the increased flux of these substances out of the sediment (when oxygen levels drop) may result in levels that are toxic or stressful to the animals. A summary of the effects of external stressors on the overall fitness of wild *N. norvegicus* is presented in Table 5.3. So far, little is known on how this cocktail of stressors affects *Nephrops*. Until now, empirical research has focused mainly on one or, at maximum, two of the stressors simultaneously. The effect of multiple stressors on *Nephrops* in the field is an important issue to address in the continuously changing environment. Traditionally, reproducing females have been studied to a lesser extent because of the perceived difficulties in interpreting the results in relation to the female reproductive cycle. Also, little is yet known about the tolerance of *Nephrops* to different stressors in the early life stages, mainly due to the fact that those life stages are hard to collect in the field and generally have to be cultured. However, many of the stressors show sex- and size-dependent differences in responses, thus emphasising the importance of studying all life stages of a species for reliable predictions of future effects of stressor(s). Sublethal measurements can be used to monitor stress in *Nephrops*, in particular, substances involved in metabolism, such as L–lactate, haemocyanin and glycogen mirroring the external conditions in a relatively short-term perspective combined with, for example, the Mn in antennae of *Nephrops* indicating a survived exposure to hypoxia. Other substances are involved

Table 5.3 Overview of possible effects on *Nephrops norvegicus* overall fitness when exposed to different external stressors

	External exposure	Growth	Reproduction	Mortality (tolerance)	Mortality (fishing)	
Temperature	21 °C			X		Upper temperature tolerance range when trawled
Salinity	25 psu			X	X	LC_{50} value (100% mortality at 21 psu), moribund; makes animals more vulnerable to both predation and fishing
Oxygen (O_2)	<15% sat	X	X	(x)	X	Stop eating, berried females emerge from burrows and become more vulnerable to both predation and fishing, lower tolerance range
Ammonia (NH_3)	300 μM	(?)				Possible energy loss due to increased irrigation
Hydrogen sulphide (H_2S)	500 μM	(x)	(?)	X	(x)	Detox of H_2S energy demanding, $LT_{50} = 22.5$ h, prone to fishery during H_2S avoidance, possible effects on externally carried embryos
Carbon dioxide (CO_2)	pH 7.7	(x)	(x)			Impaired immune defence, oxidative damage of proteins
Manganese (Mn)	>10 mg l^{-1}	(x)	(x)	(x)	X	Impaired immune defense and tracking of food increase suceptibility to infections. Impaired nerve–muscle reaction decrease swimming capacity. Animals become more vulnerable to predation and fishing
Iron (Fe)						Not known
Lead (Pb)	>1 mg l^{-1}			X		40% mortality when exposed >2 weeks

Stressor						Comments
Mercury (Hg and MeHg)	>0.1 mg l^{-1}				X	$LC_{50} < 96$ h
Cadmium (Cd)	>1 mg l^{-1}				X	$LC_{50} > 2$ weeks
Copper (Cu)	>0.1 mg l^{-1}				X	$LC_{50} = 1$ week
Zinc (Zn)	>2 mg l^{-1}				X	$LC_{50} > 3$ weeks
Cobalt (Co)	6 mg l^{-1}	(x)			X	Impaired nerve-muscle reaction decrease swimming capacity. Animals become more vulnerable to predation and fishing
Organic pollutants	300 kg tar aggregates km^{-2}				X	Decreased abundance indicating increased mortality
Plastic			(?)			Accumulates in stomach, with possible effects on feeding and growth
Fishing	Trawl	(x)	(x)	(x)	X	Direct mortality by all fishing gear, but for trawling indirect effects as well by limb loss that costs energy and egg loss that reduces fecundity, increased vulnerability to injuries
Disease	Hematodinium		(?)		X	Muscle necrosis with possible growth effects during mild infection, and lethal during severe infection

All stressors above are affecting *Nephrops* populations locally or temporarily in the wild with the exception of pH and the metals lead, mercury, cadmium, cobalt, copper and zinc, where the concentrations stated above have not been documented in *Nephrops* areas in the wild. Documented direct effects on the fitness parameters growth, reproduction or mortality (by tolerance or fishing) are denoted by an X. Indirect effects on the animals' fitness are denoted by (x). Not yet documented but possible effects are marked with a (?).

in detoxification of hazardous compounds, that is, metallothioneins and thiosulphate. The immune system appears to be a sensitive biomarker of stress and has potential for the monitoring of sublethal stress in *Nephrops* in the future. The research challenges for the future will include stressors such as emerging contaminants, nanoparticulate plastics, climate change and effects by multiple stressors. In the end, it all comes down to animal fitness, which depends on the animals' ability to survive, grow and reproduce in habitats that might be tough to tolerate at times.

ACKNOWLEDGEMENTS

The authors gratefully acknowledge comments on drafts by Roger Uglow, Catherine Bernasconi and Will Mayes.

REFERENCES

Albalat, A., Sinclair, S., Laurie, J., Taylor, A., Neil, D., 2010. Targeting the live market: recovery of Norway lobsters *Nephrops norvegicus* (L.) from trawl-capture as assessed bystress-related parameters and nucelotide breakdown. J. Exp. Mar. Biol. Ecol. 395 (1–2), 206–214.

Aller, R.C., 2004. Conceptual models of early diagenetic processes: the muddy seafloor as an unsteady, batch reactor. J. Mar. Res. 62, 815–835.

Anacleto, P., Lourenco, H.M., Ferraria, V., Afonso, C., Carvalho, M.L., Martins, M.F., Nunes, M.L., 2009. Arsenic content in seafood consumed in Portugal. J. Aquat. Food Prod. Technol. 18 (1–2), 32–45.

Aréchiga, H., Atkinson, R.J.A., 1975. Eye and some effects of light on locomotor activity in *Nephrops norvegicus*. Mar. Biol. 32 (1), 63–76.

Arnold, K.E., Findlay, H.S., Spicer, J.I., Daniels, C.L., Boothroyd, D., 2009. Effect of CO_2-related acidification on aspects of the larval development of the European lobster, *Homarus gammarus* (L.). Biogeosciences 6 (8), 1747–1754.

Aspán, A., Hall, M., Söderhäll, K., 1990. The effect of endogenous proteinase inhibitors on the prophenoloxidase activating system, a serine proteinase from crayfish haemocytes. Insect. Biochem. 20, 485–492.

Astall, C.M., Taylor, A.C., Atkinson, R.J.A., 1997. Behavioural and physiological implications of a burrow-dwelling lifestyle for two species of upogebiid mud-shrimp (Crustacea: Thalassinidea). Estuar. Coast. Shelf S. 44, 155–168.

Baden, S.P., Eriksson, S.P., 2006. Role, routes and effects of manganese in crustaceans. Oceanogr. Mar. Biol. Annu. Rev. 44, 61–83.

Baden, S.P., Neil, D.M., 1998. Accumulation of manganese in the haemolymph, nerve and muscle tissue of *Nephrops norvegicus* (L.) and its effect on neuromuscular performance. Comp. Biochem. Physiol. 119A (3), 351–359.

Baden, S.P., Neil, D.M., 2003. Manganese accumulation by the antennule of the Norway lobster *Nephrops norvegicus* (L.) as a biomarker of hypoxic events. Mar. Environ. Res. 55, 59–71.

Baden, S.P., Pihl, L., 1996. Effects of autumnal hypoxia on demersal fish and crustaceans in the SE Kattegat, 1984–91. p. 189–196. In: Andersen, J., Karup, H., Nielsen, U.B. (Eds.), Scientific Symposium on the North Sea Quality Status Report 1993, 18–21 April 1994, Ebeltoft, Denmark: Proceedings, 346 pp.

Baden, S.P., Loo, L.-O., Pihl, L., Rosenberg, R., 1990a. Effects of eutrophication on benthic communities including fish, Swedish west coast. Ambio 19 (3), 113–122.

Baden, S.P., Pihl, L., Rosenberg, R., 1990b. Effects of oxygen depletion on the ecology, blood physiology and fishery of the Norway lobster, *Nephrops norvegicus* (L.). Mar. Ecol. Prog. Ser. 67, 141–155.

Baden, S.P., Depledge, M., Hagerman, L., 1994. Glycogen depletion and altered copper and manganese handling in *Nephrops norvegicus* (L.) following starvation and exposure to hypoxia. Mar. Ecol. Prog. Ser. 103, 65–72.

Baden, S.P., Eriksson, S.P., Weeks, J.M., 1995. Uptake, accumulation and regulation of manganese during experimental hypoxia and anoxia in the decapod *Nephrops norvegicus* (L.). Mar. Pollut. Bull. 31 (1–3), 93–102.

Baden, S.P., Eriksson, S.P., Gerhardt, L., 1999. Accumulation and elimination kinetics of manganese from different tissues of the Norway lobster *Nephrops norvegicus* (L.). Aquat. Toxicol. 46, 127–137.

Baden, S.P., Håkansson, C.L.J., Spicer, J.I., 2004. Between individual concentrations during recovery from environmental hypoxia and manganese by the Norway lobster, *Nephrops norvegicus*. Mar. Biol. 143, 267–273.

Bagge, O., Munch-Petersen, S., 1979. Some possible factors governing the catchability of Norway lobsters in the Kattegat. Rapports et procès-verbaux des réunions/Conseil permanent int. pour l'exploration de la mer 175, 143–146.

Barghigiani, C., Ristori, T., Biagi, F., De Ranieri, S., 2000. Size related mercury accumulations in edible marine species from an area of the Thyrrenian Sea. Water Air Soil Pollut. 124 (1–2), 169–176.

Bateman, K.S., Tew, I., French, C., Hicks, R.J., Martin, P., Munro, J., Stentiford, G.D., 2012. Susceptibility to infection and pathogenicity of White Spot Disease (WSD) in non-model crustacean host taxa from temperate regions. J. Invertebr. Pathol. 110, 340–351.

Bell, M.C., Redant, F., Tuck, I., 2006. Nephrops species. In: Phillips, B.F. (Ed.), Lobsters: Biology, Management, Aquaculture, Fisheries Blackwell Publishing Ltd, Oxford, UK, pp. 412–461.

Bergmann, M., Moore, P.G., 2001. Survival of decapod crustaceans discarded in the *Nephrops* fishery of the Clyde Sea area, Scotland. ICES J. Mar. Sci. 58 (1), 163–171.

Bergmann, M., Wieczorek, S.K., Moore, P.G., Atkinson, R.J.A., 2001. Discard composition of the Nephrops fishery in the Clyde Sea area, Scotland. Fish. Res. 57, 169–183.

Bergmann, M., Weiczorek, S., Moore, P.G., Atkinson, R.J.A., 2002. Utilization of invertebrates discarded from the Nephrops fishery by variously selective benthic scavengers in the west of Scotland. Mar. Ecol. Prog. Ser. 233, 185–198.

Bernasconi, C.J., Uglow, R.F., 2008. Effects of emersion and re-immersion on physiological and immunological variables in creel-caught and trawled Norway lobster Nephrops norvegicus. Dis. Aquat. Organ. 81 (3), 241–247.

Bernasconi, C.J., Uglow, R.F., 2011. Purineolytic capacity response of *Nephrops norvegicus* to prolonged emersion: an ammonia detoxification process. Aquat. Biol. 11, 263–270.

Boudreau, B.P., Jørgensen, B.B., 2001. The Benthic Boundary Layer, Oxford University Press, New York p. 405.

Bridges, C.R., 2001. Modulation of haemocyanin oxygen affinity: properties and physiological implications in a changing world. J. Exp. Biol. 2 (4), 1021–1032.

Briggs, R.P., Armstrong, M.J., Dickey-Collas, M., Allen, M., McQuaid, N., Whitmore, J., 2002. The application of fecundity estimates to determine the spawning stock biomass of Irish Sea *Nephrops norvegicus* (L.) using the annual larval production method. ICES J. Mar. Sci. 59 (1), 109–119.

Bryan, G.W., 1976. Some aspects of heavy metal tolerance in aquatic organisms. In: Lockwood, A.P.M. (Ed.), Effects of Pollutants on Aquatic Organisms. Cambridge University Press, Cambridge, pp. 7–34.

Bryan, G.W., 1979. Bioaccumulation of marine pollutants. Phil. Trans. Roy. Soc. Lond. 286, 483–505.

Bryan, G.W., Ward, E., 1965. The absorption and loss of radioactive and non-radioactive manganese by the lobster, *Homarus vulgaris*. J. Mar. Biol. Assoc. UK 45, 65–95.

Butterworth, K.G., Grieshaber, M.K., Taylor, A.C., 2004. Behavioural and physiological responses of the Norway lobster, *Nephrops norvegicus* (Crustacea: Decapoda), to sulphide exposure. Mar. Biol. 144, 1087–1095.

Caldeira, K., Wickett, M.E., 2003. Anthropogenic carbon and ocean pH. Nature 425, 365.

Canli, M., Furness, R.W., 1993a. Heavy metals in tissues of the Norway Lobster *Nephrops Norvegicus*: effects of sex, size and season. Chem. Ecol. 8 (1), 19–32.

Canli, M., Furness, R.W., 1993b. Toxicity of heavy metals dissolved in sea water and influences of sex and size on metal accumulation and tissue distribution in the Norway lobster *Nephrops norvegicus*. Mar. Environ. Res. 36 (4), 217–236.

Canli, M., Furness, R.W., 1995. Mercury and cadmium uptake from seawater and from food by the Norway lobster *Nephrops norvegicus*. Environ. Toxicol. Chem. 14 (5), 819–828.

Canli, M., Stagg, R.M., 1996. The effects of in vivo exposure to cadmium, copper and zinc on the activities of gill ATPases in the Norway lobster, *Nephrops norvegicus*. Arch. Environ. Contam. Toxicol. 31 (4), 494–501.

Canli, M., Stagg, R.M., Rodger, G., 1997. The induction of metallothionein in tissues of the Norway lobster *Nephrops norvegicus* following exposure to cadmium, copper and zinc: the relationships between metallothionein and the metals. Environ. Pollut. 96 (3), 343–350.

Cannon, W.B., 1926. Physiological regulation of normal states: some tentative postulates concerning biological homeostatics. In: Pettit, A. (Ed.), A Charles Richet: ses amis, ses collègues, ses élèves. Éditions Médicales, Paris, p. 91.

Carney Almroth, B., Sköld, M., Nilsson Sköld, H., 2012. Gender differences in health and aging of Atlantic cod subject to size selective fishery. Biol. Open 1, 922–928.

Casini, M., Lövgren, J., Hjelm, J., Cardinale, M., Molinero, J.-C., Korlinovs, G., 2008. Multi-level trophic cascades in heavily exploited open marine ecosystem. Proc. R. Soc. B 275, 1793–1801.

Cerenius, L., Söderhäll, K., 2004. The prophenoloxidase-activating system in invertebrates. Immunol. Rev. 198 (1), 116–126.

Chaga, O., Lignell, M., Söderhäll, K., 1995. The haematopoietic cells of the freshwater crayfish, *Pacifastacus leniusculus*. Anim. Biol. 4, 59–70.

Childress, J.J., 1987. Uptake and transport of sulfide in marine invertebrates. In: Dejours, P., Bolis, L., Taylor, C.R., Weibel, E.R. (Eds.), Comparative Physiology: Life in Water and on Land, Fidia Research Series, vol. 9. Springer Verlag, Berlin, p. 559.

Cizewski Culotta, V., Yang, M., Hall, M.D., 2005. Manganese transport and trafficking: lessons learned from *Saccharomyces cerevisiae*. Eucariotic cell 4, 1159–1165.

Clarke, A., Fraser, K.P.P., 2004. Why does metabolism scale with temperature? Funct. Ecol. 18, 243–251.

Davies, J.M., Topping, G. (Eds.), 1997. The Impact of an Oil Spill in Turbulent Waters: The Braer, The Stationary Office Ltd, Edinburgh, p. 263.

Depledge, M., 1989. Studies on copper and iron concentrations, distributions and uptake in the brachyuran *Carcinus maenas* (L.) following starvation. Ophelia 30 (3), 187–197.

Diaz, R.J., Rosenberg, R., 1995. Marine benthic hypoxia: a review of its ecological effects and the behavioural responses of benthic macrofauna. Oceanogr. Mar. Biol. Annu. Rev. 33, 245–303.

Diaz, R.J., Rosenberg, R., 2008. Spreading dead zones and consequences for marine ecosystems. Science 321 (5891), 926–929.

Doney, S.C., Fabry, V.J., Feely, R.A., Kleypas, J.A., 2009. Ocean acidification: the other CO_2 problem. Annu. Rev. Mar. Sci. 1, 169–192.

Dunthorn, A.A., 1967. Some observations on the behaviour and development of the Norway lobster. ICES C.M. 1967/K:5 (mimeo).

Eriksson, S.P., 2000a. Temporal variations of manganese in the hemolymph and tissues of the Norway lobster, *Nephrops norvegicus* (L.). Aquat. Toxicol. 48 (2–3), 297–307.

Eriksson, S.P., 2000b. Variations of manganese in the eggs of the Norway lobster, *Nephrops norvegicus*. Aquat. Toxicol. 48 (2–3), 291–295.

Eriksson, S.P., 2006. Differences in the condition of Norway lobster (*Nephrops norvegicus* (L.) from trawled and creeled fishing areas. Mar. Biol. Res. 2, 52–58.

Eriksson, S.P., Baden, S.P., 1997. Tolerance and behaviour in juvenile *Nephrops norvegicus* (L.) during hypoxia. Mar. Biol. 128, 49–54.

Eriksson, S.P., Baden, S.P., 1998. Manganese in the haemolymph and tissues of the Norway lobster, *Nephrops norvegicus* (L.), along the Swedish west-coast, 1993–1995. Hydrobiologia 375/376, 255–264.

Eriksson, S.P., Nabbing, M., Sjöman, E., 2006. Is brood care in *Nephrops norvegicus* during hypoxia adaptive or a waste of energy? Funct. Ecol. 20, 1097–1104.

Fatt, P., Ginsborg, B.L., 1958. The ionic requirements for the production of action potentials in crustacean muscle fibers. J. Physiol. Lond. 142, 516–543.

Feely, R.A., Sabine, C.L., Lee, K., Berelson, W., Kleypas, J., Fabry, V.J., Millero, F.J., 2004. Impact of anthropogenic CO_2 on the $CaCO_3$ system in the oceans. Science 305, 362–366.

Feely, R.A., Sabine, C.L., Hernandez-Ayon, J.M., Ianson, D., Hales, B., 2008. Evidence for upwelling of corrosive "acidified" water onto the continental shelf. Science 320 (5882), 1490–1492.

Field, R.H., Chapman, C.J., Taylor, A.C., Neil, D.M., Vickerman, K., 1992. Infection of the Norway lobster *Nephrops norvegicus* by a *Hematodinium*-like species of dinoflagellate on the West Coast of Scotland. Dis. Aquat. Organ. 13, 1–15.

Fingerman, M., Devi, M., Reddy, P.S., Katyayani, R., 1996. Impact of heavy metal exposure on the nervous system and endocrine-mediated processes in crustaceans. Zool. Stud. 35 (1), 1–8.

Funch, P., Kristensen, R., 1995. Cycliophora is a new phylum with affinites to Entoprocta and Ectoprocta. Nature 378, 711–714.

Gerhardt, L., Baden, S.P., 1998. Gender- and oxygen-related irrigation behaviour of the decapod *Nephrops norvegicus* (L.). Mar. Biol. 131, 553–558.

Glasby, G.P., 1984. Manganese in the marine environment. Oceanogr. Mar. Biol. Annu. Rev. 22, 169–194.

Gornik, S.G., Albalat, A., Atkinson, R.J.A., Coombs, G.H., Neil, D.M., 2010. The influence of defined ante-mortem stressors on the early post-mortem biochemical processes in the abdominal muscle of the Norwa lobster, *Nephrops norvegicus* (Linnaeus, 1758). Mar. Biol. Res. 6, 223–238.

Graf, G., Bengtsson, W., Diesner, U., Schulz, R., Theede, H., 1982. Benthic response to sedimentation of a spring phytoplankton bloom—process and budget. Mar. Biol. 67 (2), 201–208.

Grieshaber, M.K., Hardewig, I., Kreutzer, U., Pörtner, H.-O., 1994. Physiological and metabolic responses to hypoxia in invertebrates. Rev. Physiol. Biochem. Pharmacol. 125, 43–147.

Gutteridge, J.M.C., Halliwell, B., 2000. Free radicals and antioxidants in the year 2000—a historical look to the future. In: Chiueh, C.C. (Ed.), Reactive Oxygen Species: From Radiation to Molecular Biology: A Festschrift in Honor of Daniel L Gilbert, Annals of the New York Academy of Sciences, vol. 899. pp. 36–147.

Hagerman, L., 1998. Physiological flexibility; a necessity for life in anoxic and sulphidic habitats. Hydrobiologia 375/376, 241–254.

Hagerman, L., Baden, S.P., 1988. *Nephrops norvegicus*, field study of effects of oxygen deficiency on haemocyanin concentration. J. Exp. Mar. Biol. Ecol. 116, 135–142.

Hagerman, L., Uglow, R.F., 1981. Ventilatory behaviour and chloride regulation in relation to oxygen tension in the shrimp *Palaemon adspersus* Rathke maintained in hypotonic medium. Ophelia 20, 193–200.

Hagerman, L., Uglow, R.F., 1985. Effects of hypoxia on the respiratory and circulatory regulation of *Nephrops norvegicus*. Mar. Biol. 87, 273–278.

Hagerman, L., Søndergaard, T., Weile, K., Hosie, D., Uglow, R.F., 1990. Aspects of blood physiology and ammonia excretion in *Nephrops norvegicus* under hypoxia. Comp. Biochem. Physiol. A Physiol. 1990, 51–55.

Hagiwara, S., Takahashi, K., 1967. Surface density of calcium ions and calcium spikes in the barnacle muscle fibre membrane. J. Gen. Physiol. 50, 583–601.

Harris, R.R., Andrews, M.B., 2005. Physiological changes in the Norway lobster Nephrops norvegicus escaping and discarded from the commercial trawls on the West Coast of Scotland—I. Body fluid volumes and haemolymph composition after capture and during recovery. J. Exp. Mar. Biol. Ecol. 320 (2), 179–193.

Harris, R.R., Ulmestrand, M., 2004. Discarding Norway lobster (*Nephrops norvegicus* L.) through low salinity layers—mortality and damage seen in simulation experiments. ICES J. Mar. Sci. 61 (1), 127–139.

Hart, B., 1982. Uptake of trace metals by sediments and suspended particulates: a review. Hydrobiologia 91, 299–313.

Haselmair, A., Stachowitsch, M., Zuschin, M., Ridel, B., 2010. Behaviour and mortality of benthic crustaceans in response to experimentally induced hypoxia and anoxia *in situ*. Mar. Ecol. Prog. Ser. 414, 195–208.

Hergenhahn, H.G., Aspán, A., Söderhäll, K., 1987. Purification and characterization of a high-Mr proteinase inhibitor of pro-phenol oxidase activation from crayfish plasma. Biochem. J. 248, 223–228.

Hernroth, B., Baden, S.P., Holm, K., Andrén, T., Söderhäll, I., 2004. Manganese induced immune suppression of the lobster, *Nephrops norvegicus*. Aquat. Toxicol. 70, 223–231.

Hernroth, B., Skold, H.N., Wiklander, K., Jutfelt, F., Baden, S.P., 2012. Simulated climate change causes immune suppression and protein damage in the crustacean *Nephrops norvgicus*. Fish Shellfish Immunol. 33 (5), 1095–1101.

Holmes, J.M., Gräns, A.-S., Neil, D.M., Baden, S.P., 1999. Effects of the metal ions Mn^{2+} and Co^{2+} on muscle contraction in the Norway lobster, *Nephrops norvegicus*. J. Comp. Physiol. B 169, 402–410.

Hopkinson Jr., C.S., 1987. Nutrient regeneration in shallow-water sediments of the estuarine plume region of the near-shore Georgia Bight, USA. Mar. Biol. 94, 127–142.

Hosie, D.A., Uglow, R.F., Hagerman, L., Søndergaard, T., 1991. Some effects of hypoxia and medium ammonia enrichment on efflux rates and circulating levels of ammonia in *Nephrops norvegicus*. Mar. Biol. 110, 273–279.

Intergovernmental Panel on Climate Change (IPCC), 2007. Climate Change 2007: The Physical Science Basis. Contribution of Working Group I to the Fourth Assessment Report of the Intergovernmental Panel on Climate Change. Cambridge University Press, New York.

Jansen, M., Groman, D., 1993. The effect of high concentrations of iron on impounded American lobsters: a case study. J. Aquat. Anim. Health 5, 155–156.

Jiravanichpaisal, P., Lee, B.L., Söderhäll, K., 2006. Cell-mediated immunity in arthropods: hematopoiesis, coagulation, melanization and opsonization. Immunobiology 4, 213–236.

Johansson, M.W., Söderhäll, K., 1985. Exocytosis of the prophenoloxidase activating system from crayfish haemocytes. J. Comp. Physiol. 156, 175–181.

Johansson, M.W., Keyser, P., Sritunyalucksana, K., Söderhäll, K., 2000. Crustacean haemocytes and haematopoiesis. Aquaculture 191, 45–52.

Johns, A.R., Taylor, A.C., Atkinson, R.J.A., Grieshaber, M.K., 1997. Sulphide metabolism in thalassinidean crustacea. J. Mar. Biol. Assoc. UK 77, 127–144.

Juanes, F., Smith, L.D., 1995. The ecological consequences of limb damage and loss in decapod: a review and prospectus. J. Exp. Mar. Biol. Ecol. 193 (1–2), 197–223.

Kitching, J.A., Ebling, F.J., Gamble, J.C., Hoare, R., McLeod, A.A.Q.R., Norton, T.A., 1976. The ecology of Loch Ine. J. Anim. Ecol. 45, 731–758.

Krång, A.S., Rosenqvist, G., 2006. Effects of manganese on chemically induced food search behavior of the Norway lobster, *Nephrops norvegicus* (L.). Aquat. Toxicol. 78 (3), 284–291.

Kristensen, E., 1984. Oxygen and inorganic nitrogen exchange in *Nereix virens* (Polychaeta) bioturbated sediment-water systems. J. Coastal Res. Fort Lauderdale 1, 109–116.

Kuris, A.M., Lafferty, K.D., 1992. Modelling crustacean fisheries: effects of parasites on management strategies. Can. J. Fish. Aquat. Sci. 49, 327–336.

Lee, S.Y., Söderhäll, K., 2001. Characterization of a pattern recognition protein, a masquerade-like protein, in the freshwater crayfish *Pacifastacus leniusculus*. J. Immunol. 166, 7319–7326.

Lee, S.Y., Wang, R., Söderhäll, K., 2000. A lipopolysaccharide- and β-1, 3-glucan-binding protein from hemocytes of the freshwater crayfish *Pacifastacus leniusculus*. J. Biol. Chem. 275, 1337–1343.

Lee, S.Y., Lee, B.L., Söderhäll, K., 2003. Processing of an antibacterial peptide from hemocyanin of the freshwater crayfish *Pacifastacus leniusculus*. J. Biol. Chem. 278, 7927–7933.

Lerch, F., Uglem, I., 1996. High density of *Histriobdella homari* Van Beneden, 1858 (Annelida, Polychaeta) on ovigerous female European lobsters (Decapoda, Nephropidae). Crustaceana 69, 916–920.

Liang, Z., Sottrup-Jensen, L., Aspán, A., Hall, M., Söderhäll, K., 1997. Pacifastin, a novel 155-kDa heterodimeric proteinase inhibitor containing a unique transferrin chain. Natl. Acad. Sci. U.S.A. 94, 6682–6687.

Lorenzon, S., de Guarrine, S., Smith, V.J., Ferreo, E.A., 1999. Effects of LPS injection on circulating haemocytes in crustaceans *in vivo*. Fish Shellfish Immunol. 9, 31–50.

Lourenco, H.M., Anacleto, P., Afonso, C., Martins, M.F., Carvalho, M.L., Lino, A.R., Nunes, M.L., 2009. Chemical characterisation of *Nephrops norvegicus* from Portuguese Coast. J. Sci. Food Agric. 89 (15), 2572–2580.

Lund, H.S., Wang, T., Chang, E.S., Pedersen, L.F., Taylor, E.W., Pedersen, P.B., McKenzie, D.J., 2009. Recovery by the *Nephrops norvegicus* (L.) from the physiological stresses of trawling: influence of season and live-storage position. J. Exp. Mar. Biol. Ecol. 373 (2), 124–132.

Macdonald, J.M., Shields, J.D., Zimmer-Faust, R.K., 1988. Acute toxicities of eleven metals to early life-history stages of the yellow crab Cancer anthonyi. Mar. Biol. 98, 201–207.

MacKenzie, K., 1963. *Stichocotyle nephropis* Cunningham, 1887 (Trematoda) in Scottish Waters. Ann. Mag. Nat. Hist. 6, 505–506.

Maclean, S.A., Ruddell, C.L., 1978. Three new crustacean hosts for the parasitic dinoflagellate *Hematodinium perezi* (Dinoflagellata: Syndinidae). J. Parasitol. 64, 158–160.

Magnusson, K., Ekelund, R., Dave, G., Granmo, Å., Förlin, L., Wennberg, L., Samuelsson, M.O., Berggren, M., Brorström-Lundén, E., 1996. Contamination and correlation with toxicity of sediment samples from Skagerrak and Kattegat. J. Sea Res. 35, 223–234.

Mantel, L.H., Farmer, L.L., 1983. Osmotic and ionic regulation. The Biology of Crustacea. In: Bliss, D.E., Mantel, L.H. (Eds.), Internal Anatomy and Physiological Regulation, vol. 5. Academic Press, New York, pp. 53–161.

McIntosh, A.D., Davies, I.M., Webster, L., 2001. The source and fate of polycyclic aromatic hydrocarbon (PAH) in farmed mussels and other commercial species from Loch Leven B *Nephrops norvegicus*. FRS Marine Laboratory Report No 05/01, pp. 18.

Morris, S., 1995. A new phylum from Lobster's lips. Nature 378, 661.

Mounier, Y., Vassort, G., 1975. Initial and delayed membrane currents in crab muscle fibre under voltageclamp conditions. J. Physiol. Lond. 251, 589–608.

Murray, F., Cowie, P.R., 2011. Plastic contamination in the decapod crustacean *Nephrops norvegicus* (Linnaeus, 1758). Mar. Pollut. Bull. 62 (6), 1207–1217.

Newland, P.L., Neil, D.M., Chapman, C.J., 1992. Escape swimming in the Norway lobster. J. Crustacean Biol. 12 (3), 342–353.

Nicholls, P., 1975. Inhibition of cytochrome-c oxidase by sulfide. Biochem. Soc. Trans. 3, 316–319.

Norstedt, M., 2004. Uptake of manganese through experimental feeding in Norway lobster, Nephrops norvegicus. M. Sc. Thesis in Marine Ecology, Dep. Mar. Ecol., University of Gothenburg. Sweden. Nr. 60. 27 pp. (In English).

Nyström, T., 2005. Role of protein carbonylation in quality controls and aging. EMBO J 24, 1311–1317.

Oweson, C., Hernroth, B., 2009. A comparative study on the influence of manganese on the bactericidal response of marine invertebrates. Fish Shellfish Immunol. 27, 500–507.

Oweson, C., Baden, S.P., Hernroth, B., 2006. Manganese induced apoptosis in haematopoietic cells of the lobster, *Nephrops norvegicus* (L). Aquat. Toxicol. 77, 322–328.

Perugini, M., Visciano, P., Giammarino, A., Manera, M., Di Nardo, W., Amorena, M., 2007. Polycyclic aromatic hydrocarbons in marine organisms from the Adriatic Sea, Italy. Chemosphere 66 (10), 1904–1910.

Petersen, W., Wallman, K., Pinglin, L., Schroeder, F., Knauth, H.-D., 1995. Exchange of trace elements at the sediment-water interface during early diagenesis processes. Mar. Freshw. Res. 46, 19–26.

Pihl, L., 1989. Effects of oxygen depletion on demersal fish in coastal areas of the south east Kattegat. In: Ryland, J.S., Tyler, P.A. (Eds.), Reproduction, Genetics, and Distribution of Marine Organisms, Proceedings of the 23rd European Marine Biology Symposium. Swansea, England, pp. 431–439.

Poulsen, E.M., 1946. Investigations on the Danish fishery for and the biology of the Norway lobster and the deep-sea prawn. Report Danish Biol. Station 48 (1943/45), 27–49.

Rainbow, P.S., 1996. Heavy metals in aquatic invertebrates. In: Beyer, W.N. et al., (Ed.), Interpreting Concentrations of Environmental Contaminants in Wildlife Tissues, Lewis Publishers, Boca Raton, FL, pp. 405–425.

Rainbow, P.S., 1997. Trace metal accumulation in marine invertebrates: marine biology or marine chemistry? J. Mar. Biol. Assoc. UK 77, 195–210.

Revsbech, N.P., Jorgensen, B.B., Blackburn, T.H., 1980. Oxygen in the sea bottom measured with microelectrode. Science 207 (4437), 1355–1356.

Ridgway, I.D., Taylor, A.C., Atkinson, R.J.A., Chang, E.S., Neil, D.M., 2006. Impact of capture method and trawl duration on the health status of the Norway lobster, *Nephrops norvegicus*. J. Exp. Mar. Biol. Ecol. 339, 135–147.

Rosenberg, R., 1985. Eutrophication—the future marine coastal nuisance? Mar. Pollut. Bull. 16, 227–231.

Rosenberg, R., Loo, L.-O., 1988. Marine eutrophication induced oxygen deficiency: effects on soft bottom Fauna, Western Sweden. Ophelia 29, 213–225.

Rosenberg, R., Elmgren, R., Fleischer, S., Jonsson, P., Persson, G., Dahlin, H., 1990. Marine eutrophication case studies in Sweden. Ambio 19, 102–108.

Royal Society, 2005. Ocean Acidification due to Increasing Atmospheric Carbon Dioxide, Policy Document 12/05, The Clyvedon Press, Cardiff.

Sabine, C.L., Feely, R.A., Gruber, N., Key, R.M., Lee, K., Bullister, J.L., Wanninkhof, R., Wong, C.S., Wallace, D.W.R., Tilbrook, B., Millero, F.J., Peng, T.-H., Kozyr, A., Ono, T., Rios, A.F., 2004. The oceanic sink for antropogenic CO_2. Science 305, 367–371.

Sánchez, F., Velasco, F., Cartes, J.E., Olaso, I., Preciado, I., Fanelli, E., Serrano, A., Gutierrez-Zabala, J.L., 2006. Monitoring the Prestige oil spill impacts on some key species of the Northern Iberian shelf. Mar. Pollut. Bull. 53 (5–7), 332–349.

Sánchez-Paz, A., 2010. White spot syndrome virus: an overview on an emergent concern. Vet. Res. 41, 43. http://dx.doi.org/10.1051/vetres/2010015.

Schmitt, A.S.C., Uglow, R.F., 1997. Haemolymph constituent levels and ammonia efflux rates of *Nephrops norvegicus* during emersion. Mar. Biol. 127 (3), 403–410.

Senkbiel, E.G., Wriston Jr., J.G., 1981. Haemocyanin synthesis in the American lobster, Homarus americanus. Comp. Biochem. Physiol. B 68, 163–171.

Sheldon, B.C., Verhulst, S., 1996. Ecological immunology: costly parasite defences and trade-offs in evolutionary ecology. Trends Ecol. Evol. 11, 317–321.

Shields, J.D., Stephens, F.J., Jones, B., 2006. Pathogens, parasites and other symbionts. In: Phillips, B.F. (Ed.), Lobsters: Biology, Management, Aquaculture, Fisheries, Blackwell Publishing Ltd, Oxford, UK, pp. 146–204.

Simkiss, K., 1981. Cellular discrimination processes in metal accumulation cells. J. Exp. Biol. 94, 317–327.

Simkiss, K., Taylor, M.G., 1989. Metal fluxes across the membranes of aquatic organisms. Aquat. Sci. 1, 173–188.

Small, H.J., 2012. Advances in our understanding of the global diversity and distribution of *Hematodinium* spp.—significant pathogens of commercially exploited crustaceans. J. Invertebr. Pathol. 110, 234–246.

Smith, V.J., Söderhäll, K., 1983. Induction of degranulation and lysis of haemocytes in the freshwater crayfish, *Astacus astacus* by components of the prophenoloxidase activating system in vitro. Cell Tissue Res. 233 (2), 295–303.

Smolowitz, R.M., Bullis, R.A., Abt, D.A., 1992. Pathologic cuticular changes of winter impoundment shell disease preceding and during intermolt in the American lobster, *Homarus americanus*. Biol. Bull. 183, 99–112.

Söderhäll, K., Cerenius, L., 1998. Role of the prophenoloxidase-activating system in invertebrate immunity. Curr. Opin. Immunol. 10, 23–28.

Söderhäll, I., Bangyeekhun, E., Mayo, S., Söderhäll, K., 2003. Hemocyte production and maturation in an invertebrate animal; proliferation and gene expression in hematopoietic stem cells of *Pacifastacus leniusculus*. Dev. Comp. Immunol. 27, 661–672.

Spicer, J.I., Baden, S.P., 2001. Environmental hypoxia and haemocyanin variability in Norway lobsters *Nephrops norvegicus* (L.). Mar. Biol. 139 (4), 727–734.

Spicer, J.I., Eriksson, S.P., 2003. Does the development of respiratory regulation always accompany the transition from pelagic larvae to benthic fossorial post-larvae in the Norway lobster *Nephrops norvegicus* (L.)? J. Exp. Mar. Biol. Ecol. 295 (1), 219–243.

Spicer, J.I., Hill, A.D., Taylor, A.C., Strang, R.H.C., 1990. Effect of aerial exposure on concentrations of selected metabolites in blood of the Norwegian lobster *Nephrops norvegicus* (Crustacea, Nephropidae). Mar. Biol. 105 (1), 129–135.

Stentiford, G.D., 2011. Diseases of commercially exploited crustaceans: cross-cutting issues for global fisheries and aquaculture. J. Invertebr. Pathol. 106, 3–5.

Stentiford, G.D., Neil, D.M., 2011. Diseases of Nephrops *and* Metanephrops: a review. In Grant Stentiford. "Diseases of Edible Crustaceans". J. Invertebr. Pathol. 106 (1), 92–109. http://dx.doi.org/10.1016/j.jip. 2010.09.017. PMID 21215358.

Stentiford, G.D., Shields, J.D., 2005. A review of the parasitic dinoflagellates *Hematodinium* species and *Hematodinium*-like infections in marine crustaceans. Dis. Aquat. Organ. 66, 47–70.

Sundby, B., Anderson, L.G., Hall, P.O.J., Iverfeldt, Å., Rutgers van der Loeff, M.M., Westerlund, S.F.G., 1986. The effect of oxygen on release and uptake of cobalt, manganese, iron and phosphate at the sediment-water interface. Geochim. Cosmochim. Acta 50, 1281–1288.

Svensson, A., 1975. Physical and chemical oceanography of the Skagerrak and Kattegat. Fish. Board Sweden. Inst. Mar. Res. 1, 1–88.

Taylor, H.H., Antiss, J.M., 1999. Copper and haemocyanin dynamics in aquatic invertebrates. Mar. Freshw. Res. 50, 907–931.

Taylor, A.C., Johns, A.R., Atkinson, R.J.A., Bridges, C.R., 1999. Effects of sulphide and thiosulphide on the properties of the haemocyanin of the benthic crustaceans *Calocaris macandreae* Bell, *Nephrops norvegicus* (L.) and *Carcinus maenas* (L.). J. Exp. Mar. Biol. Ecol. 233, 163–179.

Tessier, A., Turner, D.R., 1995. Metal Speciation and Bioavailability in Aquatic Systems, John Whiley and Sons Ltd, Chichester, p. 679.

Thomas, H.J., Figueiredo, M.J., 1965. Seasonal Variations in the catch composition of the Norway Lobster, *Nephrops norvegicus* (L.) around Scotland. Journal du conseil/Conseil international pour l'exploration de la mer 30 (1), 75–85.

Tyson, R.V., Pearson, T.H., 1991. Modern and ancient continental shelf anoxia: an overview. In: Tyson, R.V., Pearson, T.H. (Eds.), Modern and Ancient Continental Shelf Anoxia, Geological Society Special Publication, London, pp. 1–24.

Vaquer-Sunuyer, R., Duarte, C.M., 2010. Sulfide exposure accelerates hypoxia-driven mortality. Limnol. Oceanogr. 55, 1075–1082.

Vaquer-Sunuyer, R., Duarte, C.M., 2011. Temperature effects on oxygen thresholds for hypoxia in marine benthic organisms. Glob. Chang. Biol. 17, 1788–1797.

Viarengo, A., Nott, J.A., 1993. Mechanisms of heavy-metal cation homeostasis in marine invertebrates. Comp. Biochem. Physiol. C Pharmacol. Toxicol. Endocrinol. 104 (3), 355–372.

Visman, B., 1991. Sulfide tolerance: physiological mechanisms and ecological implications. Ophelia 34, 1–27.

Wassmann, P., 1990. Calculating the load of organic-carbon to the aphotic zone in eutrophicated coastal waters. Mar. Pollut. Bull. 21 (4), 183–187.

Weinstein, J.E., West, T.L., Bray, J.T., 1992. Shell disease and metal content of blue crabs, *Callinectes sapidus*, from the Albemarle-Pamlico Estuarine System, North Carolina. Arch. Environ. Contam. Toxicol. 23, 355–362.

White, S.L., Rainbow, P.S., 1987. Heavy metal concentrations and size effects in the mesopelagic decapod *Systellaspis debilis*. Mar. Ecol. Prog. Ser. 37, 147–151.

Zalasiewicz, J., Williams, M., Haywood, A., Ellis, M., 2011. The Anthropocene: a new epoch of geological time? Philos. Trans. R. Soc. Lond. A 369 (1938), 835–841.

Zeebe, R.E., Zachos, J.C., Calderia, K., Tyrell, T., 2008. Carbon emissions and acidification. Science 321, 51–52.

Ziino, G., Giuffrida, A., Stancanelli, A., Panebianco, A., 2002. Shell disease in *Nephrops norvegicus* from the Mediterranean Sea. Morphological and hygienic remarks. Arch. für Lebensmittelhygiene 53, 134–136.

Zuykov, M., Pelletier, E., Demers, S., 2011. Colloidal complexed silver and silver nanoparticles in extrapallial fluid of *Mytilus edulis*. Mar. Environ. Res. 71, 17–21.

Ärtebjerg, G., Carstensen, J., Axe, P., Druon, J-N., Stips, A., 2003. The 2002 Oxygen Depletion Event in the Kattegat, Belt Sea and Western Baltic. Baltic Sea Environment Proceedings No. 90, Thematic Report, Helsinki Commission, Baltic Marine Environment Protection Commission.

CHAPTER SIX

Reproduction: Life Cycle, Larvae and Larviculture

Adam Powell[*,1], Susanne P. Eriksson[†]

[*]Centre for Sustainable Aquatic Research, Swansea University, Singleton Park, Swansea, United Kingdom
[†]Department of Biological and Environmental Sciences-Kristineberg, University of Gothenburg, SE-45178 Fiskebäckskil, Sweden
[1]Corresponding author: e-mail address: a.powell@swansea.ac.uk

Contents

Abstract

Nephrops norvegicus represents a very valuable fishery across Europe, and the species possesses a relatively complex life cycle and reproductive biology across spatial and temporal scales. Insights into embryonic and larval biology, and associated abiotic and biotic factors that influence recruitment, are important since this will affect population and species success. Much of the fishery, and indeed scientific sampling, is reliant on trawling, which is likely to cause direct and indirect stresses on adults and developing embryos. We have collated evidence, including that garnered from laboratory studies, to assess the likely effects on reproduction and population. Using know-how from hatchery operations in similar species such as *Homarus* sp., we also seek to optimise larviculture that could be commercialised to create a hatchery and thus assist stock remediation. This review chapter is therefore divided into three sections: (1) general *N. norvegicus* reproductive biology, (2) life cycle and larval biology and (3) a comprehensive review of all rearing attempts for this species to date, including a likely way forward for pilot scale and hence commercial restocking operations.

Advances in Marine Biology, Volume 64
ISSN 0065-2881
http://dx.doi.org/10.1016/B978-0-12-410466-2.00006-6

© 2013 Elsevier Ltd.
All rights reserved.

201

Keywords: Norway lobster, *Nephrops norvegicus*, Langoustine, Scampi, Development, Eggs, Embryo, Zoea, Hatchery

1. INTRODUCTION

Successful reproduction is the key mechanism that individuals employ to promote a genetic influence on future populations of the species. For commercially exploited species such as the Norway lobster, *Nephrops norvegicus* (henceforth *Nephrops*), any impacts during early life history are likely to be critical for reproductive success and recruitment (Ligas et al., 2011; Sinclair, 1988); therefore research into reproduction and life history, including the biology of embryos and larvae, should be prioritised. This approach has historically assisted lobster (*Homarus* sp.) hatchery development. For the future, studies replicating ocean acidification could provide a better understanding of likely impacts of environmental change on lobsters (Arnold et al., 2009) amongst many other marine species.

Nephrops experiences a number of habitat shifts during its life cycle, where abiotic and biotic factors vary considerably, and thus adults, larvae and postlarvae need to respond to these challenges effectively. For instance, embryonic development occurs in the ambient conditions of the adult burrow over many months, requiring parental care to optimise physico-chemical parameters, particularly oxygen tension, adjacent to the brood. After hatching, this is followed by several weeks as planktonic drifters across three larval (Zoea) stages, and later metamorphosis into benthic postlarvae, where the individual again seeks the soft muddy sediment generally preferred by adults. Any aspect of the life history of an organism may change at any point in time and thereby affect its fitness, and this adds to the complexity of reproduction and reproductive success, making it challenging to quantify in the field.

While adult Norway lobsters are easy to catch, collecting juvenile or larval *Nephrops* in the field is difficult and time-consuming and the animals are easily damaged in the process. Even to date, scant data exist on the early stages of *Nephrops* beyond the first two Zoeal stages, with many studies depending upon laboratory-cultured animals. Previous efforts have shown that culturing *Nephrops* is not straightforward, and there is much scope for improvement to attain success achieved with other Nephropid lobsters (Burton, 2003; Nicosa and Lavalli, 1999).

Since at least 1958, ICES has carefully considered *Nephrops* fishing effort and has recommended studies of the species biology to safeguard the fishery (de Figueiredo and Thomas, 1967). More recently, studies have shown that *Nephrops* landings are at best historically changeable and believed by many to be declining over spatio-temporal scales (e.g. Ligas et al., 2011; Thurstan and Roberts, 2010; Chapter 7). In recent years, additional threats from disease (Stentiford et al., 2012), pollution (Baden and Eriksson, 2006; Murray and Cowie, 2011) and climate change (Findlay et al., 2011) may also threaten the fishery (for review, see Chapter 5). In addition to the ecological threats that this could pose, there is also an economic imperative: for instance, in the United Kingdom alone over 34,000 tonnes were landed in 2011, worth some £111 million and approaching 40% of the total value of total UK shellfish (Laing and Smith, 2012).

The most ambitious method of restoration, remediation and restocking is to initiate hatcheries (described in detail by Burton, 2003, for *Homarus* sp.). Recently, there has been interest in developing this technology for *Nephrops*, by adapting current lobster hatchery technology and know-how. The adoption of an aquatic species for commercial culture depends upon a detailed knowledge of the biology of early life history stages, in addition to the reproductive biology of the adult female broodstock. The reproductive cycle of *Nephrops* has been the subject of a number of investigations in the North Atlantic (Farmer, 1974; Thomas, 1964). Hence, this chapter will summarise the topic in three parts: first, a collation of literature regarding *Nephrops* reproductive biology; secondly, a review of life cycle and larval biology; finally, a comprehensive review of all rearing attempts for this species to date, including a likely way forward for pilot scale and hence commercial restocking operations.

1.1. Reproduction in *Nephrops*

The review on adult reproductive biology by Farmer (1974) remains a comprehensive work on the subject, incorporating the majority of relevant studies until the early 1970s, alongside detailed personal observations. This section aims to summarise this work and add relevant updates from the literature since the publication of this previous review.

1.1.1 External reproductive organs and sex ratio

Lobsters are dioecious and the sexes are distinguished by the position of the genital apertures (gonopores) and the shape of the first pair of pleopods. In females, the genital aperture is located on the basal segments of the third pereiopods (and in the male, on the final pair). In females, the first pair of pleopods possesses slender, short, hair-like terminals whilst in the males they are more robust and stout (termed gonopods).

One of the earliest accounts (de Figueiredo and Thomas, 1967) showed that the percentage of females caught was lowest during the period between egg extrusion and hatching; females were most abundant, relative to males, in the lowest size classes. With the onset of maturity, the growth rate of females decreased in comparison with males of corresponding size, resulting in a high percentage of females in the length classes approaching maturity size, followed by a decrease in the proportion of females to males in larger size classes (Bell et al., 2006; de Figueiredo and Thomas, 1967).

1.1.2 Internal reproductive organs and development of gametes

Male testes are elongated and H-shaped and lie ventral to the pericardium and dorsal to the digestive tract in the cephalothorax. The testes are white to transparent and thin. Paired vasa deferentia arise on the posterior arms and open on the coxopodites of the fifth pereiopods. Histological examination of lobster testes from Portuguese and Irish stocks showed that spermatogenesis occurred throughout the year, leading to the production of spermatophores (capsule of spermatozoa) in a secretory area of the vasa deferentia, where they are stored in an ejaculatory region until release (Farmer, 1974).

In females, the thelycum is a simple invagination found on the underside of the thoracic segments that accommodates spermatophores after copulation. It has no direct connection with the female internal reproductive system, which consists of an elongated H-shaped ovary in the dorsal cephalothorax with the posterior arms projecting further into the abdominal segments. Two oviducts emanate from the middle of the ovary and terminate on the coxopodites of the third pereiopods. Females also have a proportionally wider abdomen than males, thought to be an adaption to carry extruded eggs. Females are thought to mate shortly after moulting when still soft (Farmer, 1974). Relini and Relini (1989) found that in a Mediterranean population, during the summer, 'soft' males and females were both found in low proportions (2.3–2.4%). In spring, during the local mating season, the proportion of soft females increased to 7% while males remained below 3%.

1.1.3 Ovary development

It is generally believed that females have an annual cycle, although some studies for populations at the northerly part of the species range suggest a biennial cycle, that is one hatch every 2 years (e.g. Sterck and Redant, 1989). The development of ovaries and oocytes has been typically described according to five stages along a continuous series, initially by Farmer (1974), although up to nine stages have been proposed more recently (Mente et al.,

2009; Smith, 1987). The developmental stage determination is primarily attributed to the colour of the ovary, and also volume (including the distance it extends posteriorly into the abdomen). The system (summarised Table 6.1; Figure 6.1) is generally still in use for field studies and was vindicated by Tuck et al. (1997) via analysis of the ovary that identified distinct biochemical stages. Lipid, protein and water content of the ovary and ovary index (ratio of ovary: total body weight) increased significantly with maturation, while carbohydrate content decreased.

1.1.4 Onset of maturity

The onset of maturity can occur in relatively small males (2.6 cm CL; Farmer, 1974) and females (ca. 2.5 cm; Sardá, 1980). However, early studies (de Figueiredo and Thomas, 1967) found that the size of females at maturity differed in different populations. The 50% level of maturity in the 'High Adriatic' (27 mm CL) compared with 33 mm off Portugal and 22 mm in the Moray Firth. Generally, most studies since have shown that females attain sexual maturity earlier in Atlantic and UK populations compared to Portugal and the Mediterranean. For instance, Morizur (1979) recorded the smallest mated female of 19 mm with a median mature age of 23.5 mm off France, and around Scotland and the Irish Sea egg carrying females were found at 21 mm Tuck et al. (2000) and low 20's mm (Briggs et al., 2002). Around Portugal and in the Adriatic and Ligurian Seas, smallest mature females/females carrying eggs have generally exceeded 27 mm (Arrobas, 1982; de Figueiredo et al., 1982; Relini and Relini, 1985, 1989). Relini et al. (1998) investigated seven disparate populations in the Mediterranean and found that the smallest ovigerous females were between 23 and 30 mm CL, with 50% maturity size in females between 30 and 36 mm.

1.1.5 Copulation and fertilisation

Smith (1987) and Farmer (1974) both documented copulation in detail from direct observation. Copulation generally occurred at night between mature animals, within 24–48 h of a female moult. However, to our knowledge, this remains anecdotal and no detailed work has determined the precise relationship with copulation and moulting, although this is known to be variable in other related decapods. A set behavioural pattern, possibly elicited by a pheromone, includes lack of aggression or appetite, agitated searching, cleaning and digging behaviour by the male, followed by a brief courtship. The male initiates the courtship by stroking the female with his antennae for several minutes. Females are approached from behind, turned over and a

Table 6.1 Overview of female *Nephrops norvegicus* development stages

Stage	Status of female	Ovary gross description (approx. typical length)	Tissue characteristics (approx. oocyte diameter)	External eggs	Approx. duration
0	Not yet sexually mature or have undergone resorption postmoult	White and threadlike (1–1.5 cm). Previtellogenic.	Oogenesis (0.2 mm). Nuclei appear large.	Developing dark eggs of previous season	3 months unless immature
1	Initial oocyte development	Cream (2 cm). Early vitellogenic.	Oocytes undergo primary growth (0.4 mm)	Developing dark eggs of previous season	3 months
2	Intermediate oocyte development	Pale green just visible through carapace (2.5 cm). Medium vitellogenic.	Continued growth, presumably including vitellogensis (0.6 mm). Larger oocytes observed at periphery, tending to be oblong or ellipsoid in shape rather than spherical.	Developing dark eggs of previous season	3 months
3/4	Maximum oocyte development, stages could be split according to size of ovary	Dark green, visible through back of carapace (3–3.5 cm). Late vitellogenic.	Final maturation prior to release. Large oocytes with obvious nucleus surrounded by thin layer of small follicular cells and very thin 'germinal strand'. Enlarged oviduct (0.9 mm).	Generally none/ previous season's eggs for imminent release	2 months
5	Almost spent. (Additional regeneration stage.)	Mottled green/cream, similar in size to stage 1 ovary	Ovulation has occurred, few large oocytes remain unreleased, many follicular cells undergoing reabsorption (0.2–0.9 mm)	None/recently extruded dark eggs	3–4 weeks

After Symonds (1972), Farmer (1974), Bailey (1984), Smith (1987), Briggs (1988) and Mente et al. (2009).

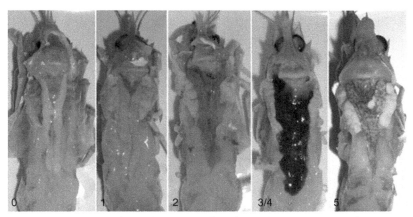

Figure 6.1 Ovary stages (0 to 5) in mature female *Nephrops norvegicus* during oocyte maturation. *Photographs courtesy of Susanne Eriksson.*

spermatophore is transferred during penetration which lasts a few seconds. The spermatophore is subsequently believed to harden inside the thelycum and is carried from copulation until the female next moults; at this time, the lining of the thelycum is shed with the spermatophore. Fertilisation of eggs probably takes place 'internally', that is as eggs pass over the surface of the thelycum prior to being extruded underneath the abdomen.

1.1.6 Fecundity

Initial studies by de Figueiredo and Thomas (1967) gave overall figures between 900 and 6000 eggs per female, with the relationship correlating with size. The relationship between fecundity and female size is typically a linear or curvilinear relationship specifically with carapace length. Several models of fecundity are summarised in Table 6.2. The abdomen width has also proved to be a correlated factor (Abelló and Sardá, 1982) but not weight (Briggs, 2002). Differences in fecundity between fishing grounds are apparent at various temporal–spatial scales (studied and reviewed by Tuck et al., 2000) including improvement over coarser sediment (Bailey, 1986), although Mori et al. (1998, 2001) found no correlation between fecundity and depth.

1.1.7 Egg loss and fecundity measurements

Studies have shown that a considerable proportion of eggs are lost between egg development in the ovary, and subsequent extrusion, incubation and spawning. Egg loss can occur in the wild or during captivity, and could be defined as acute egg loss (due to discrete stress events, potentially

Table 6.2 Fecundity models for *Nephrops norvegicus*

N	Fecundity type	Calculated average fecundity	Location	References
60	Realised	Number of eggs $= 0.526 \times$ CL (mm)$^{2.350}$	Portugal	Thomas (1964)
		Log number of eggs $= 3.0557$ log CL (mm) $- 3.7732$		de Figueiredo and Nunes (1965)
		Number of eggs $= 782 \times$ CL (cm) $- 1376$	Irish Sea	Farmer (1974)
		Log number of eggs $= 2.11$ log CL (mm) $+ 0.18$ Log number of eggs $= 3.02$ log CL (mm) $- 3.50$	Galicia	Allonso-Allende (1976)
		Log number of eggs $= 3.1635$ log CL (mm) $- 4.029$	Adriatic	Froglia and Gramitto (1979)
		Log number of eggs $= 3.194$ log CL (mm) $- 4.118$	Bretagne	Morizur et al. (1981)
76	Realised	Ln number of eggs $= 3.125 \times$ Ln CL $- 3.705$	Portugal	Abelló and Sardá (1982)
62		Ln number of eggs $= 2.488 \times$ Ln CL $- 1.559$	Catalunya	Abelló and Sardá (1982)
3–229 (1153 total)	Realised	Varied between: Number of eggs $= 0.0023 \times$ CL (mm)$^{3.835}$ Number of eggs $= 2.1959 \times$ CL (mm)$^{2.018}$	Firth of Clyde over various developmental stages	Smith (1987)
49	Potential	Number of eggs $= 0.191 \times$ CL (mm)$^{2.644}$	Firth of Clyde	Smith (1987)
33	Potential	Number of eggs $= 0.2104 \times$ CL (mm)$^{2.644}$	Jura	Smith (1987)
Several hundred (est)	Realised	Number of eggs $= 0.04609 \times$ CL (mm)$^{2.68568}$	Ligurian Sea	Relini and Relini (1989)

53	Effective	Number of eggs $= 255.392 \times$ CL (cm) $- 7426.780$	Ligurian Sea	Relini et al. (1998)
54	Effective	Number of eggs $= 120.791 \times$ CL (cm) $- 2862.498$	Adriatic	Relini et al. (1998)
50	Effective	Number of eggs $= 132.270 \times$ CL (cm) $- 3311.398$	Tyrrhenian	Relini et al. (1998)
50	Effective	Number of eggs $= 113.464 \times$ CL (cm) $- 2852.615$	Catalonian	Relini et al. (1998)
52	Effective	Number of eggs $= 90.117 \times$ CL (cm) $- 1752.364$	Euboikos	Relini et al. (1998)
45	Effective	Number of eggs $= 160.889 \times$ CL (cm) $- 3797.059$	Algarve	Relini et al. (1998)
50	Effective	Number of eggs $= 204.203 \times$ CL (cm) $- 5851.416$	Alboran	Relini et al. (1998)
17–34	Potential	Varied between: Number of eggs $= 0.1501 \times$ CL (mm)$^{2.697}$ Number of eggs $= 0.2001 \times$ CL (mm)$^{2.697}$	Firth of Clyde (7 stations)	Tuck et al. (2000)
50	Realised	Number of eggs $= 0.259 \times$ CL (mm)$^{2.566}$	W Irish Sea	Briggs et al. (2002)
Not stated but high (9–23 stations)	Potential	Number of eggs $= 0.0383 \times$ CL (mm)$^{3.1309}$	W and E Irish Sea	McQuaid et al. (2009: 2000 data)
	Realised (wild)	Number of eggs $= 0.0311 \times$ CL (mm)$^{3.1309}$	W and E Irish Sea	McQuaid et al. (2009)
	Realised (captive)	Number of eggs $= 0.0276 \times$ CL (mm)$^{3.1309}$	W and E Irish Sea	McQuaid et al. (2009)
	Effective	Number of eggs $= 0.9596 \times$ CL (mm)$^{1.9736}$	W and E Irish Sea	McQuaid et al. (2009)
	Effective (stage D counts vs. stage I larval counts)	Number of eggs $= 0.0189$ CL (mm)$^{3.170}$ versus number of eggs $= 0.00101$ CL (mm)$^{3.170}$	W and E Irish Sea	McQuaid et al. (2009)

CL=carapace length.
For definitions of fecundity type see text.

anthropogenic) and chronic egg loss (natural wastage during the protracted incubation period). A number of reasons have been suggested including anthropogenic stress from fishing (Chapman and Ballantyne, 1980) and potentially from transport. Extrusion failure, lack of fertilisation and subsequent detachment, accidental mechanical causes, infection, parasites, predation and maternal cannibalism have also been stated (de Figueiredo and Thomas, 1967; Kuris, 1991; Morizur, 1980). Some studies have also found a geographical component (Farina et al., 1999; Tuck et al., 2000).

Estimates of egg loss, however, differ from 18% to 75% loss over the incubation period (Table 6.3). Most estimates are made from females caught by trawl, and catch method seems to have an effect (Briggs et al., 2002; Tuck et al., 2000). Creeled females appear to only lose ca. 20% of their eggs (18–25%), whereas trawled females loose ca. 50% (32–75%). Similar species (*Homarus gammarus* and *Metanephrops thomsoni*) have also exhibited high egg loss after trawling (Matsuura and Hamasaki, 1987; Perkins, 1971), and other studies have shown physical and physiological damage to *Nephrops* after trawling (Milligan et al., 2009; Eriksson 2006; Ridgway et al., 2006). Females which do not normally tail flip readily lose eggs when forced to swim and the discrepancy between the two catch methods in egg loss is probably due to the escape response in combination with physical trauma and asphyxiation in the trawl (Newland, 1985). It is not known if this has any effect on recruitment in trawled areas, though berried females might very well be exposed to a trawl and manage to escape several times during the long incubation period.

Abelló and Sardá (1982) noted that fecundity measurement technique in *Nephrops* often differed, with some egg counts occurring in the ovary before extrusion and others after extrusion. Since then, a number of studies have further defined fecundity according to the stage of development to improve accuracy (Briggs et al., 2002; McQuaid et al., 2009; Tuck et al., 2000). 'Realised fecundity' is the most often applied measure of fecundity and is a simple count of the number of eggs extruded onto the pleopods at the time of fertilisation (i.e. eggs/female). This differs from the more challenging enumeration of oocytes in the ovary, or 'Potential fecundity'. This is of course a larger figure, since not all eggs are subsequently extruded. In the Irish Sea, McQuaid et al. (2009) found the egg loss between Potential and Realised fecundity was higher in captive extruded rather than wild extruded females (ca. 30% compared to 19%, respectively), while the difference between Potential and Effective fecundity was greater still and increased with female size (40–65% in 25–40 mm CL). It is possible that females can perform multiple mating or experience non-annual cycles.

Table 6.3 Reported levels of egg loss from spawning to hatching in *Nephrops norvegicus*

Area	Percentage external egg loss	Capture method	References
Portuguese Coast	75	Trawled	de Figueiredo and Nunes (1965)
Adriatic	66	Trawl fishery	Froglia and Gramitto (1979)
Adriatic	36–56	Trawl fishery	Gramitto and Froglia (1980)
Moray Firth	32–51 (11–22% due to trawling)	Trawling and creeling (different locations)	Chapman and Ballantyne (1980)
Bay of Biscay	45–50	Unknown: from 'commercial fishery' but likely trawled in this location (deep and offshore)	Morizur et al. (1981)
Portuguese Coast	68	Trawl fishery	de Figueiredo et al. (1982)
Firth of Clyde	18	Creeled	Smith (1987)
East China Sea (*Metanephrops thomsoni*)	46	Trawled	Matsuura and Hamasaki (1987)
Galicia	44 (significant geographic component, 32–68%)	Trawled	Farina et al (1999)
North Tyrrhenian Sea	49.5	Trawled	Mori et al. (1998)
Firth of Clyde	Not stated (significant geographic component, 33% difference between highest and lowest)	Trawled	Tuck et al. (2000)
Irish Sea	35 (11.5% due to trawling)	Trawled (short duration) and some creeling	Briggs et al. (2002)
Irish Sea	47	Trawled (short duration)	McQuaid et al. (2009)

After Smith (1987) and Briggs et al. (2002).

In addition, eggs are additionally lost during incubation (i.e. between extrusion and before hatching). Mori et al. (1998) measured egg loss at different developmental stages according to egg colouration, and confirmed that egg loss increased as incubation progressed—suggesting that for accuracy, egg quantification is required as close to hatching as possible. However, the majority of studies have captured animals via trawling and it is difficult to assume that this has had no effect on eggs loss. The count of eggs remaining just prior to hatching is termed 'Effective fecundity'. Furthermore, 'Effective fecundity' measured using counts of released stage I Zoea larvae (in comparison to 'ready to hatch' eggs) found that the latter measurement was an overestimate, with about half the number of larvae produced than would be anticipated (McQuaid et al., 2009). Indeed, some larviculture studies went yet further and only considered the most 'active' hatched larvae to be likely to thrive (de Figueiredo, 1971; Smith, 1987). Figure 6.2 summarises the types of fecundity measurements used in the literature. In the present study, the average calculated loss of eggs from 'Realised' to 'Effective' fecundity of a 30 mm female was ca. 40% (from average 1000 to 600 eggs; see Table 6.2). This is in concordance with the egg loss data found in Table 6.3, where creeled specimens lost on average 20% and trawled 50%. The average calculated loss of eggs from potential to effective fecundity was extrapolated from Table 6.2, and decreased from 1600 to 600 eggs; thus, a ca. 60% egg loss from oocyte to hatched larvae.

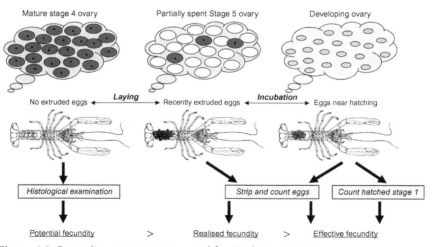

Figure 6.2 Fecundity measurements used for *Nephrops norvegicus*.

1.1.8 Spawning

Farmer (1974) described spawning in detail. Females spawn at night, lying on their back resting on the substrate. During spawning, the oocytes (unfertilised eggs) pass from the ovary through the oviduct and are extruded externally from the gonopore. The oocytes pass over the thelycum and become fertilised, and eggs attach on the abdominal segments held by the pleopods. As with other decapods, extruded eggs are placed in an egg mass underneath the abdomen, ranging from between the fourth pereiopods at the anterior-most extent, posteriorly to the margin between the abdomen and the telson. The egg mass is held against the female by assistance from all four pairs of pleopods, via setae emanating from exopod and endopods. Abelló and Sardá (1982) noted that females have a proportionally wider abdomen than males. Experiments by Farmer (1974) showed that egg laying in captivity in communal tanks was more successful with substrate than without, although this may also have afforded lesser disturbance from conspecifics. Not all oocytes are released at spawning. Some remain and are reabsorbed, referred to as gonad stage 5, which gives the gonad a mottled appearance of white germinal cells and green reabsorbing mature oocytes (Table 6.1; Figure 6.1).

1.1.9 Incubation and egg development

The duration of the embryonic development from fertilised egg to hatched Zoea larvae varies from 6 to 10 months, depending on latitude and habitat (Farmer, 1974; Mori et al., 1998; Sardá, 1995). Embryonic development is sustained throughout that time by the large yolk, which mainly consists of the protein lipovitellin (for review, see Subramoniam, 2011). Yolk cleavage occurs within a few days of spawning, depending on temperature and then development is arrested for several months over the winter; warmer temperatures reduce the period of arrested development. Subsequent development through the later embryonic stages is relatively rapid and is accompanied by a change in egg colour from dark green through to paler green and orange/pink, as protein bound carotenoids in the yolk are free and become incorporated into developing chromatophores.

Egg development was detailed initially to IV–V stages by Figueiredo and Barraca (1963), de Figueiredo (1971), and de Figueiredo and Vilela (1972). Initially, the research was observational and assigned developmental stages to the eggs, namely based on fraction of yolk (D_1–D_4 stages describe 1/2, 1/4, 1/8 and 1/16 of the yolk remaining) and size (ca. 1.5–1.8 mm diameter during development). However, a more detailed nine-stage categorisation was suggested by Dunthorn (1967), summarised in Table 6.4 and Figure 6.3.

Table 6.4 Egg development in *Nephrops norvegicus*

Stage	Egg mass colour	Description	Proximate analysis	
			Increase	Decrease
1	Dark green	Yolk segmented. No embryo		
2	Dark green	First appearance of embryo		
3	Dark green	Optic lobes and appendages present		
4	Dark green	Eye pigments present as thin crescentic shaped areas. Triangular shaped ocellus		
5	Dark green/greyish green	Well formed embryo occupying about 1/2 egg capsule. Eye pigment comma-shaped. Ocellus sometimes present	Wet weight, water, ash, protein	Dry weight, lipid, carbohydrate, energy
6	Greyish green/olive	Chromatophores present on limbs		
7	Olive	Embryo occupying about 2/3 capsule. Chromatophores on limbs and telson		
8	Light orange/brownish orange	Yolk almost all absorbed. Eye pigment oval. Chromatophores on carapace		
9	Pale orange/ brownish orange	Formation of complete larva within capsule prior to hatching		

After Dunthorn (1967) and Smith (1987).

Other obvious developmental landmarks include the early appearance and filling of the dark pigment of the eyes during stage IV (Dunthorn, 1967). Heart beats have been documented in early stage IV *Nephrops* embryos (Eriksson et al., 2006). The exact timing onset of heartbeats is unknown for *Nephrops* but in the closely related *Homarus americanus* heartbeats are initiated at 14% of embryonic development (Helluy and Beltz, 1991) which is equivalent to stages I–II in *Nephrops* (Dunthorn, 1967).

Many animals show some form of parental care, often seen as an evolutionary response to harsh environments, high predation and intense competition for resources during early life stages (Clutton-Brock, 1991). The degree and type of parental care vary greatly amongst different taxa and many marine crustaceans display indirect development where only the female protects the embryos. *Nephrops* show active brooding type behaviour (elevated total pleopod activity) when carrying eggs, showing increase in effort over the incubation period and can be initiated by abiotic stress (Eriksson et al., 2006). The increased active brood care over egg development is possibly a result of mass-specific increase in oxygen consumption of the eggs.

1.1.10 Egg and embryo developmental changes and biochemistry

Eggs increase in size from ca. 1 to 2 mm diameter during development (Figueiredo and Barraca, 1963; Mori et al., 1998; Smith, 1987). Mori et al. (2001) found an inverse relationship between egg volume and depth (possibly due to reduced food): in the Tyrrhenian Sea, eggs averaged

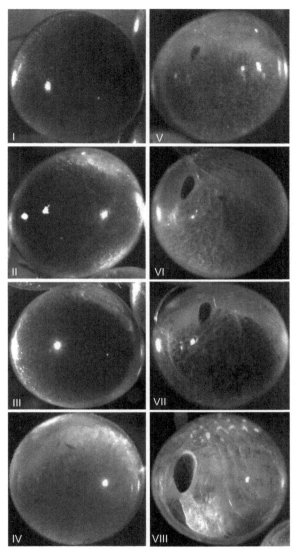

Figure 6.3 Egg developmental stages I–VIII for gravid female *Nephrops norvegicus* (see Table 6.4). *Photographs courtesy of Susanne Eriksson.*

1.4 mm^3 at greater than 500 m depth but at less than 450 m depth, they increased to 1.55 mm^3. No correlation has, however, been found between the female size and the egg volume or size (McQuaid et al., 2009; Mori et al., 2001; Smith, 1987), and McQuaid et al. (2009) suggested that the local stocks invested in quantity rather than size, as a strategy against high egg loss.

Egg diameter throughout development was detailed according to egg stage by Smith (1987) who found an increase of 1.19–1.56 mm in two populations from around the Forth of Clyde, Scotland. For these animals, proximate biochemical analysis was performed from fresh spawned eggs to hatching: The increase in size with development was found to coincide with an increase in protein, water (and wet weight) and salts (ash), with the latter two predominating in the later stages. Dry weight, lipid, carbohydrate (and correspondingly energy content) decreased during development to Zoea I, leading the author to suggest that larvae that hatch late in a brood on the same female could suffer from less energy content. Pochelon et al. (2011a) found high inter- and intra-brood differences in embryo fatty acid content, although there was no obvious pattern along an anterior-posterior plane or either side of the midline. Differences were suggested to relate to food quality and quantity and/or differential lipid catabolism during the incubation period of developing embryos. Further disparities in larval composition were found between geographic locations; for example, Rotllant et al. (2004) found higher absolute values of C, H and N in Mediterranean larvae typically hatching in the winter, compared to those from the Irish Sea hatching in the spring with more favourable conditions.

More detailed analysis has been performed to understand the early larval protein and lipid requirements and to potentially improve broodstock nutrition and hence success. Rosa et al. (2003) also found an increase in egg volume and water content. Total and free amino acids increased, the latter potentially from active transport of nitrogen containing compounds from seawater. Essential amino acids were conserved, while non-essential and free non-essential amino acids increased during embryonic development. The decrease of several lipid classes was attributed to energy storage/liberation, membrane, cell component and hormone production during development.

1.1.11 Hatching

Hatching, and the more specific events of eclosion (the detailed processes that occur during emergence of a larva from the egg), occurs similarly to other Astacidae. Hatching occurs at night over several successive evenings (5–20 days) using similar behavioural postures; in captivity, this occurs in the evening or overnight (onset of darkness) with dozens to several hundred of Zoea released per night (Farmer, 1974; Möller and Branford, 1979; Smith, 1987). The female posture is similar to *Homarus* sp., whereby the female stands on the tips of the pereiopods with the abdomen raised well clear of the substrate in a characteristic pose. The telson and uropods are spread to produce a fan shape, and the pleopods are vigorously agitated to detach and waft larvae posteriorly away from the female. In the wild, the female probably ventures out of the burrow to allow dispersal, and in captivity females have been observed to leave shelter. This process may take minutes to hours (Farmer, 1974; Smith, 1987).

The events of eclosion are described in detail (Farmer, 1974; Smith, 1987, who remain the only authors, to our knowledge, to suggest a transient pre-Zoeal stage in the later stages of eclosion). Generally, the egg margin in decapod crustaceans typically consists of a trichromatic outer membrane and one inner chitinous membrane (Cheung, 1966). Smith (1987) and Farmer (1974) also describe late stage embryos as surrounded by at least two membranes, a thicker outer and a discrete inner layer, and a third membrane, probably a pre-Zoeal exoskeleton. The calcium content of this pre-Zoeal exoskeleton is not known, but an elevated increase in calcification occurs later on when pelagic larval stages metamorphose into postlarvae. From this time, the proportion of calcium remains similar, although the thickness of the exoskeleton increases with growth (Spicer and Eriksson, 2003).

It is surmised that mechanical stress during hatching and an increase in internal osmotic pressure, evidenced by increased egg water and ash content, causes collapse of a breakage plane in the membranes. Egg membranes of the lobsters *H. gammarus* and *H. americanus* have been shown to dramatically increase their permeability to water and minerals just before hatching (Pandian, 1970a,b) and this has also been indirectly shown in *Nephrops* where embryonic metal uptake increases dramatically during the final embryonic stages (Eriksson, 2000).

Smith (1987) found that a minority of released embryos remain encased in both layers or are not completely liberated from the inner membrane and remain bound over the head or appendages. In the latter case, some had to be artificially removed allowing normal development to occur, or else, they eventually died. The majority of embryos naturally escape from these membranes and within seconds swim upward in the water column while concurrently moulting the pre-Zoeal exoskeleton (third layer). Animals are typically curled in a ball at this stage, although during this process, the appendages and abdomen move and flex to a greater degree, allowing the Zoea I to emerge. The entire process lasts between 5 min and several hours for a particular individual.

1.2. Seasonality of the reproductive cycle

Researchers have generally found that the *Nephrops* reproductive cycle, especially egg laying and hatching (and therefore incubation period) varies throughout the year depending on population. In the wild, this appears to be proportional to water temperature, and potentially light intensity/photoperiod, and thus approximates to the location of populations according to geographic distribution, depth, and probably also temporally between years according to annual weather patterns and climate.

Farmer (1974) outlined the typical sequence and seasonality of internal and external development in females in the Irish Sea. In warmer regions such as the Mediterranean, development rate and incubation period are shorter (~6 months; Mori et al., 1998) than in the North Atlantic where it can be around 10 months (Farmer, 1974). The relationship between incubation period and temperature was experimentally proven by Farmer (1974), who halved this duration by raising the temperature by 10 °C ('degree day' development has also since been recorded). Relini et al. (1998) found that seven Mediterranean populations (Latitude 36–44 N) were generally similar, but at lower latitudes and shallower depths, the onset of brooding and hatching was expedited. In general, southerly populations hatched in winter whilst northerly populations hatched in spring.

While larvae appear in the water column between March and June in the Western Irish Sea (e.g. Nichols et al., 1987), in more southerly latitudes they appear earlier (e.g. January–April off Portugal; dos Santos and Peliz, 2005). Tables 6.5 and 6.6 give an overview of anticipated monthly cycles for spawning, hatching and incubation period, while Figure 6.4 describes a likely latitudinal relationship.

2. LIFE CYCLE AND LARVAL DEVELOPMENT

2.1. Larval nomenclature

Nomenclature of *Nephrops* larvae differs in the literature, with between 3 and 5 separate larval stages suggested. Three free-swimming pelagic stages have been historically documented (Andersen, 1962; Farmer, 1974; Jorgenson, 1925; Karlovak, 1955; Santucci, 1926; Sars, 1884) and have been described as: Stages I–III; Zoea I–III; Zoea I, Mysis I and Mysis II. Stage IV is generally considered to be the first postlarval stage, since the animal is similar in appearance to adults, is benthic and is believed to have completed metamorphosis. Rotlland et al. (2001) considered the first benthic stage IV to be a postlarva or decapodid, so that stage V, achieved after a further successful moult, was a true 'juvenile'. To add to the confusion, Farmer (1974) and Smith (1987)

Table 6.5 General annual reproductive cycle for female *Nephrops norvegicus* in the Irish Sea

Time	Stage	Comments
August–September	Egg laying	Appearance of 'berried' females. Ovaries spent, that is high ratio of stage V to stage IV ovaries.
September–April	Carrying eggs	Eggs remain attached to females and undergo incubation and development. Typically early ovarian developmental stages.
April–June	Eggs hatch and pelagic larvae released	Initial and intermediate oocyte development.
May–August	Copulation swiftly after moulting	Majority of eating and foraging occurs. Spermatophore carried until eggs are extruded. Intermediate and mature oocyte development.

document a very brief pre-Zoea (stage I) between release from the maternal abdomen and an initial moult to Zoea I (stage II), with Zoea II and III becoming stages III and IV respectively, before a first benthic postlarva, effectively stage V. Irrespective of the observations and precise biological accuracy, most recent papers have maintained stages I–IV to equal ZI–III and the first benthic postlarval stage, PL1. Figure 6.5 outlines the basic life cycle.

2.2. Life cycle

The life cycle of *Nephrops* is summarised below from diagrams and observations on the topic by Smith (1987; Figures 6.5 and 6.6). Zoea I released from females ascend through the water column via active swimming showing positive phototaxis towards 400–600 nm, and high barokinesis. Experiments showed that larvae could actively swim at around 20–30 mm per s increasing with stage. The larvae undergo a diel vertical migration, with many Zoea larvae residing within the top 40 m, occasionally rising to within 20 m between dusk and dawn (Hill, 1990; Hillis, 1975). Larval depth preference usually coincided just below the pycnocline and with light intensity, although occasionally concentrations were found at depth (potentially synchronous hatching). Lateral dispersal also occurs during the pelagic larval stages, which could last many weeks depending on local temperature, and may reduce recruitment onto suitable substrates if prolonged (Hill, 1990).

Table 6.6 Spatial changes in spawning, egg hatching and incubation period from north to south latitudes for *Nephrops norvegicus*

Approx. laying	Approx. hatching	Average incubation period (days)	Location	References
May–June	May–July	380	Iceland	Eiriksson (1970)
Jun–Aug	May–Aug	320	Faeroes	Andersen (1962)
Aug–Nov	May–Jun	270	Scottish waters	Thomas and Figueiredo (1965)
Sept	May	250	Scotland and Ireland	de Figueiredo and Thomas (1967)
Aug–Sep	May–Aug	300	NE England	Symonds (1972)
Aug–Sept	Apr–Jun	240	Irish Sea	Farmer (1974)
Sept	May	270	Irish Sea	O'Rioridan (1964)
Jun–Jul	Jan–Feb	210	Adriatic	Karlovak (1955)
June–Jul	Jan	210	Adriatic	de Figueiredo and Thomas (1967)
Aug–Sept	Feb–Mar	180	Portugal	Figueiredo and Barraca (1963)
Aug–Sept	Feb–Mar	195	Portugal	de Figueiredo and Thomas (1967)
Jul	Jan–Mar	ca. 200	Greece	Mente et al. (2009)
Aug–Nov	Dec–Mar	ca. 180	Ligurian Sea	Relini and Relini (1989)
Sept–Nov	Feb–Mar	180 (est)	Ligurian Sea	Relini et al. (1998)
Oct–Nov	Dec–Jan	120 (est)	Adriatic	Relini et al. (1998)
Sept–Nov	Dec–Jan	120 (est)	Tyrrhenian	Relini et al. (1998)
Aug; Dec	Dec–Jan	120 (est)	Catalonian	Relini et al. (1998)
Jul; Dec	Dec–Jan	120 (est)	Euboikos	Relini et al. (1998)
Sept; Dec	Dec–Jan	120 (est)	Algarve	Relini et al. (1998)
Sept; Dec	Nov–Dec	120 (est)	Alboran	Relini et al. (1998)

Some data is estimated from graphs.
After Farmer (1974) and Mente et al. (2009).

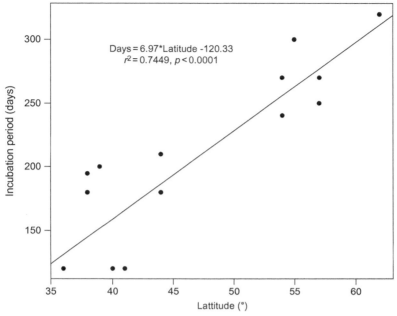

Figure 6.4 Relationship between latitude and incubation period for *Nephrops norvegicus*. Values estimated from the literature.

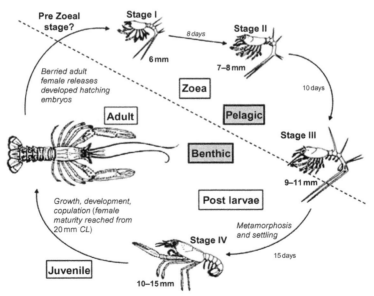

Figure 6.5 *Nephrops norvegicus* life cycle. Larval stages and duration at 15 °C. *After Santucci (1926).*

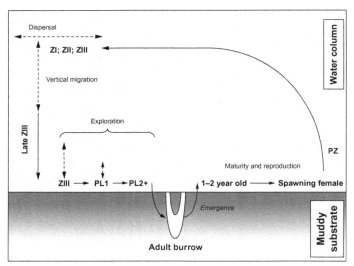

Figure 6.6 *Nephrops norvegicus* life cycle. Larval and adult habitat and distribution. *After Smith (1987).*

During Zoea III, larvae become increasingly negatively phototactic and reside at depth in a demersal (hyper-benthic) habitat, although in experiments they showed little interest in preference for substratum types (Smith, 1987). After metamorphosis, first stage postlarvae (PL1) retained some swimming and exploratory behaviour, although this declined at PL2+ and substratum preference increased. In experiments, mud > muddy sand > sand > gravel, burrows in larger sized substrate were poorer, while there was evidence to suggest intermoult duration was shorter in mud. Occasional catches of juvenile *Nephrops* indicate that they settle in fine-grained mud substrata in the same areas as the adults (Cobb and Wahle, 1994). Less interest was observed in hand constructed burrows although previous adult burrows were explored, suggesting a chemotactic influence. In captivity, PL were able to construct small burrows (U-shaped, T-shaped or Y-shaped) although in the wild reside amongst adult burrows. This observation is supported by data collected in the field (Rice and Chapman, 1971) and in the laboratory (Eriksson and Baden, 1997) that indicate that the juveniles (PL2 or older) build complex burrow systems. The older the postlarvae the faster they start to burrow, already significantly quicker between PL1 of age 2 and 4 days after metamorphosis (Eriksson and Baden, 1997). There is some indication that juvenile *Nephrops* are preyed upon at settlement by demersal fish (Andersen, 1962; Brander and Bennett, 1989), which means that their survival, and indirectly recruitment, may increase with their ability to hide in a burrow.

Occasional catches of juvenile *Nephrops* indicate that they rarely emerge from their burrows during their first year (Cobb and Wahle, 1994). There is some evidence that they not only burrow for refuge but also to search for food (McIntyre, 1973). After several years sub-adults emerge more frequently, attain maturity and mate; the smallest females found with eggs are approximately 20 mm CL (e.g. Briggs et al., 2002).

2.3. Larval characteristics from ZI to III and PL

Larvae and larval morphology has been observed and recorded by Sars (1884), Jorgenson (1925), Santucci (1926) and latterly by de Figueiredo and Thomas (1967). Figure 6.7 provides visual representations of the stages,

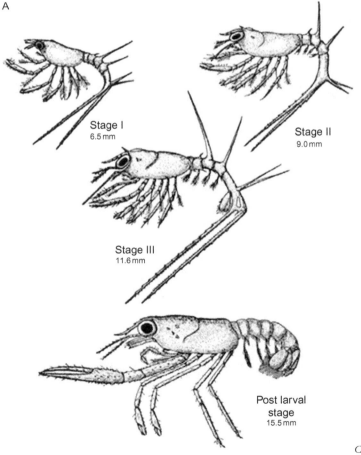

A

Stage I
6.5 mm

Stage II
9.0 mm

Stage III
11.6 mm

Post larval
stage
15.5 mm

Continued

B

Figure 6.7—Cont'd *Nephrops norvegicus* larvae, Zoeal stages I–III and postlarval stage I. (A) Drawings by Santucci (1926) and (B) photomicrographs by the authors.

and Table 6.7 gives a written account of the morphological development using detailed written accounts from these early studies.

3. *NEPHROPS* LARVICULTURE

3.1. Current hatchery techniques for Nephropidae/ Astacidae family

Globally, decapod crustaceans are produced via the capture fishery and aquaculture operations (total production for the latter is ca. 1.4 billion tonnes/year, about 20% of total production) and consists predominantly of marine penaeid shrimps, but also include significant numbers of freshwater prawns, crayfish and marine and freshwater crabs (Wickins and Lee, 2002). Clawed and spiny lobsters arise almost entirely from the capture fishery (ca. 300,000 million tonnes/year).

The adoption of an aquatic species for commercial culture depends upon an in-depth knowledge of the species life cycle, especially that involved in reproduction and early development. The knowledge is species-specific, though some general knowledge can be used for animals with a similar life history or tolerance. Crayfish or crawfish *Procambrus* sp. are a related species that have also been cultured for several decades, although they contrast in terms of greater robustness, extensive and large-scale methods of culture, and they occur in warm freshwater systems which experience seasonal dry out (McClain and Romaire, 2007). However, this extensive, subtropical, freshwater system has little cross over with *Nephrops*. Knowledge exists for successful attempts to artificially propagate Nephropid clawed lobsters, although these have been mainly restricted to species within the genus *Homarus* (Provenzano, 1985) for restocking into the wild as small juveniles, and hence the annual tonnage is negligible.

3.1.1 Homarus *sp. culture*

Currently, hatcheries represent the most ambitious and technology-driven method of restoration, remediation and restocking measures. Initial attempts to culture both the European (*H. gammarus*) and American (*H. americanus*) lobsters were begun independently over 100 years ago (Nicosa and Lavalli, 1999; Van Olst et al., 1980) and these have been developed since this time. There are currently two main hatcheries in the United Kingdom and several other operations in NW Europe. Initial evidence has shown measures of success using scientific methodology such as tagging (Schmalenbach et al., 2011), alongside anecdotal evidence of increased juvenile recruitment by fishers.

Table 6.7 Characteristic developmental changes in *Nephrops norvegicus* larvae

Larval stage	Approx. size (mm)	Cephalothorax (CT) characteristics	Rostrum	Antennae	Appendages	Abdomen	Tail
Stage I / Zoea I	5.5–7	Median, dorsal, longitudinal groove	Devoid of teeth; half length of CT	First and second consist of two joints	All pereiopods bear setose exopodites. First two anterior pairs are chelate. Pleopod buds appear under abdomen	Six segments, first hidden under CT. Three to six segment bare median dorsal spines.	Segment 6 bares two divergent terminal projections (acaudal fork) almost as long as the larva.
Stage II / Zoea II	7.5–10	Additional transverse groove divides CT into cephalic and cardiac regions. Two supraoccular spines.	Three small outgrowths along lateral margins	Bases of both antennae divided into three segments	Pleopods become biramous (not segment 1)	As stage I	As stage I
Stage III / Zoea III	10–12	As stage II	3 pairs of dorso-lateral teeth	Flagellum of second antennae becomes highly segmented	Pleopods developed (not segment 1)	As stage I	Large spines have obvious teeth. Simple Uropod developed.
Stage IV / PL1	12–15	Additional small spines present	Additional pairs of teeth present	Endopodite of second antenna very elongated	Pleopod development of segment 1 occurs at maturity	No large spines. Segment 1 visible.	No caudal fork. Uropod developed

Many UK byelaws currently prevent fishermen from landing egg-carrying female lobsters (berried hens; e.g. http://www.cornwall-ifca.gov.uk/); however, some fishermen are permitted to land berried hens for the hatchery. These are maintained in a broodstock facility until the eggs hatch, with the broodstock being returned to the fishers. The wild gravid females release several thousand larvae per animal in tanks, with larvae collected and reared separately in cylindro-conical 'Paxton' hoppers. In Norwegian and UK hatcheries, larvae are reared at high density throughout the ca. 3-week pelagic larval stage in a 'clear water' system—namely a high energy upwelling tank fed copious brine shrimp (*Artemia* sp.) and/or copepods (*Calanus* sp.) in an effort to reduce cannibalism. Green water systems, that is slightly larger volume microcosms including microalgae are typically favoured in the United States, Spain and Ireland. After metamorphosis, postlarval lobsters are removed, housed in individual rearing 'cells' in raceways, and after several weeks, juveniles are released into an appropriate sub-littoral habitat (background and technical details given in detail by Burton, 2001, 2003).

Hatchery-driven research has sought to optimise their operations and reduce their cost via sustainable techniques to improve survival and growth. Recent interest in commercial farming of *H. gammarus* at industrial scales, due to an increased worldwide market demand for lobsters (Prodohl et al., 2007) has made these goals more imperative. Since initial investigations a century ago, industrial development with academic expertise has slowly improved survival success by health management (Daniels et al., 2010; Scolding et al., 2012); nutrition (e.g. Schmalenbach et al., 2009) and continuing investigation of rearing containers and overall method (Beal and Protopopescu, 2012; http://aquahive.co.uk/).

3.1.2 Comparison of Nephrops and Homarus larviculture

Lobster *H. gammarus* hatcheries continue to produce juveniles in the United Kingdom and Ireland (Scolding et al., 2012) and there has been commercial interest in developing this technology for *Nephrops*, using current lobster knowledge as a starting point. The life cycle and reproduction biology are similar to *Nephrops* and thus there are likely to be transferrable techniques. However, after consideration of species-specific reproductive biology and author personal observation, as well as learning from recent trials (see below), technically there are additional considerations:

1. Broodstock animals are smaller and less robust, with adult *Nephrops* less tolerant to changes in abiotic factors such as temperature and salinity compared to *Homarus* sp. This means that greater care must be taken

in sourcing animals from creel fisheries and in transporting and introducing animals to holding tanks.

2. *Nephrops* have a lower fecundity than *Homarus*, with the latter species
   carrying up to 12,000 eggs for a specimen of 85-mm carapace width
   (Burton, 2003).
3. Larvae are more fragile, need to be handled less and be kept in lower densities.
   The Zoeal tail spine is also proportionally longer, which adds a potential
   threat of larvae colliding and becoming caught in nets, etc. (A. Powell and
   S. Eriksson, pers. obs.). The more convoluted structure may also result in
   greater moult associated mortality/fouling/abnormalities, which has been
   observed in tropical lobster larvae (Ritar et al., 2006).
4. The larval stage is around 50% longer than *Homarus* (approximately
   30+ days rather than approaching 20 days at 15 °C as found in
   Homarus). However, *Nephrops* broodstock and larvae appear less canni-
   balistic than *Homarus* sp., and at low densities broodstock do not appear
   to show aggression and do not require claw banding (A. Powell and
   S. Eriksson, pers. obs.).

3.2. Larviculture, 1971: Present

In the first half of the twentieth century and beyond, initial observations of
larvae were performed on individuals captured using plankton trawls (e.g.
Jorgenson, 1925). These specimens were often damaged and any further
attempt at onwards culture or experimentation was usually hampered
(e.g. Hillis, 1975). Recent attempts at rearing *Nephrops* larvae from captured
gravid broodstock began in the early 1970s. Laboratory scale work has been
attempted a number of times in recent decades until the present date, with
approximately 15 studies published to our knowledge. The approaches are
summarised in detail (Table 6.8) with an accompanying overview below.
Generally, the species appears less robust compared to other species, with
high mortality between successive larval stages.

3.2.1 Broodstock

Although some initial studies procured larvae from trawls (Hillis, 1975) or
removed eggs from females (de Figueiredo, 1971), the majority of studies
have worked with larvae released naturally by captive broodstock.

Smith (1987) acknowledged that removal of eggs was sub-optimal due to
a lack of maternal care (maintenance and water agitation) in the final stages of
development. While some females have been removed by trawling with
little information regarding onwards husbandry, more recent papers

Table 6.8 Overview of *Nephrops norvegicus* larviculture to date

References	Broodstock	Temp.	Salinity	Water change	Aeration	Rearing container	Larval density	Feed	Feeding ration	Miscellaneous husbandry	Summary of larviculture results
de Figueiredo (1971) and de Figueiredo and Vilela (1972)	See 1979(a). Described egg stages according to declining yolk fractions and embryo development, D_1–D_4	7–10, 11–14, 13–17 °C.	33, 35, 38, 40 ppt	Every 2 days	Low (150 bubbles per min), strong (>150) or very strong (constant)	250, 400 and 1000 mL incubating units, constructed with mesh 0.3, 0.6 and 1.2 mm	Various but not detailed; apparently high according to stated optima (eggs no greater than 100 per unit).	Variety: *Artemia* nauplii, *Crangon crangon* eggs; cockle/mussel tissue; conspecifics.	See 1979(a)	Authors investigated differences between rearing success with inactive, active and very active larvae (latter more successful). Mortality also thought to arise from bacteria and fungus.	Eggs; removing from female not optimum but best if taken at late stage, 38 ppt, 400-mL pots, 0.6 mm diameter at lower temperature range at relatively low density. Larvae; shrimp eggs, low aeration; warmer temperature range; max. $15 \times ZI$, $10 \times ZII$ in 400 mL, individual ZIII in 100 mL.
Hillis (1975)	Larvae taken from plankton trawls; small numbers of broodstock taken (little discussion of husbandry)	Ranged 16–22 °C and 11–13 °C.	Not stated	Cages no, but kept in aquaria with filtered water. Beakers, yes	Cages yes; beakers, no	Small cages of plastic framework and mesh, or 250-mL beakers	Individual	*Artemia salina*, supplemented in stage III and replaced in the postlarval phase by pieces of mussel *Mytilus edulis* gonad and small squid *Alloteuthis* sp. flesh.	Daily; quantities not given	Larvae moved in vacuum flasks to maintain temperature in transit	Wild caught larvae at stage III more successful in further development. Incomplete moulting observed and believed to primary cause of mortality; tended to occur more often with protracted development. Cannibalism and possible infection/ciliate infestation observed.

Continued

Table 6.8 Overview of *Nephrops norvegicus* larviculture to date—cont'd

References	Broodstock	Temp.	Salinity	Water change	Aeration	Rearing container	Larval density	Feed	Feeding ration	Miscellaneous husbandry	Summary of larviculture results
de Figueiredo (1979a)	Portuguese broodstock transported 30 km in small containers immediately after landing. Eggs gently removed from females.	Eggs, 11–14 °C or 14–17 °C. Larvae, 15–19 °C.	Eggs, 37 ppt Larvae, 37 ppt, reduced to 33 ppt.	Daily but was reduced during later stages of development for eggs; Larvae, daily change.	Yes for groups of eggs or larvae but not individual larvae.	Eggs—In small nets (0.6 mm) in water filled beakers 400 mL. Larvae, in 1000-mL beakers (group) in 100-mL beakers (individually)	Eggs in batches ca. 200; larvae, in groups (7–19 per L stage I; 10 per L stage II; 5 larvae per L stage III) or individually (1 larvae per 100 mL)	Effectively a green water microcosm: *Artemia salina*, *Crangon crangon* eggs, small pieces *Fucus spiralis*; *Nannochloris* sp.	Not stated, presume daily in excess.	Eggs and larvae: Millipore water and antibiotics used. Hatched larvae selected according to strong swimming behaviour.	Eggs removed at later developmental stages and at higher temperatures had greatest success (up to 30% larvae). Young eggs more susceptible to infection; later eggs more sensitive to handling. Increased duration of emersion responsible for low hatching rate (?) Larvae held individually in 100-mL beakers had a greater survival than larvae held in groups (13 per L reduced to 5 per L).
de Figueiredo (1979b)	Paper covers PL reared in previous study outlined above	16–20 °C.	32–33 ppt	Daily then every 2 days; filtered seawater used.	Yes	400-mL beakers until PL3; then 1000 mL	Individual (1–2 per L)	PL1, *Crangon crangon* eggs; PL2, small amphipods and isopods <2 mm; PL3+ live large amphipods >3 mm. All stages had small fronds of *Fucus spiralis* and 2–3 mL *Nannochloris* sp.	Daily, then every second day for PL3+.	N/A	Additional live prey and algae were a positive addition; some similar observations recorded for length increase, claw regeneration and feeding. Survival between PL stages was poor (PL4, 12% of population). Lethal limit suggested to be >20 °C (accidental increase killed majority of animals).

Reference	Source	Temperature	Salinity	Water management	Aeration	Container	Larval density	Food	Food density	Husbandry	Observations
Anger and Puschel (1986)	Trawled Easterly North Sea, overwintered in flow through; spawned early May	12 °C	32 ppt	Entire, batch, daily	Slight	3-L beakers	10 per beaker (i.e. 3.3 per L)	*Artemia* sp. (San Francisco brand) reared on *Dunaliella tertiolecta* body length equal to larval CL	100 per beaker (i.e. 33 per L, 10 times larval density)	Mortality and moult stage checked daily; stock managed to keep similar stages separate	Despite destructive sampling ($n = 12$ per stage), very high survival rates achieved for stages I and II; stage II and a few stage IV achieved. Development time for stages I and II approx. 8–10 days, stage III 10–12 days.
Smith (1987) Thesis	Creel caught very locally. Kept in filtered and UV water, individually, in dim light with aeration. Food withheld. Larvae taken from late stage females with a net.	8–20 °C	Not stated but likely >33 ppt	Initially flow through, then static. Complete change every 2 days	Yes, airstones. Intensity not stated.	Various (see text). General rearing in receptacles 200 mL to 4 L	Individual and singly, between 3 and 23 per L	Non-supplemented instar 1 *Artemia* (ZI); 4-day-old *Artemia* supplemented with microalgae (ZII, ZIII); mysis and brown shrimp for PL.	Not stated (in excess?)	Very low light intensities used.	Development time for successive Zoeal stages inversely proportional with temperature and 16 °C seemed to be the overall 'optimum' temperature for development. Density of 3 Larvae per L was optimum, with better success for individual culture. Suggested ZI, ZII mass culture and individually for ZIII onwards. Starvation not conducive to development.

Continued

Table 6.8 Overview of *Nephrops norvegicus* larviculture to date—cont'd

References	Broodstock	Temp.	Salinity	Water change	Aeration	Rearing container	Larval density	Feed	Feeding ration	Miscellaneous husbandry	Summary of larviculture results
Thompson and Ayers (1989)	Caught Isle of man in September; pairs placed in 48-L tanks with limited flow (batch? From beach wells), 10–12 °C, 12:12 photoperiod; spawned April–June	[Experimental treatment: 7, 9, 11, 13 and 15 °C]	33–34 ppt, reduced to 30 ppt for a significant time	A few litres exchanged every day to maintain salinity (ca. 5% per day)	Apparently high (induced significant evaporation)	Cylindrical chambers 103 mm × 80 mm (lengths of pipes with drilled holes, covered in mesh) 16 suspended in 50-L tanks	1 per container, containers ca. 250 cm³ (effective density 4 per L)	*Artemia salina.* Occasional 'wild' plankton ≤270 μm and microencapsulated feed 90–250 μm diameter.	Not recorded	Condition/developmental stage of larvae were checked every day. Water samples were taken at regular intervals to determine salinity, nitrate, nitrite and ammonia levels (maintained close to natural oceanic concentrations).	Relatively few larvae produced in relation to number of spawning females (33 over 3 years). Temperature reduced developmental duration, and correlated with increased survival between stages and developmental stage attained. No difference in results between years.
Eriksson and Baden (1997) and Spicer and Eriksson (2003)	Females creel caught off SW Sweden, May, and maintained individually in flow-through aquaria	Strictly 12 or 15 °C	12 or 34 ppt	Flow though, water flow ca. 2 L per min. All seawater used was filtered through an 80–1m mesh and then exposed to UV. Originated locally from depth (35 m)	Not stated (?)	280 L cylindro-conical vessels, central 60 μm filter. Semi-benthic Zoea III were placed individually (115 × 75 × 80 mm = just under 700 cm³), compart-mented aquaria under RAS	Not stated, but broods from different females were kept in separate vessels	Freshly hatched *Artemia* sp. nauplii (<36 h old)	*Ad libitum*	Dim light in an 18-h light:6-h dark regime. Larvae were washed weekly for 15–60 min in a 200 mg per L Streptomycin solution	Mean cumulative survival across three broods was ca. 16% from ZI (Zoea I) to ZII; 5%, ZI to ZIII; and 2.5% ZI to PL1 (juvenile). The first ZII were observed day 8, and the first moult into ZIII started on day 18. The development from hatched larva to metamorphosed juvenile took an average of 33 days at 15 °C. Animals actively leave low oxygen environments.

Dickey-Collas et al. (2000)	Trawled Irish Sea in August; overwintered individually in chambers in continuous upwelling system with filtered water. Fed mussels every 2 days. Near hatching Females bathed (0.5 ppm malachite green) and moved to RAS, water 16 °C, filtered and UV sterilised.	16 °C [*Experimental treatment: ranged from 8 to 16 °C*]	Not stated but regularly checked	Batch sterilised seawater, every 2 days	Apparently none for individual larval rearing	Previous mass culture failed; 500-mL beakers used containing 300 mL liquid	Individual (effective density 3.3 per L)	Effectively a green water microcosm: *Tetraselmis suecia* and newly hatched *Artemia* nauplii (variety AF480)	*Artemia* (ca. 8 per mL); microalgae (17% of overall volume, 50 mL unknown cell density added to 250 mL culture water).	Highly active larvae used. Malachite Green used for 3 min, *Artemia* after hatching, larvae and beaker every 2 days.	Broadly similar (after Thompson and Ayers, 1989) developmental trends with increased temperature regimes.
Rotland et al. (2001) and Rotlant et al. (2004)	Caught off Barcelona, January–March from 400 m depth; maintained individually in cylindro-conicals at 15 °C, 12:12 photoperiod, fed mussels	15 °C	Not stated	Not stated, but agitated ('running')	Not stated	Connected boxes	1 per box from mixed batch	[*Experimental treatment: Live enriched Artemia; Frozen Artemia; 200 µm shrimp feed or mussels*].	*Ad libitum*, twice per day	Treatments performed in randomised block	Live enriched *Artemia* superior (advanced stage reached, development rate, growth and survival).

Continued

Table 6.8 Overview of *Nephrops norvegicus* larviculture to date—cont'd

References	Broodstock	Temp.	Salinity	Water change	Aeration	Rearing container	Larval density	Feed	Feeding ration	Miscellaneous husbandry	Summary of larviculture results
Pochelon et al. (2009, 2011a,b)	Potted in deep water off West Portugal. No further information on husbandry	15 °C	35 ppt	Not stated but RAS system. Water filtered, passed through activated carbon and UV irradiated.	Not stated for stock animals, slight during experiments.	20-L cylindro-conical vessels	Not stated for stock rearing. Short experiments were individual/40-mL beaker (25 per L)	Newly hatched *Artemia* sp. (San Francisco brand). [*Experimental treatment included Nauplii; larger Metanauplii; enriched Metanauplii with two different dried algal products*].	5 per mL for stock; [*Experimental treatment included 0.5, 1, 3 and 5 per mL (1.25–12.5 times larval density) including investigation of feeding history*].	Generally reared in darkness [*Experimental treatments included 24, 12 and 0 h of light*]. For experiments 40-mL beakers were used for freshly hatched or recently moulted stage II animals	12:12 photoperiod and highest prey density resulted in higher prey ingestion. More Nauplii were consumed by stage I and more Metanauplii were consumed by stage II. Metanauplii enrichment had little effect on prey ingestion by stage II but reduced ingestion in stage I animals. Stage II consumed more prey than stage I.

(e.g. Spicer and Eriksson, 2003) have specified a need to reduce stress via creeling, appropriate short-term transport, and providing satisfactory water quality during captivity. This has also been suggested in related species *Metanephrops japonica* (Okamoto, 2008).

3.2.2 Rearing vessels and physical conditions

de Figueiredo (1971), de Figueiredo and Vilela (1972), de Figueiredo (1979a,b), and Hillis (1975) used receptacles constructed from net or mesh around a framework, typically at quite a high density. Later work has generally opted for beakers/containers with a larval density of typically 2–4 per L, typically finding that low or individual culture has promoted better survival between stages since cannibalism and other physical damage can occur. Rearing vessels have been usually supplemented with modest additional aeration. Optimisation of temperature has been frequently studied and appears to be a critical factor in successful development. Within an upper tolerance limit of about 18–20 °C, warmer incubation expedites development and improves survival between developmental stages and metamorphosis, with thermal optima between about 14 and 16 °C. Steady salinity and low light intensity also tend to improve culture success.

3.2.3 Hygiene management

Bacteria and fungi have caused mortalities in eggs and larvae (de Figueiredo, 1971, 1979a) and some studies have used antibiotics and antibacterials in static or flow-through systems (Dickey-Collas et al., 2000; Eriksson and Baden, 1997) or frequent water changes in small green water containers (e.g. Anger and Puschel, 1986). The small biomass suggests that in larger systems, a satisfactory flow rate with appropriate hygiene management such as filtration, UV treatment (e.g. in RAS; Pochelon et al., 2009, 2011a,b) also reduces the incidence of infection.

3.2.4 Nutrition

As with many other crustacean species, enriched brine shrimp *Artemia* sp. have been used as a standard larval feed in almost all of the studies, either alone or in conjunction with other food items. Ration is typically daily and in excess, with some evidence that increasing prey size with larval size is beneficial (Anger and Puschel, 1986; Pochelon et al., 2009). Some studies have employed 'green water' systems, using microalgae and *Artemia* sp. (e.g. Dickey-Collas et al., 2000), although the majority of studies have been 'clear water' systems, incorporating pieces of macroalgae, live isopods/amphipods, marine invertebrate tissue/eggs or microencapsulated feed. The precise feeding regime is peculiar to particular studies and there is little evidence to show if a particular regime is superior. Smith (1987) investigated starvation, which reduced the number of larvae reaching successive stages and increased development time.

3.3. Likely future directions for *Nephrops* hatchery

Patchy research over the past 40 years has highlighted the challenges of culturing significant quantities of *Nephrops* postmetamorphosis, while there is scant information regarding ongrowing of postlarvae. The topic is ripe for significant research to optimise current protocols for laboratory scale culture, and also for upscaling to pilot and commercial sized operations. This is likely to demand consideration of larval ecology, reproductive biology, previous laboratory attempts and using *Homarus* sp. hatchery technology which is likely to be transferable. However, mortality due to 'no known cause' has often been observed, either from moult complications, spontaneously after metamorphosis or with obvious differences in batch success between females. It seems likely that a large quotient of rearing success is dependent on genetic quality of the brood or conversely, congenital abnormalities yet to be established.

It is likely that current scientific projects will produce a hatchery handbook within the current decade, requiring consideration of previous work since the early 1970s, present refinements and, potentially, continuous informal refinements by industry as seen in *Homarus* sp. hatcheries. It is also likely that improving rearing success, by reducing mortality and improving growth and development, will be achieved in four ways—attentive husbandry, improving rearing systems, appropriate and sufficient nutrition and optimal hygiene.

3.3.1 Broodstock husbandry and handling

Since the species is not particularly robust (compared to many other related decapods) it seems sensible to collect, transport and maintain captive broodstock with great care to avoid mortality and excessive loss of eggs during incubation. Creeling (or perhaps even diving) rather than trawling for females and maintaining animals individually in well aerated viviers (or at least damp conditions for short journeys) with hatchery operations situated close to stocks to avoid long periods of transport will undoubtedly prove optimal. Maintaining a constant temperature regime throughout the supply chain, from inhabited water temperature, in vivier and in receiving tank will also reduce stress. Once in captivity, low light intensity and dark-coloured tanks, with appropriate photoperiod will be required, and high water quality and aeration would also avoid stress in the brood and eggs. Claw banding should not be required in communal holding tanks if density is low and refuges are provided, although the species does show clear behavioural elements to ascertain dominance in aquaria (Katoh et al., 2008); it may be wise to consider individual holding units if space allows. Disinfection of female abdomen using antibacterial/fungal agents (and perhaps quarantining) is generally performed with *H. gammarus* although nothing is known of the effects on *Nephrops*.

Larval manipulation should be avoided where possible—for example, females likely to spawn within days should be removed to spawn in rearing tanks (also assists pre-Zoea/unhatched) or via a low impact larval collection system, with netting abolished unless absolutely necessary. Temperature control of ovarian development should be possible to manage and prolong the overall spawning period. In the NE Atlantic, this can occur between April and July. With a gradual increase in temperature, overwintered brood caught in the autumn could start spawning in February; cooler water applied to spring caught animals could delay spawning until September, provided changes are very gradual and do not exceed thermal tolerance limits.

3.3.2 Rearing systems

Appropriate rearing systems and hydrodynamics are likely to play a major part in providing a 'quantum leap' in larval rearing success—since nutrition, hygiene and physico-chemical water parameters appear generally satisfactory. Many studies have reared larvae individually (occasionally in 'green water' systems) and have enjoyed better survival than communally reared larvae. However, this requires a large resource investment in terms of husbandry and time, and thus a tradeoff is required between the effort applied and eventual rearing success. It is possible to culture larvae communally in relatively large, 'clear water' hoppers (currently practiced for *H. gammarus* hatcheries in United Kingdom and Norway), and this is likely to be a way forward since it economises resources, an important consideration for any industrial scale hatchery.

Optimisation of vessel shape, aeration, hydrodynamics and density is required to achieve as closely as possible an individual environment without the added cost of 'individual room service'. Hoppers with minimal dead spaces, promoting a circular cell-like movement could be one way forward, for example 'Paxton' hoppers with improved architecture. Larval density for communal systems seems to be between 3 and 10 per L, although this seems to decline within weeks due to mortality. While the precise rearing container size, shape and water hydrodynamics have yet to be optimised, it seems that excellent water quality is key for success (i.e. previous studies with fluctuating temperature and salinity have reduced progression to later stages), with a temperature of 15–17 °C critical to drive development.

Commercially workable systems for ongrowing benthic postlarvae do not exist, with sediment-filled raceways used (Eriksson and Baden, 1997; Spicer and Eriksson, 2003), although 'Orkney cells' for *H. gammarus* are worthy of investigation and currently appear satisfactory (A. Powell and S. Eriksson, pers. obs.). Furthermore, recent developments such as 'Aquahives' for late larval and postlarval stages have been proven for European lobsters (http://aquahive.co.uk/), and it may be possible to employ them for *Nephrops*. At least two different inserts exist, hive-like plates for *Homarus* ZIII onwards, and stacks of larger cup-like cells which may be suitable for earlier larval stages of other decapods. These culture larvae separately, providing the benefits of individual rearing without the added resource burden. It seems possible that Smith (1987) pre-empted a successful and economical way forward by suggesting batch culture in Kriesel-like tanks for initial stages, and individual culture for stage III onwards.

3.3.3 Nutrition

Live brine shrimp *Artemia* sp. appear to have been investigated across all previous larval rearing, and in general appear to be optimal or at least satisfactory, with evidence throughout the literature, specifically for *Nephrops* but also for a range of other species (Sorgeloos et al., 1998). While other feeds (e.g. live isopods or other locally wild caught material) have been comparable, the use of these items in commercial operations would be challenging. *Homarus* hatcheries in the United Kingdom have recently moved away from *Artemia*, favouring preserved zooplankton consisting mostly or entirely of copepods *Calanus* sp. which appear satisfactory and require less preparation. Larval penaeid shrimp feeds, which have known composition and sterility, have not yet been investigated to our knowledge.

Although approximately 3 *Artemia* per mL appears to be satisfactory, there is evidence that feeding strategy could be optimised further via a range of practices such as enrichment, disinfection, probiotics and optimised prey density and size (e.g. Pochelon et al., 2009; Scolding et al., 2012). In later stages, the typically semi-benthic habitat of Zoea stage III and PL1 and the pelagic (or surface) habit of *Artemia* may effectively reduce prey availability (suggesting that enclosed individual systems and/or a change to alternative/larger prey such as copepods or mysids may be beneficial). Constant feeding systems, rather than daily feeding, may also be possible with prototype systems under development. Postmetamorphosis, weaning onto dry pellet or 'dust' feed will be advantageous to ascertain composition, improve energetic and growth, and maintain water quality and ease of cleaning. Little work on diet composition has been performed in the species in captivity, although in adults, consideration of the amino acid profile is likely to be required (Mente, 2010).

3.3.4 Hygiene management

Infection has been observed in previous studies and maintaining high water quality is essential. For *H. gammarus* hatcheries, disinfectants such as Chloramine T (Pharmaq, UK) are regularly used during larval rearing although the fragility of *Nephrops* prevents the possibility of regular sieving and rinsing. One potential method, not tested in the species but becoming increasingly utilised for crustaceans, is a constant low level direct supply of ozone in rearing systems to manage bacterial loading of the culture water, larvae and live feeds. This has been performed with some success on *H. gammarus* larvae and postlarvae (Scolding et al., 2012). Probiotics and use of antibacterial enrichments are other methods and have also been implemented with similar species (Daniels et al., 2010).

4. CONCLUSIONS/SUMMARY

Nephrops is less robust than its closest relatives of the Nephropids, *Homarus* spp. The smaller size also reduces individual and species fecundity (both potential and realised) and possible reproductive success. The *Nephrops* fishery mainly consists of mass-capturing via trawling and there is no restriction to land berried female *Nephrops* (in comparison to almost total creeling for *Homarus* fishery, in addition to strict regulations across most of its range). This makes the species more susceptible to fishing-induced mortality than the other Nephropids.

To maintain a sustainable stock of *Nephrops*, many believe creeling should be promoted, in particular to reduce adult stress and thereby stress-related egg loss, and enable better fitness of discards. The longer larval stage in *Nephrops* suggests a possibility for spread over larger geographic areas, especially since the muddy substrate they prefer also widespread along their geographic range. In the future, ranching or seeding natural stocks, enabling the possibility of establishing new grounds, will demand techniques to mass-culture and recruit postlarvae.

Compared to *Homarus* sp. which has a higher individual unit price, it is probable that potential *Nephrops* hatcheries will require a different business model that is embedded inside local community fisheries management (specifically encouragement of space-based management; see Chapter 7, Torridon case study). It could also improve the relationship between trawlers and creelers, perhaps by using undersize or over-quota animals within hatchery, or even nursery, areas.

ACKNOWLEDGEMENTS

The authors wish to thank Dr. Sara Barrento, Swansea University for assistance with obtaining and translating Portuguese texts and informal discussions with staff and colleagues at the Orkney Lobster Hatchery, Orkney, UK.

REFERENCES

Abelló, P., Sardá, F., 1982. The fecundity of the Norway Lobster *Nephrops norvegicus* (L.) off the Catalan and Portuguese coasts. Crustaceana 43, 13–20.

Alonso-Allende, J.M., 1976. Notes on the biology of *Nephrops norvegicus* (L.) from the North West of Spain. ICES CM, 1976/K:5.

Andersen, F.S., 1962. The Norway lobster in Faeroe waters. Meddelelser fra Kommissionen for Danmarks Fiskeri-Og Havundersögelser 3, 265–326.

Anger, K., Puschel, C., 1986. Growth and exuviation of Norway lobster (*Nephrops norvegicus*) larvae reared in the laboratory. Ophelia 25, 157–167.

Arnold, K.E., Findlay, H.S., Spicer, J.I., Daniels, C.L., Boothroyd, D.P., 2009. Effect of CO_2-related acidification on aspects of the larval development of the European lobster, *Homarus gammarus* (L.). Biogeosciences 6, 1747–1754.

Arrobas, 1982. Some aspects on the biology and fishery of *Nephrops norvegicus* (L.) from the south Portuguese coast. ICES C.M. 1982/K.

Baden, S.P., Eriksson, S.P., 2006. Role, routes and effects of manganese in Crustaceans. Oceanogr. Mar. Biol. 44, 61–83.

Bailey, N., 1984. Some Aspects of Reproduction in *Nephrops*. Shellfish Committee Document CM 1984/K:33.

Bailey, N., Howard, F.G., Chapman, C.J., 1986. Clyde *Nephrops*: biology and fisheries. Proc. Roy. Soc. B 90, 501–518.

Beal, B.F., Protopopescu, G.C., 2012. Ocean based nurseries for cultured lobster *(Homarus americanus* Milne Edwards) Postlarvae: field experiments off the coast of Eastern Maine to examine effects of flow and container size on growth and survival. J. Shellfish Res. 31, 177–193.

Bell, M., Redant, F., Tuck, I., 2006. *Nephrops* species. In: Phillips, B. (Ed.), Lobsters: Biology Management, Aquaculture and Fisheries. Blackwell Publishing, Oxford, pp. 420–469.

Brander, K.M., Bennett, D.B., 1989. Norway lobsters in the Irish Sea. Modelling one component of a multispecies resource. In: Caddy, J.F. (Ed.), Marine Invertebrate Fisheries: Their Assessment and Management, John Wiley and Sons Inc., Somerset, NJ, pp. 183–204.

Briggs, R.P., 1988. A preliminary analysis of maturity data for Northeast Irish Sea. *Nephrops*. ICES CM 1988/K, 21 p.

Briggs, R.P., Armstrong, M.J., Dickey-Collas, M., Allen, M., McQuaid, N., Whitmore, J., 2002. The application of fecundity estimates to determine the spawning stock biomass of Irish Sea *Nephrops norvegicus* (L.) using the annual larval production method. ICES J. Mar. Sci. 59, 109–119.

Burton, C.A., 2001. The role of lobster (*Homarus* spp.) hatcheries in ranching, restoration and remediation programmes. Hydrobiologica 465, 45–48.

Burton, C.A., 2003. Lobster hatcheries and stocking programmes: an introductory manual. Sea Fish Industry Authority Aquaculture Development Service. Seafish Report, SR552.

Chapman, C.J., Ballantyne, K.A., 1980. Some observations on the fecundity of Norway lobsters in Scottish waters. International Council for the Exploration of the Seas Council Meeting Papers, C.M.1980/K:25.

Cheung, T.S., 1966. The development of egg-membranes and egg attachment in the shore crab, Carcinus maenas, and some related decapods. J. Mar. Biol. Assoc. U.K. 46, 373–400.

Clutton-Brock, T.H., 1991. Monographs in Behavior and Ecology—The Evolution of Parental Care. Princeton University Press, Princeton, NJ XIII + 352 pp.

Cobb, J.S., Wahle, R.A., 1994. Early life history and recruitment processes of clawed lobsters. Crustaceana 67, 1–25.

Daniels, C.L., Merrifield, D.L., Boothroyd, D.P., Davies, S.J., Factor, J.R., Arnold, K.E., 2010. Effect of dietary *Bacillus* spp. and mannan oligosaccharides (MOS) on European lobster (*Homarus gammarus* L.) larvae growth performance, gut morphology and gut microbiota. Aquaculture 304, 49–57.

Dickey-Collas, M., McQuaid, N., Armstrong, M.J., Allen, M., Briggs, R.P., 2000. Temperature-dependent stage durations of Irish Sea Nephrops larvae. J. Plankton Res. 22, 749–760.

dos Santos, A., Peliz, I.A., 2005. The occurrence of Norway lobster (*Nephrops norvegicus*) larvae off the Portuguese coast. J. Mar. Biol. Assoc. U.K. 85, 937–941.

Dunthorn, A., 1967. Some observations on the behaviour and development of the Norway lobster. ICES (K) 5, 1–11.

Eriksson, S.P., 2006. Differences in the condition of Norway lobsters (*Nephrops norvegicus* (L.)) from trawled and creeled fishing areas. Mar. Biol. Res. 2, 52–58.

Eriksson, S.P., 2000. Variations of manganese in the eggs of the Norway lobster, *Nephrops norvegicus* (L.). Aquat. Toxicol. 48, 291–295.

Eriksson, S.P., Baden, S.P., 1997. Behaviour and tolerance to hypoxia in juvenile Norway lobster (*Nephrops norvegicus*) of different ages. Mar. Biol. 128, 49–54.

Eiriksson, H., 1970. On the breeding cycle and fecundity of the Norway lobster in south west Iceland. International Council for the Exploration of the Seas Council Meeting Papers, C.M.1970/K:6.

Eriksson, S.P., Nabbing, M., Sjoman, E., 2006. Is brood care in Nephrops norvegicus during hypoxia adaptive or a waste of energy? Functional ecology 20, 1097–1104.

Farmer, A.S.D., 1974. Reproduction in *Nephrops norvegicus* (Decapoda: Nephropidae). J. Zool. 174, 161–183.

Farina, A.C., Freire, J., Gonzalez-Gurriaran, E., 1999. Fecundity of the Norway lobster *Nephrops norvegicus* in Galicla (NW Spain) and a review of geographical patterns. Ophelia 50, 177–189.

Findlay, H.S., Wood, H.L., Kendall, M.A., Spicer, I.J., Twitchett, R.J., Widdicombe, S., 2011. Comparing the impact of high CO_2 on calcium carbonate structures in different marine organisms. Mar. Biol. Res. 7, 565–575.

de Figueiredo, M.J., 1971. Sobre a cultura de crustaceaos decapodes em laboratorio: *Nephrops norvegicus* (Lagostim) e *Penaeus kerathurus* (Camarao). Bol inform. Inst. Biol. Marit. Lisboa 1, 1–17.

de Figueiredo, M.J., 1979a. Artificial culture of *Nephrops norvegicus* (L.). I—further studies on larval culture of *Nephrops norvegicus* (L.) reared from the egg. Bol. Inst. Nac. Invest. Pescas 1, 5–12.

de Figueiredo, M.J., 1979b. Artificial culture of *Nephrops norvegicus* (L.). II—some studies on the growth of early post-larvae of *Nephrops norvegicus* (L.) reared from the egg. Bol. Inst. Nac. Invest. Pescas 1, 13–23.

de Figueiredo, M.J., Barraca, I.F., 1963. Contribuicao para o conhecimento da pesca e da biologica do lagostim (*Nephrops norvegicus* L.) na costa portuguesa. Notas e Estudos Inst. Biot. Marit. Lisboa 28, 1–45.

de Figueiredo M.J., Margo, O., Franco, M.G., 1982. The fecundity of *Nephrops norvegicus* (L.) in Portuguese waters. International Council for the Exploration of the Sea C.M., 1982/K, 29, 14 pp.

de Figueiredo, M.J., Nunes M.C., 1965. The fecundity of the Norway lobster *Nephrops norvegicus* (L.) in Portuguese waters. Conseil International pour l'Exploration de la Mer CM 1965, Contr. No. 34, 5 pp.

de Figueiredo, M.J., Thomas, H.J., 1967. *Nephrops norvegicus* (Linnaeus, 1758) Leach—a review. Oceanogr. Mar. Biol. Annu. Rev. 5, 371–407.

de Figueiredo, M.J., Vilela, M.H., 1972. On the artificial culture of *Nephrops norvegicus* reared from the egg. Aquaculture 1, 173–180.

Froglia, C., Gramitto, M.E., 1979. An estimate of the fecundity of Norway lobster (Nephrops norvegicus) in the Adriatic Sea. Rapp. Comm. Int. Mer. Médit. 25/26 (4), 227–229.

Gramitto, M.E., Froglia, C., 1980. Osservazioni sul potenziale riproduttivo dello scampo (*Nephrops norvegicus*) in Adriatico. Mem. Biol. Mar. Oceanogr. (Suppl.) 10, 213–218.

Helluy, S.M., Beltz, B.S., 1991. Embryonic development of the American Lobster (*Homarus americanus*)—quantitative staging and characterization of an embryonic molt cycle. Biol. Bull. 180, 355–371.

Hill, A.E., 1990. Pelagic dispersal of Norway lobster *Nephrops norvegicus* larvae examined using an advection—diffusion-mortality model. Mar. Ecol. Prog. Ser. 64, 217–226.

Hillis, J.P., 1975. Captive rearing of larvae of the Dublin Bay prawn *Nephrops norvegicus* (L.). Irish Fish. Investig. B 16, 3–12.

Jorgenson, O., 1925. The early stages of *Nephrops norvegicus*, from the Northumberland plankton, together with a note on the post-larval development of *Homarus vulgaris*. J. Mar. Biol. Assoc. U.K. 13, 870–879.

Karlovak, O., 1955. A summary of our present knowledge regarding the Norway Lobster (*Nephrops norvegicus* L.) in the Adriatic. General Fisheries Council for the Mediterranean. Proc. Tech. Pap. 3, 417–421.

Katoh, E., Johnson, M., Breithaupt, T., 2008. Fighting behaviour and the role of urinary signals in dominance assessment of Norway lobsters, *Nephrops norvegicus*. Behaviour 145 (10), 1447–1464.

Kuris, A.M., 1991. A review of patterns and causes of crustacean brood mortality. In: Wenner, A., Kuris, A. (Eds.), Crustacean Issues, 7: Crustacean Egg Production. A.A. Belkema, Rotterdam, pp. 117–141.

Laing, I., Smith, 2012. Shellfish production in the UK in 2011. In: Laing, I., Smith, D. (Eds.), Shellfish News 34, 39–40. Available on line http://www.cefas.defra.gov.uk/publications-and-data/shellfish-news.aspx.

Ligas, A., Sartor, P., Colloca, F., 2011. Trends in population dynamics and fishery of *Parapenaeus longirostris* and *Nephrops norvegicus* in the Tyrrhenian Sea (NW Mediterranean): the relative importance of fishery and environmental variables. Mar. Ecol. Evol. Persp. 32, 25–35.

Matsuura, S., Hamasaki, K., 1987. Loss of Eggs Attached to the Pleopods in *Metanephrops thomsoni* (Bate, 1888) (Crustacea, Decapoda, Nephropidae). J. Fac. Agric. Kyushu Univ. 31, 405–410.

McClain, W.R., Romaire, R.P., 2007. Procambarid Crawfish: Life History and Biology. Southern Regional Aquaculture Centre, Publication no. 2403, 6 pp.

McIntyre, J.D., 1973. Meiobenthos. Proc. Challenger Soc. 4, 1–9.

McQuaid, N., Briggs, R.P., Roberts, D., 2009. Fecundity of *Nephrops norvegicus* from the Irish Sea. J. Mar. Biol. Assoc. U.K. 89, 1181–1188.

Mente, E., 2010. Survival, food consumption and growth of Norway lobster (*Nephrops norvegicus*) kept in laboratory conditions. Integr. Zool. 5, 256–263.

Mente, E., Karapanagiotidis, I.T., Logothetis, P., Vafidis, D., Malandrakis, E., Neofitou, N., Exadactylos, A., Stratakos, A., 2009. The reproductive cycle of Norway lobster. Journ. Zoo. 278, 324–332.

Milligan, R.J., Albalat, A., Atkinson, R.J.A., Neil, D.M., 2009. Effects of trawling on the physical condition of the Norway lobster *Nephrops norvegicus* in relation to seasonal cycles in the Clyde Sea area. ICES Journal of Marine Science 66, 488–494.

Möller, T.H., Branford, J.R., 1979. A circadian hatching rhythm in *Nephrops norvegicus* (Crustacea: Decapoda). In: Naylor, E. (Ed.), Cyclic Phenomena in Marine Plants and Animals: Proceedings of the 13th European Marine Biology Symposium, Isle of Man, 27 September–4 October 1978, pp. 391–397.

Mori, M., Biaga, F., Ranieri, S.de, 1998. Fecundity and egg loss during incubation in Norway lobster (*Nephrops norvegicus*) in the North Tyrrhenian Sea. J. Nat. Hist. 32, 1641–1650.

Mori, M., Modena, M., Biaga, F., 2001. Fecundity and egg volume in Norway lobster (*Nephrops norvegicus*) from different depths in the northern Tyrrhenian Sea. Sci. Mar. 65, 111–116.

Morizur, Y., Conan, C., Guenole, A., Onnes, M.H., 1981. Fecondite de *Nephrops norvegicus* dans le golfe de Gascogne. Mar. Biol. 63, 319–324.

Morizur, Y., 1980. Evaluation de la perte d'oeufs lors de l'incubation chez nephrops norvegicus (L.) dans la région Sud-Bretagne, France. Crustaceana 41, 301–306.

Morizur, Y., 1979. Evaluation de la perte d'oeufs lors de l'incubation chez *Nephrops norvegicus* dans la region de Sud-Bretagne. Int. Count. Explor. Sea Comm. Meet. (Shellfish Benthos Comm.) K 45, 1–9.

Murray, F., Cowie, P.R., 2011. Plastic contamination in the decapod crustacean *Nephrops norvegicus* (Linnaeus, 1758). Mar. Pollut. Bull. 62, 1207–1217.

Nichols, J., Bennett, D., Symonds, D., Grainger, R., 1987. Estimation of the stock size of adult Nephrops norvegicus (L.) from larvae surveys in the western Irish Sea in 1982. J. Nat. Hist. 21 (6), 1433–1450.

Nicosa, F., Lavalli, K., 1999. Homarid lobster hatcheries: their history and role in research, management and aquaculture. Mar. Fish. Rev. 61 (2), 1–57.

Newland, P.L., 1985. The control of escape behaviour in the Norway lobster, *Nephrops norvegicus* (L.). PhD Thesis, University of Glasgow, Scotland.

Okamoto, K., 2008. Use of deep seawater for rearing Japanese scampi lobster (*Metanephrops japonica*) broodstock. Rev. Fish. Sci. 16, 391–393.

O'Rioridan, C.E., 1964. *Nephrops norvegicus* the Dublin Bay prawn in Irish waters. Sci. Proc. R. Dublin Soc. 1B, 131–157.

Pandian, T.J., 1970a. Ecophysiological studies on the developing eggs and embryos of the European lobster Homarus gammarus. Mar. Biol. (Berl.) 5, 154–167.

Pandian, T.J., 1970b. Yolk utilization and hatching time in the Canadian lobster Homarus americanus. Mar. Biol. (Berl.) 7, 249–254.

Perkins, H.C., 1971. Egg loss during incubation from offshore northern lobsters (Decapoda, Homaridae). Fish. Bull. U.S. 69, 451–453.

Pochelon, P.N., Calado, R., dos Santos, A., Queiroga, H., 2009. Feeding ability of early zoeal stages of the Norway lobster *Nephrops norvegicus* (L.). Biol. Bull. 216, 335–343.

Pochelon, P.N., da Silva, T.L., Reis, A., dos Santos, A., Queiroga, H., Calado, R., 2011a. Inter-individual and within-brood variability in the fatty acid profiles of Norway lobster, *Nephrops norvegicus* (L.) embryos. Mar. Biol. 158, 2825–2833.

Pochelon, P.N., Queiroga, H., Rotllant, G., dos Santos, A., Calado, R., 2011b. Effect of unfavorable trophic scenarios on amylase and protease activity of *Nephrops norvegicus* (L.) larvae during their first vertical migration: a laboratory approach. Mar. Biol. 158, 2079–2085.

Prodohl. P.A., Jorstad, K.E., Triantafyllidis, A., Katsares, V., Triantaphyllidis, C., 2007. European lobster—*Homarus gammarus*. In: Svåsand, T., Crosetti, D., García-Vázquez, E., Verspoor, E. (Eds.) *Genimpact—Evaluation of Genetic Impact of Aquaculture Activities on Native Populations*. A European network (EU contract no. RICA-CT-2005–022802). Final scientific report, pp 91–98.

Provenzano Jr., A.J., 1985. The Biology of Crustacea. Vol. 10. Economic Aspects, Fisheries and Culture. In: Provenzano Jr., A.J. (Ed.), Academic Press Inc., Orlando, FL, 331 pp.

Rice, A.L., Chapman, C.J., 1971. Observations on the burrows and burrowing behaviour of two mud-dwelling decapod crustaceans, *Nephrops norvegicus* and *Goneplax rhomboides*. Mar. Biol. 10, 330–342.

Ritar, A.J., Smith, G.G., Thomas, C.W., 2006. Ozonation of seawater improves the survival of larval southern rock lobster, *Jasus edwardsii*, in culture from egg to juvenile. Aquaculture 261, 1014–1025.

Relini, L.O., Relini, G., 1985. Notes on the distribution, reproductive biology and fecundity of *Nephrops norvegicus* in the Ligurian Sea. FAO Fish. Rep. 336, 107–111.

Relini, L.O., Relini, G., 1989. Reproduction of *Nephrops norvegicus* L. in isothermal Mediterranean waters. In: Ryland, J.S., Tyler, P.A. (Eds.), 23rd European Marine Biology Symposium, pp. 153–160.

Relini, L.O., Zamboni, A., Fiorentino, F., Massi, D., 1998. Reproductive patterns in Norway lobster *Nephrops norvegicus* (L.), (Crustacea Decapoda Nephropidae) of different Mediterranean areas. Sci. Mar. 62 (Suppl. 1), 25–41.

Ridgway, I.D., Taylor, A.C., Atkinson, R.J.A., Chang, E.S., Neil, D.M., 2006. Impact of capture method and trawl duration on the health status of the Norway lobster, *Nephrops norvegicus*. J. Exp. Mar. Biol. Ecol. 339, 135–147.

Rosa, R., Morais, S., Calado, R., Narciso, L., Nunes, M.L., 2003. Biochemical changes during the embryonic development of Norway lobster, *Nephrops norvegicus*. Aquaculture 221, 507–522.

Rotlland, G., Charmantier-Daures, M., Charmantier, G., Anger, K., Sarda, F., 2001. Effects of diet on *Nephrops norvegicus* (L.) larval and postlarval development, growth, and elemental composition. J. Shellfish Res. 20, 347–352.

Rotllant, G., Anger, K., Durfort, M., Sarda, F., 2004. Elemental and biochemical composition of *Nephrops norvegicus* (Linnaeus, 1758) larvae from the Mediterranean and Irish Seas. Helgoland Mar. Res. 58, 206–210.

Santucci, R., 1926. Lo sviluppo e l'ecologia post-embrionali dello "Scampo" *Nephrops norvegicus* (L.) nel Tirreno e nei mari nordici. Mem. R. Com. Talassogr. Ital. 125, 1–36.

Sardá, F., 1980. Contribución al conocimiento de la biología de *Nephrops norvegicus* (L.). Estudio del ciclo de intermuda. Tesis doctoral, Universidad de Barcelona, unpublished.

Sardá, F., 1995. A review (1967–1990) of some aspects of the life history of Nephrops norvegicus. ICES Mar. Sci. Symp. 199, 78–88.

Sars, G.O., 1884. Bidrag til Kundskaben om Decapodernes Forvandlinger. I. *Nephrops, Calocaris, Gebia*. Arch. Math. Naturvidenskab 9, 155–204.

Scolding, J.W.S., Powell, A., Boothroyd, D.P., Shields, R.J., 2012. The effect of ozonation on the survival, growth and microbiology of the European lobster (*Homarus gammarus*). Aquaculture 364–365, 217–223.

Schmalenbach, I., Buchholz, F., Franke, H.D., Saborowski, R., 2009. Improvement of rearing conditions for juvenile lobsters (*Homarus gammarus*) by co-culturing with juvenile isopods (*Idotea emarginata*). Aquaculture 289, 297–303.

Schmalenbach, I., Mehrtens, F., Janke, M., Buchholz, F., 2011. A mark-recapture study of hatchery-reared juvenile European lobsters, *Homarus gammarus*, released at the rocky island of Helgoland (German Bight, North Sea) from 2000 to 2009. Fish. Res. 108, 22–30.

Sinclair, M., 1988. Marine Populations: An Essay on Population Regulation and Speciation Washington Sea Grant Program, University of Washington Press, Seattle, 252 pp.

Smith, R.S.M., 1987. The biology of larval and juvenile *Nephrops norvegicus* (L.) in the firth of Clyde. Ph.D. Dissertation, Glasgow University, Glasgow.

Sorgeloos, P., Coutteau, P., Dhert, P., Merchie, G., Lavens, P., 1998. Use of brine shrimp, *Artemia* spp., in larval crustacean nutrition: a review. Rev. Fish. Sci. 6, 55–68.

Spicer, J.I., Eriksson, S.P., 2003. Does the development of respiratory regulation always accompany the transition from pelagic larvae to benthic fossorial postlarvae in the Norway lobster *Nephrops norvegicus* (L.)? J. Exp. Mar. Biol. Ecol. 295, 219–243.

Sterck, W., Redant, F., 1989. Further evidence on biennial spawning of the Norway Lobster *Nephrops norvegicus*, in the central North Sea. ICES CM 1989/K2, 9 p.

Stentiford, G.D., Neil, D.M., Peeler, E.J., Shields, J.D., Small, H.J., Flegel, T.W., Vlak, J.M., Jones, B., Morado, F., Moss, S., Lotz, J., Bartholomay, L., Behringer, D.C., Hauton, C., Lightner, D.V., 2012. Disease will limit future food supply from the global crustacean fishery and aquaculture sectors. J. Invertebr. Pathol. 110, 141–157.

Subramoniam, T., 2011. Mechanisms and control of vitellogenesis in crustaceans. Fish. Sci. 77 (1), 1–21.

Symonds, D.J., 1972. The fishery for the Norway Lobster *Nephrops norvegicus* (L) off the north east coast of England. In: Fishery Investigations Series II, vol. 27. Ministry of Agriculture, Fisheries and Food, London, pp. 1–35.

Thomas, H.J., 1964. The Spawning and fecundity of the Norway lobsters (*Nephrops norvegicus* L.) around the Scottish coast. J. Conseil Int. Explor. Mer 29, 221–229.

Thomas, H.J., Figueiredo, M.J., 1965. Seasonal variations in the catch composition of the Norway lobster, *Nephrops norvegicus* (L.) around Scotland. Journ. Conseil 30, 75–85.

Thompson, B.M., Ayers, R.A., 1989. Laboratory studies on the development of *Nephrops norvegicus* larvae. J. Mar. Biol. Assoc. U.K. 69, 795–801.

Thurstan, R.H., Roberts, C.M., 2010. Ecological meltdown in the Firth of Clyde, Scotland: two centuries of change in a coastal marine ecosystem. PLoS One 5 (7), e11767. http://dx.doi.org/10.1371/journal.pone.0011767.

Tuck, I.D., Taylor, A.C., Atkinson, R.J.A., Gramitto, M.E., Smith, C., 1997. Biochemical composition of Nephrops norvegicus: changes associated with ovary maturation. Mar. Biol. 129, 505–511.

Tuck, I.D., Atkinson, R.J.A., Chapman, C.J., 2000. Population biology of the Norway lobster, *Nephrops norvegicus* (L.) in the Firth of Clyde Scotland II: fecundity and size at onset of sexual maturity. ICES J. Mar. Sci. 57, 1227–1239.

Van Olst, J.C., Carlberg, J.M., Hughes, J.T., 1980. Aquaculture. In: Cobb, J.S., Phillips, B.F. (Eds.), The Biology and Management of Lobsters: Ecology and Management, vol. 2. Academic Press Inc, London, pp. 333–384.

Wickins, J.F., Lee, D.O'.C., 2002. Crustacean Farming, Ranching and Culture, second ed. Blackwell, Oxford.

CHAPTER SEVEN

Nephrops Fisheries in European Waters

Anette Ungfors[*,1], Ewen Bell[†], Magnus L. Johnson[‡], Daniel Cowing[‡], Nicola C. Dobson[‡], Ralf Bublitz[‡], Jane Sandell[§]

[*]Department of Biological and Environmental Science Kristineberg, Gothenburg University, Fiskebäckskil, Sweden
[†]Centre for Environment, Fisheries and Aquaculture Science (CEFAS) Lowestoft, United Kingdom
[‡]Centre for Environmental and Marine Sciences, University of Hull, Scarborough, United Kingdom
[§]Scottish Fishermen's Organisation Ltd., Peterhead, United Kingdom
[1]Corresponding author: e-mail address: anette.ungfors@bioenv.gu.se

Contents

Abstract

This review focuses on the Norway lobster (*Nephrops norvegicus*) as a resource, describing how the fishery has developed from the 1960s to the present day to become one of the most economically important fisheries in Europe. In 2010, the total landings were 66,500 tonnes, of which UK fishers landed a significant part (58.1%). The *Nephrops* fishery is also important for countries such as Ireland (11.7% of the total) and Sweden (1.9%)

Advances in Marine Biology, Volume 64
ISSN 0065-2881
http://dx.doi.org/10.1016/B978-0-12-410466-2.00007-8

© 2013 Elsevier Ltd.
All rights reserved.

247

where it is of regional importance. Some are also taken in the Mediterranean, where Italian, Spanish and Greek fishers together take approximately 7% of the total landing. More than 95% of *Nephrops* are taken using single- or multi-rig trawlers targeting *Nephrops* or in mixed species fisheries. In regions such as Western Scotland and the Swedish West Coast, creel fisheries account for up to a quarter of the total landings. Across the range, a small proportion (<5%) is taken using traps in a fishery characterised by larger sized animals that gain a higher price and have lower discard and by-catches of ground fish with low mortalities. The trawling sector, however, is reducing the by-catches of ground fish with the aid of technical measures, such as square-mesh panels and grids and national systems of incentives. Assessments for *Nephrops* are operated via the 34 functional units (FUs) regarded as stocks. Changes in management procedures have arisen as a result of the advisory input from underwater TV fishery-independent stock surveys. The total allowable catch does not follow FUs but is agreed upon per management area.

Keywords: Norway lobster, Trawl evolution, Creel fisheries, Fisheries management, TAC, Functional units, Stock assessment, UWTV

1. FISHERY HISTORY

The fishery for the Norway lobster, *Nephrops norvegicus* (hereafter referred to as *Nephrops*), has increased significantly in the North Atlantic and the Mediterranean over the past five decades. *Nephrops* is marketed as scampi, langoustine, Dublin Bay prawn or Cigalas and, unlike 50 years ago, is now regarded as a delicious shellfish by chefs and consumers throughout Europe. Landings rose steadily and sharply until 1985 but have been more or less stable since then (Figure 7.1A, FAO). Recent landing statistics of *Nephrops* from all countries show that 66,544 tonnes were landed in 2010. Most of this came from the North-East Atlantic where 38,600 tonnes (58.1%) were taken by the United Kingdom, 7800 tonnes (11.7%) by the Irish, 4800 tonnes (7.2%) by the French, 4300 tonnes (6.5%) by the Danish, 2500 tonnes (3.8%) by Icelandic fishers and 1200 tonnes (1.9%) by the Swedish. Landings in Spain and Portugal have decreased significantly since the 1970s (Figure 7.1A). Some fishing occurs in the Mediterranean, a more deep-sea fishery compared to the North-East Atlantic shelf and slope fishery, where Italian, Spanish and Greek fishers take 3300 tonnes (5%), 700 tonnes (1%) and 500 tonnes (0.7%), respectively (FAO, 2010 landings).

However, as indicated in the country statistics and shown by landings per region (Figure 7.1B), landings by UK and Irish fishers and those in the North Sea (IV and IIIa) and the English Channel, and the Irish Sea and the Celtic Sea (VII) have increased during the past 20–30 years, while landings by

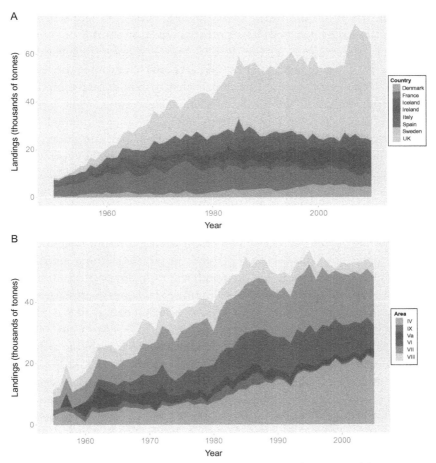

Figure 7.1 Total Norway lobster (*Nephrops norvegicus*) landings given (A) per country (tonnes) from 1950 to 2010, summarised for the North-East Atlantic and the Mediterranean. The landings rose sharply up to 1985 but since then have been more or less stable at around 60,000 tonnes. The United Kingdom is taking nearly 60% of the total landing, of which Scottish vessels are capturing around 80%. Ireland, France and Denmark otherwise contribute, but landings in Spain and Portugal have decreased. Countries with landings below 50 tonnes per year have been ignored, and the Spanish landing includes both the North-East Atlantic and the Mediterranean region (www.fao.org; FishStatJ database). (B) The total landings per fishing sea region from 1955 to 2005 (excluding the Mediterranean). It is to be noted that landings in the North (IV, VI, VII) have increased, but they have decreased to the South (IX). *Redrawn from ICES 2010 Climate report, Engelhard and Pinnegar (2010).* (For the colour version of this figure, the reader is referred to the online version of this chapter.)

Spanish fishers and those from the Portuguese coast (IX) and the Bay of Biscay (VIII) have decreased. The potential effect of climate on *Nephrops* abundance is reviewed in the ICES climate report (Engelhard and Pinnegar, 2010), along with other factors that could have caused the regional change in landing patterns: (1) The climate change hypothesis is tempting, as southern populations at the species border are decreasing, while northern stocks are increasing. This is a pattern noted in other exploited stocks (Heath et al., 2012). However, 13 LPUE (landings per unit effort) series from the northern functional units (3–16) were investigated for correlations with surface seawater temperature (SST) and the North Atlantic Oscillation (NAO; Zuur et al., 2003). Neither SST nor NAO was shown to correlate significantly with the LPUE of *Nephrops*, but six groups with similar fluctuations were found. The three other hypotheses for the South–North change in landings are as follows: (2) differences in fishing pressure related to the regional maturity of the fishery, (3) trends in targeted fishing effort with decline in multi-gear fisheries in the South and (4) indirect effects of fishing through reduction in natural predators leading to increases in crustaceans as a result of their main predator cod (*Gadus morhua*) being over-exploited (Brander and Bennett, 1989). Interestingly, a fifth suggestion relating to the decrease in *Nephrops* in the South is that the decrease in discards after the multi-gear crash has led to reduced feeding opportunities contributing to reduced stocks (see Grabowski et al., 2010). Cheung et al. (2012) review the potential effects of climate change on the future UK and Irish fisheries, and suggest that changes in temperature, ocean pH and oxygen levels are likely to affect marine ecosystems and their associated fisheries. Simulations of high scenario CO_2 levels on another commercially important crustacean, the European lobster (*Homarus gammarus*), have suggested that carapace mass may be lowered at the final larval stage (Arnold et al., 2009). Impacts from ocean acidification on, for example, UK and Irish fisheries have yet to be detected, but estimations of potential losses are huge.

The United Kingdom receives the majority of the EU quota allocations for the North Sea (IV, IIIa) and Western Scotland (VIa). Scottish vessels take 75–80% of the UK landings, and Northern Ireland and England take around 10% each. *Nephrops* from Scotland can have a higher value, especially larger whole animals, which can be sold at a higher price per kilogram, equivalent to that of European lobster, *Homarus gammarus* (D.R. Collin & Son Ltd., personal communication; Sandberg et al., 2004). Around 50% of the UK landings are exported to Spain, France and Italy. One-third of the total *Nephrops* landings are taken from Scottish waters and have risen from a few tonnes in the

1960s to over 31,000 tonnes in 2009 with a value of £78.3 million, making it the second most valuable fishery in Scotland after mackerel (*Scomber scombrus*). The first-hand sale value for Irish fishers selling *Nephrops* (€33.4 million) is ranked second in Ireland, behind the lower value but higher bulk of Atlantic mackerel (42.3 tonnes, €44.7 million; Marine Institute, 2012).

Nephrops fisheries occur on muddy bottoms with specific silt and clay content (Farmer, 1975), a sediment-type necessary to fulfil the requirement of a burrowing behaviour (see Chapter 2). Often, this habitat and, therefore, fishery are deeper than for other commercial crustacean fisheries in Europe, such as brown crab (*Cancer pagurus*) and European lobster. The depth varies in Northern Europe from 20 m in the sea lochs of Scotland to over 500 m on the shelf ridge west of the Hebrides. The fishery within the Mediterranean is notably deeper compared to the North-East Atlantic, mainly located on the upper and middle continental slope (300–600 m depth) or at even 500–800 m depth East of Corsica and NW and W of Sardinia (MEDITS trawl surveys 1994–1999 in Abelló et al., 2002), and also at a shallower 200–300 m depth in the Central Adriatic Sea (Morello et al., 2009).

There are variations in the biological parameters of *Nephrops* from different grounds in the North Atlantic East and in the Mediterranean, which have received a lot of attention over the years (e.g. Abelló et al., 2002; Bell et al., 2006; Farmer, 1975; ICES, 2004b, 2006; Ulmestrand and Eggert, 2001). There has been a particular focus on defining the parameters of importance for assessment and management, such as growth and maturation size as well as size frequencies at different depths and sediment types. In the management section, we report on the size at onset of maturity (SOM), and in Section 4.1, we discuss the shortcomings of growth data as a parameter for stock assessments. For a detailed review of the distribution of *Nephrops* by sediment type, see Chapter 2.

Here, we give a geographical overview of the different fisheries within five main areas: (1) the North Sea, (2) Western Scotland, (3) the Celtic Sea, the Irish Sea, and W Ireland, (4) the Iberian Peninsula, and (5) the Mediterranean (Table 7.1, Figure 7.2). This is mainly to give a brief insight into where the main fisheries are conducted, and the specific characteristics of the fishery for each region, including the current stock status. Each of these sea areas has separate and discrete fishing grounds, referred to as functional units (FUs) and assigned numbers. FUs are treated in more detail in Section 3.

Nephrops are found around the southern coast of Iceland (FU 1) with 10 discrete grounds identified at depths between 100 and 300 m (Eiriksson, 1999). There are no populations of *Nephrops* reported along the northern coast and this

Table 7.1 Functional units, sub-divisions and TAC areas for Norway lobster
(*Nephrops norvegicus*)

Functional unit	Name	ICES sub-division	TAC area
1	Iceland south coast	Va	
2	Faroes	Vb	
3	Skagerrak	IIIa	IIIa; EC waters of IIIb
4	Kattegat	IIIa	IIIa; EC waters of IIIb
5	Botney Gut—Silver Pit	IVb,c	EC waters of IIa and IV
6	Farn Deeps	IVb	EC waters of IIa and IV
7	Fladen Ground	Iva	EC waters of IIa and IV and Norwegian waters of IV
8	Firth of Forth	IVb	EC waters of IIa and IV
9	Moray Firth	Iva	EC waters of IIa and IV
10	Noup	Iva	EC waters of IIa and IV
32	Norwegian Deep	Iva	Norwegian waters of IV
33	Off Horns Reef	IVb	EC waters of IIa and IV
34	Devil's Hole	IVb	EC waters of IIa and IV
11	North Minch	Via	VI; EC waters of Vb
12	South Minch	Via	VI; EC waters of Vb
13	Clyde	Via	VI; EC waters of Vb
14	Irish Sea East	VIIa	VII
15	Irish Sea West	VIIa	VII
16	Porcupine Bank	VIIb,c,j,k	VII
17	Aran Grounds	VIIb	VII
18	Ireland NW coast	VIIb	VII
19	Ireland SW and SE coast	VIIg,j	VII
20	NW Labadie	VIIg,j	VII
21	Baltimore	VIIg,j	VII
22	Galley	VIIg,j	VII

Table 7.1 Functional units, sub-divisions and TAC areas for Norway lobster (*Nephrops norvegicus*)—cont'd

Functional unit	Name	ICES sub-division	TAC area
23	Bay of Biscay North	VIIIa	VIIIa, VIIIb, VIIId and VIIIe
24	Bay of Biscay South	VIIIb	VIIIa, VIIIb, VIIId and VIIIe
25	North Galicia	VIIIc	VIIIc
31	Cantabrian Sea	VIIIc	VIIIc
26	West Galicia	IXa	IX and X; EC waters of CECAF[1]
27	North Portugal (N of Cape Espichel)	IXa	IX and X; EC waters of CECAF[1]
28	South-West Portugal (Alentejo)	IXa	IX and X; EC waters of CECAF[1]
29	South Portugal (Algarve)	IXa	IX and X; EC waters of CECAF[1]
30	Gulf of Cadiz	IXa	IX and X; EC waters of CECAF[1]

[1]Fishery Committee for the Eastern Central Atlantic.

division is thought to result from the difference in oceanic conditions. The south coast of Iceland is influenced by the warmer Atlantic Gulf Stream, whereas the northern coast is within the Arctic waters. Experimental fisheries for *Nephrops* started in 1939, but a targeted commercial fishery did not commence until 1951, which took another 7 years to develop fully. Until the extension of territorial waters excluded foreign vessels in the early 1970s, these grounds were subjected to significant fishing activity by other nationalities.

Faroe Islands (FU 2)—A limited creel fishery exists around Faroe Islands (FU 2). In the past 10 years, a single Faroese vessel has on one occasion landed 1.55 tonnes of *Nephrops* to Scotland from area Vb (Caroline Cowan, personal communication).

Skagerrak (FU 3) and Kattegat (FU 4)—The landings in Skagerrak and Kattegat for 2010 were taken by Danish (73%), Swedish (24%) and Norwegian fishers (2%). Danish fishers have the majority of the total allowable catch (TAC), which is 5200 tonnes for 2013. The Danish fishery is a trawl fishery, whereas an increasing, and now significant, part of 26% of the Swedish landings of 1386 tonnes in 2012 was taken by the creel sector (Figure 7.3). There is also a small creel fishery in the Hvaler area along the Norwegian Skagerrak coast. However, for the Skagerrak and Kattegat as a

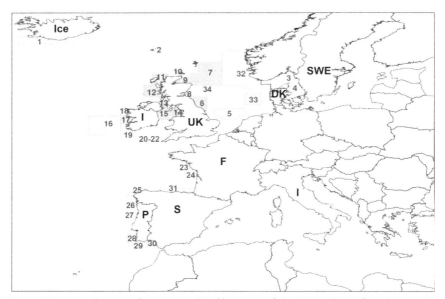

Figure 7.2 Map showing the geographical location of the 34 *Nephrops* functional units (1–34) in the North Sea (ICES sub-division IVa–c, IIIa), Western Scotland (ICES sub-division VIa), and in the Celtic Sea, Irish Sea and around Ireland (ICES sub-division VIIa–j), the Iberian Peninsula (sub-division VIIIa–c, IX) and in the northern area around Iceland and the Faroes (sub-division Va–b; see Table 7.1). (For the colour version of this figure, the reader is referred to the online version of this chapter.)

whole, only around 6% of the total landings (total landings were 8500 tonnes in 2010, ICES, 2012b; IIIa) is taken by creels. The Swedish creel fishery is described further in Section 1.6. The discard problem in *Nephrops* trawls and the decline of cod in the North Sea and not least in Kattegat led to the mandatory use of square-mesh panels (SMPs) in 2004, in line with the Cod Recovery Plan (EU regulation 423/2004). Further action was taken in 2009 when the centralised allocation of days at sea was changed to a kW-day system. This is discussed further in Section 3. Trawl selection improvements and the future of the region are discussed in Madsen and Valentinsson (2010).

1.1. North Sea

The main fisheries are Fladen Ground and Farn Deeps, and also Firth of Forth, Moray Firth, and, to a lesser extent, the Noup and Devil's Hole.

Fladen Ground (FU 7)—Within the Fladen Ground, substrates suitable for *Nephrops* burrows are distributed more or less continuously over a very large area (approximately 30,000 km^2). The distribution is slightly patchier

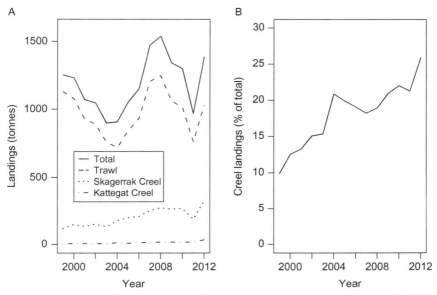

Figure 7.3 Swedish Norway lobster (*Nephrops norvegicus*) landings in 1999–2012. (A) Trawl in Skagerrak and Kattegat and creel landings (tonnes) in Skagerrak and Kattegat, respectively, within the ICES sub-division IIIa and (B) the proportion of creel to total landing (%) based on same data. *Data from SwAM, Fisheries statistics.*

towards the SW of the ground and sediments are patchy and coarse towards the North. The hydrographical conditions in this area are well suited to the retention of larvae due to a large-scale seasonal gyre that develops in the late spring over a dome of colder water. Because of the distance between fishing grounds and landing ports, the fishery in the Fladen is restricted to multi-day trips and is, therefore, fished by larger boats than the more inshore grounds. The fleet includes modern purpose-built boats from 12 up to 35 m, fishing mainly with 80-mm mesh size twin-rigs. Fishing trips are between 3 and 9 days, with the larger boats staying longer at sea. The Fladen fishery generally follows a similar pattern every year, with different areas producing good fishing at different times of the year. Abundance (according to the underwater television (UWTV) surveys) is generally higher on the soft and intermediate sediments located in the centre and south-east of the ground. Males consistently make the largest contribution to the landings although the sex ratio does vary. The Fladen is the largest unit in the North Sea and contributes just <50% of the overall TAC. The ICES advice for 2013 (ICES, 2012b) recommended that catches from the area should be no more than 10,000 tonnes, a reduction of

4100 tonnes from the 2012 advice. The more than significant reduction in the advised outtake from the area created a significant decrease in the overall TAC for the North Sea for 2013. As with all of the FUs within the northern North Sea, fishers are reporting increasing levels of predatory fish, such as cod, hake (*Merluccius merluccius*) and saithe (*Pollachius virens*) on the *Nephrops* grounds and are linking this with the decline in the abundance of *Nephrops*.

Farn Deeps (FU 6)—The Farn Deeps fishery is predominantly a winter trawl fishery (October–March), with only a small level of creel activity occurring in more isolated areas. The ground supports a fishery of 2000–3000 tonnes per year, taken principally by English trawlers, although visiting Scottish vessels account for between 20% and 30% of the landings (ICES, 2012b). The ground extends from Teesside up to around the Scottish border and ranges from around 50 to 110 m in depth. There has been an active *Nephrops* fishery off the North-East English coast since the end of the nineteenth century, albeit punctuated by the two World Wars. Prior to the First World War, landings peaked around 1911 at around 700 tonnes and did not reach these levels again until after 1960. It is hypothesised that the reduction in landings between these two periods is a result of lower effort, with many boats having been lost in the conflicts. The modern fishery comprises a wide range of vessel sizes with the local fleet typified by small vessels 15–20 m in length towing single-rig gears, a migrant fleet from Scotland towing single and twin rigs and one from Northern Ireland towing twin, triple and quadruple rigs. Since 2000, there has been an increase in the effort of vessels targeting *Nephrops* using multi-rig trawls. In 2004, they accounted for about 10% of the landings by weight and for about 20% by 2006. Since 1990, a moderate proportion (20–40%) of the landings from this fishery has been as tails (as opposed to whole). Until the mid-2000s, it was common practice to sort the catch while it was tied up in harbour and there are anecdotal reports of a population of *Nephrops* becoming established in the mouth of the river Tyne as a result of discarded animals. The major part of the sorting now takes place while the rig is steaming back to port. Fishing is usually limited to trips of 1 day with 2 hauls of 3–4 h being carried out. ICES (2012b) reports that the fishing effort is too high in the unit and that the biomass is, consequently, below the relevant reference point for a sustainable fishery.

Firth of Forth (FU 8)—The Firth of Forth ground is located close inshore to the Scottish coast and extends far up into the river estuary. The substrate composition is mainly muddy sand and sandy mud with only a small amount of the softest mud. In the late 1950s, the seine netters started to see big catches of *Nephrops* but discarded them as they were of no value.

This developed and by the early 1960s, the *Nephrops* were landed to be tailed for scampi. The fleet in the area targeted whitefish when it was abundant and *Nephrops* when it was less so. As the whitefish fishery developed further, and the use of the pair trawl became popular, the *Nephrops* fishery declined, until the mid-1990s, by which point the entire fleet was dependent on *Nephrops* alone. In the last 5 years, the fleet has decreased significantly and a number of vessels have started to pot for crab and lobster. During 2006/2007, the number of vessels regularly fishing in the Firth of Forth was estimated at 40 (23 under 10 m and 19 over 10 m), but in 2012, there were only 16 or 17 (Pittenweem, F.M.A., personal communication). The fishery provides income for many of the very small vessels working from the ports of the Firth of Forth, and many of them operate from the under-10 m pool quota, managed by the UK fisheries administrations, instead of the Producer Organisation allocations, which can sometimes restrict the fishing opportunities in the area. Some vessels, normally active in the Farn Deeps, come north from Eyemouth and South Shields, but, in turn, Firth boats sometimes move to other grounds when catch rates drop during the late spring moulting period. The area saw an influx of the larger trawlers from the North-East of Scotland in mid-2012 and this, unfortunately, coincided with significant quantities of fresh water going into the estuary, which may have influenced the reduced landings by the local fleet (Pittenweem, personal communication). Although the fleet has historically used single rigs and actively resisted the use of twin-rigged gear, a number of vessels have made the transition during 2012. There are very few, if any, *Nephrops* creelers in the area, but there are a significant number of crab and lobster pots and gear conflict has been reported. Landings have been variable throughout the time series (since 1981) and effort has reflected this, but it has been above F_{MSY} (the theoretical level of fishing effort that would result in a maximum sustainable yield) in the most part (ICES, 2012b). Despite this, the biomass of the stock is good and consistently well above B_{trig} (i.e. the biomass at which ICES recommends that further management action should be triggered to prevent further decline; ICES, 2012b), possibly as a result of larval recruitment from outside the area. The fishing pattern is highly seasonal with night fishing occurring predominantly during the summer and daylight fishing during the winter. The area is characterised by catches of smaller *Nephrops*, and discard rates are sometimes high.

Moray Firth (FU 9)—The Moray Firth is a relatively sheltered inshore area that supports populations of juvenile pelagic fish and relatively high densities of squid at certain times. The Moray Firth borders the Fladen functional unit (FU 7) and there is some evidence of *Nephrops* populations lying across this boundary. The fishery occurs the year round, supporting a variety of vessels using both single and twin-rigged gear. The *Nephrops* fishery in the area commenced in earnest in the late 1960s/early 1970s and was predominantly prosecuted by relatively small (under 21 m) local vessels using single trawls that had previously been fishing for species such as cod and sprat in this inshore zone. *Nephrops* had previously been caught in the area by smaller vessels, mainly from Buckie, working up and down the Moray Firth coast in the inshore waters. The fishery gradually evolved to include the twin rig in the late 1980s/early 1990s. Today, the FU provides not only a year-round fishery for smaller local boats based in and around the Moray Firth coast, but also a refuge for the larger vessels when the weather is bad, restricting fishing in the Fladen area. The more nomadic and larger vessels are also known to visit the area when the fishing is particularly good. The local fleet does not generally have a particular discard issue, although it is known that some haddock (*Melanogrammus aeglefinus*) may be discarded in the summer months. There is very little, if any, creeling for *Nephrops* in the area, although some gear conflict issues have arisen between the *Nephrops* trawlers and the lobster and crab creelers. Landings throughout the lifetime of the fishery have been fairly stable but, as suggested previously, the effort in the area has been variable. That said, abundance has been healthy throughout the period for which information is available, indicating some resilience, particularly as the landings prior to 2007 are likely to have been under-reported. In recent years, a squid fishery has been seasonally important in the Moray Firth leading to a switch in the target fishery of the local fleet. The effort in the area increased in 2011 (ICES, 2012b) but may have decreased in 2012 due to a change in fishing patterns. Effort within the unit has been variable throughout the time series with biomass remaining relatively stable, although reducing more recently.

Botney Gut (FU 5)—A deep channel cutting east–west across the central North Sea, known as the Botney is actually a submerged river bed that has supported a moderate *Nephrops* fishery since the early 1970s with small landings reported from the late 1940s. The fishing area covers 1850 km^2 and spans both UK and Dutch territorial waters. This fishery is the most internationally diverse in the North Sea and has at times been used by vessels from Belgium, the Netherlands, Denmark, Germany and the United Kingdom. There is no creeling for *Nephrops* within the Botney Gut area.

Off Horns Reef (FU 33)—West of the Horns Reef lies a mid-sized *Nephrops* ground (5700 km^2). The landings from this FU were marginal for many years before a Danish fishery developed between 1993 and

2004, taking over 1000 tonnes. Since then, other countries have exploited the area and annual landings have been over 1400 tonnes. Relatively little is known regarding the distribution of muddy sediments in the area, which has not been surveyed by UWTV.

Norwegian Deeps (FU 32)—The spatial extent of commercial *Nephrops* activity within the Norwegian Deeps area is the largest in the North Atlantic at over 55,000 km^2 but is subjected to relatively little fishing activity, which has never taken more than 1200 tonnes. The fishery is commercially prosecuted by Swedish, Danish and Norwegian vessels and supports some recreational creeling as well. The commercial activity is based on trawl fisheries (both directed and mixed fisheries) as well as some by-catch in the northern shrimp (*Pandalus borealis*) fishery. North of 60 °N, the fishing activity is predominantly creeling.

Noup (FU 10)—The Noup fishery started at around the same time as the Moray Firth fishery and was primarily used by the same vessels but has since reduced in importance, as supported by the low, but variable, landings since 1997. As with the Moray Firth, the area was originally exploited by single-rigged trawlers, with a move to twin rigging late in the 1980s. At present, the fishery is prosecuted by around four vessels. The quantity landed from the area is now very small, accounting for <1% of the overall North Sea landings. There is very little information available on the stock, but it was thought to be relatively stable when the last survey was undertaken in 2007 and effort has reduced significantly since then (ICES, 2012b).

Devil's Hole (FU 34)—This unit was assessed individually for the first time in 2012 and the first UWTV survey was carried out in the same year. In advice before this date, the FU, although the largest of such, was thought of purely as part of the component of the stock that was outside of the defined FUs. The fishery started in the 1990s and was, arguably, opened up by vessels from Buckie. These vessels were relatively small in size and capacity and initially operated single-rigged gear, with some changing to twin rig later in the decade. The fishery currently supports 15–20 Scottish vessels with variable, relatively low, landings.

1.2. West of Scotland

On the west coast of Scotland, there are *Nephrops* fisheries in the North Minch (FU 11), South Minch (FU 12), the Firth of Clyde (FU 13) and, to a lesser extent, at Stanton Bank and in more offshore areas on the shelf edge. Small inshore trawlers targeting *Nephrops* take most of the catch, but some are caught by larger twin-rig vessels. Creel fishing accounted for 18% and 21%, respectively of landings in the North and South Minch in 2009 (including Loch Torridon, which is described in more detail in Section 1.6). Creel-caught *Nephrops* are generally larger and in better condition than those caught by trawling (Ridgway et al., 2006). They attract high prices in the live export market and provide an important source of income for small local boats. Creels are used mainly in inshore areas and sea lochs, where trawler access may be limited by the seabed or legislation. In some areas, both fishing methods are used and gear conflicts sometimes occur. Effort within the West of Scotland as a whole was anticipated to be increased during 2012 with an increased number of vessels working on the grounds.

North Minch (FU 11)—The North Minch extends from Kinlochbervie and covers most of the coast of Lewis and the northern part of Skye. As in many other areas, the seine vessels began catching *Nephrops* in the mid-1960s but either discarded them or, to a lesser extent, tailed them, as no real market was accessible. As the market began to develop, vessels from the east coast began fishing in the North Minch with trawl gear and were out-fishing the local vessels. As the west coast vessels began to change to trawling, significant fleets developed in towns such as Gairloch, Lochinver and Stornoway, with the numbers peaking in the early 1970s. In 2011, 79% of the landings from the North Minch were landed by trawlers, and quota was not restrictive. Effort restrictions operate in the area and until 2012 have not been restrictive. In 2012 and, to a lesser extent, 2011, significant numbers of east coast vessels fished in the area, leading to effort issues and the need for the Fisheries Administration to take action. Fishing mortality has fluctuated over time but does not appear to have had an adverse impact on the biomass. The ICES advice for 2012 recommends that landings should be restricted to 4200 tonnes, indicating that the fishery is exploited at sustainable levels. In recent years, fishermen have not reported the same high levels of predatory species as have been seen in the North Sea, but significant quantities of haddock were seen on the grounds in 2012. It is thought that this is due to vessels working with different fishing gear, rigged to catch a fish by-catch, and, to a lesser extent, increased abundance of haddock.

South Minch (FU 12)—The South Minch extends from the northern part of Skye and encompasses Muck and part of Jura at its southern end. The western extent encompasses Barra and South Uist and extends some way seaward from the west coast of South Uist. Fishing activity in the South Minch concentrated mainly on pelagic species and creeling until the 1960s when fishermen from the Clyde migrated north bringing their dedicated gear with them. The fishery has many characteristics in common with the North Minch, although the abundance appears to be a little more variable. Many vessels operating in the North Minch also operate in the south. Seventy eight percent of the landings in 2011 were made by trawlers, with the fishery being mainly Scottish, although very small landings are made by Northern Irish and, to a lesser extent, English vessels. ICES recommend catches of not more than 5800 tonnes for 2013, suggesting that the fishery is being exploited sustainably as the recommended number is well below the numbers of recent landings (ICES, 2012b).

Firth of Clyde (FU 13)—The Firth of Clyde (FU 13) is located off the west coast of Scotland near Ayrshire and is one of the three FUs in division VIa. FU 13 fishing grounds extend from Ayrshire westward into the Atlantic waters around Arran, Kintyre and Jura, stopping just south of Mull. The fishery has existed in these waters since the 1960s and can be split into two sub-areas, the first being the Firth of Clyde and the second, the Sound of Jura. These waters were used as fishing grounds for species such as herring until the 1980s when *Nephrops* became the dominant species (Combes and Lart, 2007). *Nephrops* are now considered to occur in high densities in both the Firth of Clyde and Sound of Jura (more than 0.8 burrowper m^2; ICES, 2012b). In 2011, fishing in FU 13 produced 6431 tonnes of *Nephrops* from these grounds. The catch was a combined effort from both trawls (accounting for 97% of the catch) and creels (accounting for 3% of the catch; ICES, 2012b), with vessels earning approximately £2000 per tonne of trawled *Nephrops* (Combes and Lart, 2007) with a minimum landing size (MLS) of 20-mm carapace length (CL; ICES, 2012b). Discards from trawls equated to 556 tonnes; while no data are available for discards from creels, the number is expected to be very low (ICES, 2012b). Approximately 92% of *Nephrops* landed in FU 13 are from Scottish vessels with the remaining 8% landed by other UK and Irish vessels. The area used to trawl for Nephrops is 3000 km^2 (Wieczorek et al., 1999) with the by-catch including cod, haddock and whiting (*Merlangius merlangus*). However, temporal and area bans that are enforced in FU 13 have led to a rise in the number of creel boats as trawlers are unable to fish over the weekend (ICES, 2012b). It is estimated that approximately one-third of the creel boats operate round the year. As a result of the findings and advice from ICES, it is suggested that landings should not exceed 5600 tonnes for the Firth of Clyde and 800 tonnes for the Sound of Jura (total of 6400 tonnes for FU 13). Anecdotally, the general size profile of Clyde landings decreased in 2012. This may have been purely due to localised increases in landings. Observations from early 2013 indicate that effort may again be high in the area as the visiting fleet appeared in January as opposed to the more normal March.

1.3. Celtic Sea, Irish Sea and West Ireland

Within the ICES sub-division VII (the fishing grounds around Ireland), 7 FUs are used to assess *Nephrops* stocks within three broader grounds, the Celtic Sea, the Irish Sea and West Ireland (ICES, 2012a). Trawling is the primary means of *Nephrops* fishing within these grounds, yet this may not be the most efficient method for landing. ICES (2012a) stated that *Nephrops* could be seen in grounds that were newly trawled, suggesting that at least some of these animals may be impervious to trawling efforts. This phenomenon of *Nephrops* grounds has been hinted at previously (Vergnon and Blanchard, 2006).

1.3.1 Celtic Sea FUs 20–22

The Celtic Sea is located to the south of Ireland. It is bordered in the north-east by St. George's Channel and in the east by the Bristol Channel and the English Channel and contains three FUs. Two small FUs (20 and 21, 'Labadie') are always considered together, and FU 22 Celtic Sea—the Smalls. Both single and twin trawlers are employed in FUs 20–21with the fishery dominated by French and Republic of Ireland vessels (98%). Although this fishery was thought to be stable, landings have decreased considerably over the past few years, with the French catch decreasing dramatically in this area from 90–95% to 45–50%. The FU 20–21 fleet in 2011 landed 1237 tonnes. Although discards were not calculated because of scarce observations, discards in the FU are thought to be high. Anecdotal evidence suggests that some of the Irish fleet are making the transition to multi-rigged gear, with the potential for increasing the overall pressure on the units in the Celtic Sea. The Irish fleet also saw a general increase in the amount of tails landed in 2012 as a whole, mainly as a result of the high price that the product was attracting.

FU 22—Smalls has existed as a fishery since the 1960s and is currently thought to be a relatively stable and productive stock. *Nephrops* landings in FU 22, an area covering 2881 km^2, are dominated by vessels from the Republic of Ireland (more than 95%) with a moderate density of *Nephrops* (around 0.5 burrows per m^2). In 2011, 617 tonnes of *Nephrops* were landed from these grounds, all of which were by otter trawls, with only 9% discards. Whiting and cod and occasionally haddock and hake dominate any by-catch. ICES advise that landings of *Nephrops* in FU 22 should not exceed 3100 tonnes in 2013 (ICES, 2012b).

1.3.2 Irish Sea

In the Irish Sea, *Nephrops* are exploited mainly in the waters to the west of the Isle of Man (FU 15—Irish Sea West) and most landings are by Northern Ireland (6000 tonnes per annum) and the Republic of Ireland (2000 tonnes). These landings give a combined first-sale annual value of about €14.7 million, which makes it the most valuable fishery in these waters with *Nephrops* occurring in very high densities (around 1.1 burrows per m^2), resulting in smaller sized animals (ICES, 2012b). However, among the Western Irish fisheries, there has been a prevalence of a dinoflagellate parasite, *Hematodinium* that has been present since 1994 with an estimated annual mean infection prevalence of 18% in 1996. Peak infection is believed to occur in April/May, with juveniles being more susceptible to the parasite. Significantly infected muscle tissue is reportedly bitter in taste, resulting in the meat being inedible (Briggs and McAliskey, 2002). A severe outbreak of the disease could have devastating consequences for one of the largest fisheries in the area in terms of stock recruitment and fishery revenue (Field et al., 1992). In 2011, FU 15 landed 10,162 tonnes and discarded a further 2700 using otter trawls from 5289 km^2 of fishing grounds. The trawls used are a combination of single and twin gears and 70–99 mm meshes, with an MLS of 20-mm CL. However, in order to reduce the by-catch (mainly juvenile whiting, cod and haddock), separator trawls and Swedish grids were used by approximately 45% of the Irish vessels (ICES, 2012b). Landings for the 2013 season should not exceed 9300 tonnes, similar to what the *Nephrops* stock in FU 15 has sustained for several years. However, within the domain of the Irish Sea, *Nephrops* are also fished off the West England coast from the north of Wales to the south-west border of Scotland (FU 14—Irish Sea East). This FU actually comprises two grounds and there is a small population fished in Wigtown Bay, just north of the main ground, although *Nephrops* in this area are in fairly low densities (around 0.3 burrows per m^2). In 2011, the *Nephrops* catch in FU 14, where the fishery operates mainly in spring and early summer and is male dominated, was 561 tonnes with around 28% discards, with the by-catch of this area comprising primarily plaice (*Pleuronectes platessa*), but cod and whiting as well. The number of vessels responsible for this catch numbered in the region of 60 vessels, with the fleet consisting of approximately 25 English and 35 Northern Irish vessels, using 70–99 mm meshes. The English used predominately single otter trawls, while over 70% of the Northern Irish vessels used multi-rig trawls (summary—38% twin: 62% single otter trawls; ICES, 2012b). Based on reports by ICES, landings within this FU should not exceed 880 tonnes in 2013. The Northern Irish fleet working in the Irish Sea reported a steady year in 2012, with 'good fishing, good prawns and good prices'. The good price reduced the overall landings and kept the overall quality to a high standard by reducing the economic pressure on fishermen to land inferior prawns or to fish harder. Effort may have increased in the Irish Sea in 2012 due to displacement of the Northern Irish fleet from VIa to VII.

1.3.3 West Ireland

The Western Ireland grounds comprise three FUs. The Porcupine Bank stock (FU 16) is located off the west of Ireland in the North Atlantic and the fishery operates mainly between April and July. *Nephrops* are taken by the Spanish fleet as by-catch in a multi-species trawl fishery together with demersal fish, comprising mainly hake, anglerfish (*Lophius piscatorius*) and megrim (*Lepidorhombus whiffiagonis*). However, *Nephrops* landings from Spanish trawlers fishing on this bank have decreased by about 90% in the last 25 years, from 3873 tonnes in 1982 to 483 tonnes in 2007. González-Herraiz et al. (2009) found a negative correlation with the NAO index with a time lag of 6.5 years. *Nephrops* density in this area is considered extremely low (<0.3 burrows per m^2). Although the stock size is thought to have increased over recent years, this may be only as a result of reduced fishing pressure since from 2010, FU 16 has employed a seasonal closed area during female emergence (first 3 years May–July but from 2013 only in May). This regulation has been adhered to well by the fleets allowing protection for a large area of the stock. In 2011, 1187 tonnes of *Nephrops* were landed in these waters using both single and twin rigs, with exploitation rates of females being lower than that of males. Discards are thought to be nominal, although the actual discard rates for 2011 are uncertain (ICES, 2012b). Landings in FU 16 should not exceed more than 1800 tonnes as advised by ICES, managed by a separate quota. This separation of quota has led to displacement of effort into other areas and has seen the larger offshore vessels typical of the FU 16 fishery on some of the more inshore grounds within divisions VI and VII. This is discussed further in Section 3.

The second of the West Ireland FUs, FU 17—Aran Grounds extend north from the South coast of County Clare to the North coast of County Galway and are approximately 909 km^2. These grounds are considered to have a fairly high density of small *Nephrops* (0.9 burrows per m^2), yet fluctuations have been observed in the stock size. In addition, the Aran Grounds have a catch sex ratio that varies greatly with a high percentage of males landed during the autumn. In previous years, FU 17 has been fished by French, English and the Republic of Ireland vessels; however, over the past few years, landings in these grounds have been solely by the Republic of Ireland. Currently, 90% of the fishery employs twin-rigged vessels with the total landings being 600 tonnes in 2011, all of which were from otter trawls. The by-catch of these grounds consists of hake, megrim and monkfish (ICES, 2012b). Recent surveys have shown the Aran Grounds stock in significant decline and harvest rates may be reconsidered. Data from 2004 to 2007 show that previously landings on these grounds had been restricted. ICES, 2012b has advised that *Nephrops* catch in the Aran Grounds should not exceed 590 tonnes in 2013.

FU 18—The *Nephrops* grounds to the northwest of Ireland have not been officially assessed, like the other FUs around Ireland, and ICES (2012b) states that there is currently no major *Nephrops* fishery in the area. Any *Nephrops* caught in this area are believed to be minimal in number and are not expected to exceed 200 tonnes. The recorded landings from 2011 are over 20 tonnes.

FU 19—Ireland SW and SE coast grounds extend along the entire Irish South coast from the north coast of County Kerry on the west to County Wexford on the east, spanning 1653.26 km^2. *Nephrops* in FU 19 are found to occur in moderate densities (around 0.5 burrows per m^2) with higher landing rates of males in comparison to females. In 2011, 608 tonnes of *Nephrops* were landed from these grounds. All landings were from otter trawls employing either single or twin rigs with 80–99 mm cod-end mesh size. The majority of vessels landing in these waters were from the Republic of Ireland (responsible for 96% of total landings), with a few French (3%) and English (1%) vessels. Discards were in the region of 18% with by-catch species including hake, megrim, anglerfish and monkfish (ICES, 2012b). Landings in FU 19 for 2013 should not exceed 820 tonnes, as advised by ICES.

1.4. Iberian Peninsula

The southern region of the North Atlantic distribution consists of nine FUs, namely, the Bay of Biscay (FUs 23–24); North and West Galicia (25–26); the Cantabrian Sea (FU 31); North, South and South-West Portugal (FUs 27–29) and the Gulf of Cadiz (FU 30). These southern *Nephrops* populations are distributed in much deeper waters than their northern counterparts (from about 80 to about 700 m) and here the distribution is patchy depending mainly on the sediment type. In the northern Bay of Biscay, this species occurs at depths ranging from 80 to 120 m (Conan et al., 1994); in Galicia, from 90 up to more than 600 m of depth (Fariña, 1996); and in South Portugal and the Gulf of Cadiz, between 200 and approximately 700 m (Figueiredo and Viriato, 1989; Ramos et al., 1996).

It is uncertain when *Nephrops* began to be exploited commercially in these southern areas, but De Buen (1916) refers to a commercial fishery for *Nephrops* in Spain and indicates its presence along the entire Iberian Peninsula coast. Since then, *Nephrops* fisheries have expanded considerably, in particular, over the last 50 years. During the past few decades, the trawling fleet have undergone considerable technological improvements (i.e. gear development, GPS) resulting in an increase in fishing effort. Landings of *Nephrops* in southern areas have been decreasing since the mid-1980s. Currently, the largest *Nephrops* catches are obtained in the Bay of Biscay with a total of 3398 tonnes in 2010, followed by the Iberian Atlantic coast with 275 tonnes, and the Cantabrian Sea and North Galicia division with 42 tonnes (Engelhard and Pinnegar, 2010). In this last division, the fishing effort is not specifically directed to *Nephrops* but to a mixed fishery that targets a number of different demersal species.

In the Bay of Biscay and the Iberian Peninsula Atlantic waters, *Nephrops* is mainly caught in a mixed bottom-trawling fishery and can be either a target species or a by-catch of finfish/crustacean trawlers. Mixed finfish trawlers target *Nephrops* together with hake, black anglerfish (*Lophius budegassa*), monkfish, megrim, horse mackerel (*Trachurus trachurus*), mackerel (*S. scombrus*) and blue whiting (*Micromesistius poutassou*). The crustacean fleet in the southwest and South Portugal targets, besides *Nephrops*, deep-water shrimps such as the rose shrimp (*Parapenaeus longirostris*) and also the red shrimp (*Aristeus antennatus*). During the last two decades, these vessels, traditionally targeting predominantly *Nephrops*, transferred most of their fishing effort towards the rose shrimp. Today, the latter species is the most important species for the crustacean trawling fleet in South Portugal. The decrease in *Nephrops* landings in the South-West/South Portuguese coast can also be related to this shift of fishing effort, but no specific studies exist that enable an assessment of effort changes. The same shift of effort happened in the Gulf of Cadiz where the rose shrimp also achieves a remarkably high first-sale price and where the fishing grounds are closer to the coast and easier to reach (90–380 m of depth).

Historically, *Nephrops* populations have been exploited along these coasts by trawling, but nowadays creels also account for a small proportion of the total landings. Creels were used locally on the west coast of Portugal, but only in areas unavailable to trawlers because of either legislative restrictions on access or unfavourable bathymetry. Although creels are very selective for *Nephrops* and proved to be financially profitable in the Portuguese west coast (mean first-sale price in 2007–2009 was €40 per kg), the expansion of this fishing gear is unlikely, mainly because of likely conflict with the trawling activity (Leocádio et al., 2012).

Nephrops represents a very valuable resource in these southern countries. One peculiar case is Portugal where, despite the relatively insignificant importance of landings in the context of global European figures, local first-sale values are considerably higher than elsewhere in Europe; it was €21.02 per kg in 2011, while the average prices per kilogram for the United Kingdom and France in 2011 were €5.60 and €9.25, respectively. These high prices achieved in Portugal are due to the quality of the product as *Nephrops* are sold fresh, refrigerated (from the trawling fleet) or alive (from the artisanal creeling fleet). The catch per unit effort (CPUE) in Portuguese waters has decreased from about 120 kg per day (per vessel) in 1989 to about 50 kg in 1995, and has remained more or less stable since then (ICES, 2004b). Based on the recommendations made by the WGNEPH, a zero TAC was set for 2003, 2004 and 2005 to let the stock recover.

1.5. The Mediterranean

1.5.1 Balearic Islands

The fishery in this part of the Mediterranean is operated by 40 trawlers working on the upper slope between 350 and 600 m depth with average annual catches of 10 tonnes (3.5 tonnes for females and 6.5 tonnes for males) between 2002 and 2009. Typical by-catch species in this fishing area are hake, megrim, blue whiting and anglerfish. The stock of *Nephrops* in this area is considered overexploited and for the period of 2002–2009, a pseudo cohort of 25,000 and a population biomass of 35 tonnes had been estimated (Guijarro, 2010).

1.5.2 Adriatic Sea (GSA 18)

In the Adriatic Sea, one-third of the total Italian landings are taken off the West coast and *Nephrops* ranks first among all the crustacean species exploited in terms of value. There have, however, been declines in catches since 1993 (Vrgoč et al., 2004). *Nephrops* can be found at depths of 30 m in the Northern Adriatic to 400 m in the southern area (Marano et al., 1998), with higher population densities in the north. Important fishing grounds for demersal trawlers are at 70 m depths off Ancona (Central Italy on the Adriatic coast) and at 220 m in the Pomo pit further off in the Middle Adriatic basin. The *Nephrops* fishery in Italy has been managed and regulated since 1980 by limited numbers of fishing licences, area limitations, mesh size restrictions, MLSs (EC 1967/06[1]) and seasonal fishing bans. The fishing effort and landings have declined since 2004 and annual landings have gone down from 1300 tonnes (2007) to 865 tonnes (2011) (Cardinale et al., 2012).

1.5.3 The Ligurian and Northern and Central Tyrrhenian Sea (GSA 9)

Nephrops in this area are targeted by bottom trawlers with a fleet of about 80–100 to date, fishing at 300–500 m depths along muddy slopes. Most trawlers targeting *Nephrops* also catch other commercial species (hake, deep-sea pink shrimp, horned octopus (*Eledone cirrhosa*), squids (*Todaropsis eblanae*)) and non-commercial ones. The *Nephrops* fishery in this area is the most valuable, with total annual landings between 248 tonnes in 2005 and 228 tonnes in 2008 (Cardinale et al., 2011).

1.6. Socio-economically interesting fishery regions

1.6.1 Loch Torridon

A creel fishery has existed for approximately 40 years on the North-West coast of Scotland, especially around Loch Torridon (FU 11, North Minch). In 1984, the Inshore Fisheries (Scotland) Act removed the ban on using mobile gear closer than three nautical miles from the shore. This resulted in an increased and intense conflict between trawlers and creelers. Creelers sought an exclusion zone where trawlers were banned and they could co-manage the fishery. In November 2000, this was recognised and legislated for (Bennett and Hough, 2007). At the same time, an exclusive trawl area of similar size was established solely for the use of trawlers. The creel fishermen in Loch Torridon set up a company, Shieldaig Export Ltd., to collectively supply the live *Nephrops* market. The *Nephrops* captured are mainly live-stored and exported to the Spanish market by air freight. Internal rules were also established for the closed area under the Torridon Management Plan (Bennett and Hough, 2007). The measures included a maximum number of creels for one (2 × 400) or two or more (2 × 800) fishermen per vessel, only one set of gear to be hauled per day, a maximum of 200 fishing days

---

[1] Council Regulation (EC) No. 1967/2006 of 21 December 2006 concerning management measures for the sustainable exploitation of fishery resources in the Mediterranean Sea, amending Regulation (EEC) No. 2847/93 and repealing Regulation (EC) No. 1626/94. Official Journal of the European Union, L409/11.

per year, berried females to be released and escape gaps to be of 22 mm allowing specimens up to 40-mm CL to escape, although the MLS is 20-mm CL. The Loch Torridon creelers take approximately 100–150 tonnes per year of a TAC for the management area C (FUs 11–13) of ca. 11,000 tonnes. The Loch Torridon creel fishery became the first *Nephrops* fishery to be eco-labelled under Marine Stewardship Council (MSC), underpinning the voluntary Loch Torridon Management Plan and otherwise sustainable fishery managed under TAC in management area C (FU 11; Mason et al., 2002). However, in 2011, the eco-label was withdrawn. The MSC eco-label, which has been under fire recently (Froese and Proelss, 2012), does not seem to have been missed by the creelers who suggested that it was of little value for marketing, although it is likely that this is because of the nature of the continental market rather than a general observation.

1.6.2 The Swedish coastal area

A creel fishery exists on the Swedish Skagerrak coast and 90% of the creel landing of 359 tonnes for Swedish *Nephrops* in 2012 came from here (Figure 7.3A). There is a smaller creel fishery prosecuted by Swedish fishermen in the Kattegat also. The creel fishery was introduced in the mid-1980s (Eggert and Ulmestrand, 1999) and the landings were 10–15% of the total in 1986–1999 (Swedish National Board of Fisheries, 2010). After the trawl border was moved out from 2 to 3–4 nautical miles, it gradually increased and has now reached 26% (Figure 7.3B). In 2006, borders for trawling and creeling were regulated further to give more space to the creelers, and from 2008, a permit for creeling was required to minimise the risk of overestablishment of creelers (Swedish National Board of Fisheries, 2006). Few new permits are currently given. For the year 2013, 109 vessels (102 skippers) have a permit to use creels in Sweden in vessels from 5 to 12.7 m, with most vessels 8–10 m in length (48%). The total TAC for Skagerrak and Kattegat in 2013 is 5200 tonnes, of which 1367 tonnes are for Swedish fishermen with 25% for creeling only. The first-hand sale value for the Swedish *Nephrops* landings in 2012 was €12.8 million, which ranked second in landings from the West coast after the northern shrimp (€14.0 million), and fourth in the country after herring (*Clupea harengus*; €17.0 million) and cod (€15.2 million). However, in Sweden, prices for *Nephrops* are between 8% and 23% (mean 18%) higher, about €3 per kg, for creel captured *Nephrops* compared to trawl caught (Swedish National Board of Fisheries, 2010). This is probably because creeled *Nephrops* are larger and deemed to be of higher quality. The *Nephrops* landing is sold shortly after landing mainly at the Gothenburg, Smögen or Strömstad fish auction, and then redistributed further. *Nephrops* are sold sorted or in different size categories. Swedish regulations state that *Nephrops* creels must be set in depths deeper than 30 m (SwAM, 2004:36), and the fishery often operates at 40–80 m depth. The fishers shoot 300–400 creels per day on average but only up to 800 creels (single fishermen) or 1400 (two or more fishers) are allowed (SwAM, 2004:36). It takes approximately 30 min to lift, empty, re-bait and shoot a fleet of 50 creels (Ungfors, A., personal observation). Soaking time for the gears is 2.5 days on average. The Swedish *Nephrops* creelers also target European lobster and brown crab (using lobster or crab pots).

2. CAPTURE METHODS

Trawling is the main gear used for capturing *Nephrops*. However, in some areas, particularly in Western Scotland and in the Swedish Skagerrak coastal region, the proportion caught by creeling is of significance and has increased markedly over the last 40 and 28 years of use, respectively. Nowadays, around 20% in the North and the South Minches and the Clyde, and 26% of the Swedish TAC are taken by creels, which means an even larger percentage in value terms. The length frequency differs in catches taken by trawls and creel captures (Jansson, 2008; Leocádio et al., 2012; Figure 7.4). The higher value of creel-caught animals means that a lower quantity can be landed to generate the same income (Table 7.2). The cost for fuel for creel fishers is also less. Ziegler and Valentinsson (2008) took a holistic view of the *Nephrops* industry using life-cycle analyses (LCA) and found that 2.2 l of diesel were used per kilogram of creel-captured *Nephrops* landed, which can be compared to selective trawl fishery using 4.3 l, mainly used by the fishing operations. Ziegler (2006) showed that the conventional trawl fishery uses 9.0 l of diesel per kilogram of *Nephrops* landed, the discrepancy due mainly to a cleaner catch composition (higher % of *Nephrops*) in the species-selective trawl compared to the conventional one.

Stocks with larger MLS where higher frequencies of the larger individuals are taken, are more suitable for creel capture. The highest MLS is in Skagerrak and Kattegat (40-mm CL), making this a promising area for creel fishers. The ecological foot print, for example, the benthic effect on the ecosystem using trawl versus creel has been evaluated and it is suggested that 1 h of trawling results in the same impacted area on the seafloor as during 1 year of the entire Swedish *Nephrops* creel fishery (Hornborg et al., 2012; Ziegler, 2006; Ziegler and Valentinsson, 2008). However, the consequences of

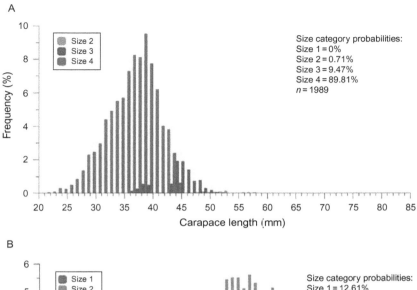

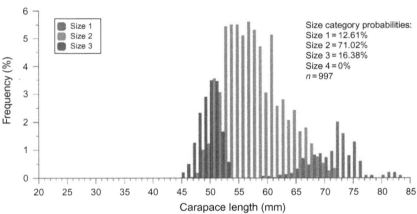

Figure 7.4 Size distribution in trawl (A) versus creel (B) catches shows that larger sizes are landed with creels for Norway lobster (*Nephrops norvegicus*). Size at first capture is larger and length distribution of the catch is greater in creels. Data are shown for the Portuguese coast (Leocádio et al., 2012). (For the colour version of this figure, the reader is referred to the online version of this chapter.)

Nephrops trawling activity on the sea bottom ecology are complex and there is scarcity of information on comparable control areas without a trawl fishery. Surveys using, for example, wreck areas as non-fished areas (Ball et al., 2000) or closures (Tuck et al., 1998) suggest that it is the fishing intensity *per se* that impacts the benthic ecology. Ball et al. (2000) found some indications of disturbance at the inshore trawling grounds used for less

Table 7.2 Examples of size-sorting categories in use for Norway lobster (*Nephrops norvegicus*) (up to four categories, XL-S, regional differences may occur), and the approximate first-hand sale value (€ per kg) for these in the three given areas

	Sorting category	Extra large (XL)	Large (L)	Medium (S)	Small (S)	Reference
UK waters	Numbers per kg	1–9	10–14	15–20	21–26	Don fishing company[a], Peterhead
	Approx. value (€ per kg)[a]	10	8	7	5	
Sweden (FUs 3 and 4)	Numbers per kg	1–5	6–10	11–15	16–20	
	Total length[b] (cm)		>16	14.5–16	13–14.5	
	Approx. value (€ per kg)	24.3–29.4–33.6	17.4–30	9.3–17.4	7–8.1	Gothenburg[b]
			7.0–15.5–39.1	5.9–10.9–27.8	5.7–8.6–25.5	Smögen[c]
			13.3	9.9		Strömstad[d]
Portuguese waters	CL (mm)	>60	47–72	36–53	>44	Leocádio et al. (2012)
	Approx. value (€ per kg)	47–130	23–67	7–24	2–10	

[a]Price to fishers.
[b]Gothenburg Fish auction mainly use two sorting categories, over 16 cm and 13–16 cm total length, but sell according to fishers' own sorting; main value changes during the year are given (lower spring–higher summer season).
[c]Smögen Fish auction use three to four sorting categories (min—average—max for year 2012 is given).
[d]Strömstad Fish auction use two sorting categories (over 16 cm and 13–16 cm, average for 2012 is given), the average value for creel captured in 2012 is 13.3 (no size sorting).
The size categories are affected by the minimum landing size (Figure 7.11 for more information). The first-hand sale is higher for the larger *Nephrops* (more than double), but the values differ during the year.

intensive fishing (e.g. the disappearance of echinoderm *Brissopsis lyrifera*) but that the trawling grounds still contained large molluscs, and the number of species did not differ significantly between untrawled and trawled areas. However, in deeper and more intensively trawled areas, there is a complete absence of large benthic infauna except *Nephrops*. In the Bay of Biscay, a community indicator (the ABC method—abundance–biomass comparison) has been used to evaluate trawling impacts on benthic invertebrates (Vergnon and Blanchard, 2006). In theory, disturbed habitats will have a lower average biomass than abundance compared to less disturbed areas where the biomass is larger (comparing curves for biomass and abundance for ranked species). The results of the ABC method were inconsistent with the theoretical expectation for these communities and the measured levels of fishing intensity, as the biomass of macro benthos (not least of *Nephrops* and *Liocarcinus* sp.) remained relatively high in the heavily trawled areas.

The sex ratio in the capture during the year differs, perhaps more in trawl landings than creeling. This is a result of the reproductive behaviour of *Nephrops*, where females are ovigerous for months during which time they are more prone to stay in their burrows. Males dominate trawl captures in the egg-bearing season. However, that egg-bearing females do not feed is apparently a 'truth with some modification'. Aguzzi et al. (2007) analysed the empty stomachs of both berried and non-berried females from captures taken at different times of the day and night. From this and laboratory studies in burrow emergence behaviour, they concluded that both berried and unberried females feed outside of their burrows (see Chapter 3). This can in fact be seen in creeling, and it can be regarded as a negative effect of creeling that a larger proportion of berried females are seeking food from the baited creel, which seems to attract even egg-bearing females into the cages. Aguzzi et al. (2007) hypothesise that the egg-bearing stage only modifies the range and duration of emergence. Although Jansson (2008) does not present data on berried versus non-berried females, the sex ratio in the Swedish creel fishery in 2 years of observations (64% males:36% females) is similar to the trawling data (65% males:35% females) from the region.

2.1. Net design

A commercial fishery for *Nephrops* was started in the late 1950s. Before this, the *Nephrops* was considered as an unwanted product caught as by-catch in finfish trawl fisheries. However, as the market demand increased, especially in the south of Europe, the fishing industry started to use seine nets of 55-mm mesh size and wooden otter doors to increase the opening width (ICES, 2004a). Eventually, a conventional trawl came into use, using mesh sizes of 70–80 mm with a rather low headline. This trawl evolved into a so-called scraper, also with a low headline but with wider wings and a heavier construction (ICES, 2004a). Initially, trawls were manufactured using natural fibres and used on muddy areas, but as harder grounds were explored, the net was protected using disc structures of rubber thread through lead rings.

Another improvement in the trawl was the development of the twin-rig trawl by Danish fishermen in the mid-1980s. This came from the tropical shrimp fisheries in the Gulf of New Mexico and was first used in European waters for *Pandalus* prawns but was quickly adopted for *Nephrops* across Europe. The net and, therefore, net drag are smaller in a twin rig compared to a single rig with a similar entrance width (Figure 7.5), with less engine effort required as a result, and the speed can be increased, leading to a dramatic increase in capture efficiency (Sangster and Breen, 1998). Eggert and Ulmestrand (1999) used a factor of 1.7:1 for conversion of twin-rig versus single-rig LPUE in their *Nephrops* economic modelling. There is some dispute within the fishing industry as to the environmental impact of the twin rig compared to the single rig as, in order to maximise contact with the seabed, a 'clump' of chain is positioned on the footrope between the two sets of gear. Single-rig fishers claim that the clump is causing a greater impact upon

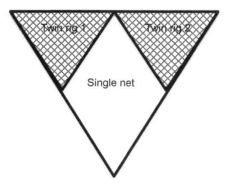

Figure 7.5 Twin-rig set-up reduces net drag. The trawl opening area (illustrated as the horizontal line) is theoretically similar for a single and a twin-rigged trawl, but advantage in reduced net drag is reached as the twin-rigged trawls are shorter why less water friction. *Modified from ICES (2004b).*

the seabed than their gear. To date, there has been no scientific evaluation of this claim.

The discarded portion of catches from *Nephrops* trawl fisheries is a widely recognised problem. The North–East Atlantic trawl fishery for *Nephrops* has the fifth highest discard ratio in the world (Catchpole et al., 2006a, 2008). The problematic high discard rates of under-sized *Nephrops* and by-catches of vulnerable finfishes (North Sea; Catchpole et al., 2005, 2008) have been the main driver for the many improvements in trawl design and the more stringent regulation. Reviews of trawl improvements and the resulting change in the selection pattern of *Nephrops* size and by-catch of other species have already been compiled (ICES, 2004a; Catchpole and Revill, 2008; Madsen and Valentinsson, 2010) and are not repeated herein. The approaches taken are square-mesh panels (SMPs) at various distances from the cod-end (Briggs, 2010; Catchpole et al., 2006b; Drewery et al., 2010; Krag et al., 2008), other 'windows' increasing the mesh size, often in the upper panel (Ingólfsson, 2011; Madsen et al., 2012), structures separating the trawl into horizontal layers using different species-specific escape behaviour within the trawl (Main and Sangster, 1981), and the use of a firm grid to physically sort out larger finfishes through an opening and let the smaller *Nephrops* be sieved in the cod-end (Catchpole et al., 2006b; Drewery et al., 2010; Valentinsson and Ulmestrand, 2008; Figure 7.6A and C). Due to the large amount of by-catch and discard of groundfish, especially cod, in the conventional *Nephrops* trawling (90-mm diamond mesh), the use of a species-selective grid, that is, trawls equipped with a Nordmøre-type firm sorting grid with a 35-mm bar distance and a 70-mm square mesh cod-end has become mandatory in Swedish national waters since 2004 (SwAM, 2004:36). The grid in the upper part of the trawl lets the fish escape through it, while the target species is retained on the bottom of the trawl.

The vulnerability of the *Nephrops* individuals to trawling is highly dependent on their emergence behaviour (Aguzzi and Sarda, 2008; Aguzzi et al., 2003, 2005). Their emergence behaviour from the burrow for feeding or mating is triggered at a specific light intensity, and impacted by depth, season and water parameters resulting in emergence patterns covering diurnal (deep-living stocks), nocturnal (shallow living) or crepuscular (medium depths), and is consistent with the idea that *Nephrops* is a visual feeder. Even within a specific region, the emergence pattern can vary following changes in light and oceanography, which has a profound impact on when the fishers trawl (see Chapter 3).

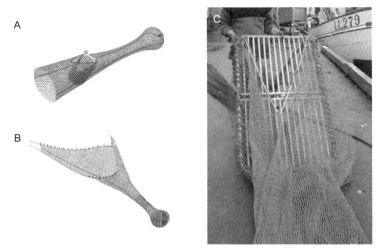

Figure 7.6 Different technical gear solutions to reduce discards (under-sized *Nephrops*) and by-catches (mainly gadoids) in *Nephrops* trawls. (A) A schematic trawl with the Nordmøre sorting grid device and escape window, (B) a schematic trawl with a square-mesh panel (SMP) in the cod-end and (C) photo of the Swedish species-selective trawl with the 35-mm Nordmøre sorting grid with 70-mm square mesh cod-end on a pier in Lysekil, Swedish West coast (photo by Therese Jansson). *Panels A and B from ICES (2004a), produced by FRS, Crown copyright.* (For the colour version of this figure, the reader is referred to the online version of this chapter.)

2.2. Creel types

Creel fisheries for *Nephrops* have now been established for three to four decades in regions such as Western Scotland and the Swedish Skagerrak with a few vessels also in the Kattegat. In the Faroe Islands, creeling for *Nephrops* is the only capture method as trawling is banned. A commercial and recreational creeling fishery occurs along the Norwegian coast, but the literature on these minor fisheries is scarce. Creeling also occurs in the north–eastern part of the Adriatic Sea and pilot creeling operations comparing trawl and creel-capture potentials have been performed for the Central Adriatic Sea (Morello et al., 2009) and the Portuguese coast (Leocádio et al., 2012). Creel trials have also been tested by fishers' organisations, for example, in Northern Ireland (ANIFPO, 2007). The typical creel design is either a one- or a two–chamber-type with a (1) half round transection (flat bottomed with a circular top, D-shaped, Figure 7.7A), which has advantages on board the vessel such as securing the creel pile while the fisher is standing on the deck between hauling and shooting, and staying put in strong currents as the drag is less, or is (2) square–shaped (box type, Figure 7.7B), increasing the inner volume. One or, most often,

Figure. 7.7 Photos of Norway lobster (*Nephrops norvegicus*) creel fishery in Skagerrak. (A) Top left, a single half round creel (D shaped), (B) top right square (box type) single creel, (C) below left, a double chamber square (parlour) creel and (D) below right, creel baited with salted herrings. Photo: Anette Ungfors. (For the colour version of this figure, the reader is referred to the online version of this chapter.)

two entrances with rings of 70–75 mm inner diameter lead towards the bait enclosed within the main chamber (Figure 7.7A and B). The size and air weight differs among the creels with weights ranging from 2.6 to 7.5 kg depending on the diameter of the steel frame (e.g. Carapax Ltd.). Scottish and Swedish creels have entrances on the longest sides, but creels in use in the Adriatic Sea (Croatian or Italian types) have entrances on the shorter side (Morello et al., 2009). Creels of the two-chambered type have an additional chamber at one end (or 'parlour', Figure 7.7C) where the *Nephrops* and by-catch such as crabs gather. A two-chambered creel increases the capture efficiency as seen, for example, for European lobster and brown crab (Lovewell et al., 1988), as the escape potential from the creel is reduced, which is of great value to the fishermen, especially when a longer soaking time, for example, owing to bad weather or weekends is needed. *In situ* filming of the American lobster (*Homarus americanus*) in and around a trap revealed that 94% escaped (Jury et al., 2001). Inter and intra-specific social interactions strongly affect captures. Effects of both sex and size have been demonstrated, that is, where the opposite sex is either attracted or reversed depending on the first individual entering the creel, and also that smaller sizes can be scared away when larger ones have already entered (Watson and Jury, 2013). By-catches of aggressive swimming crabs (*Liocarcinus* spp.), the spider crab (*Hyas araneus*) and the brown crab are often found in creels on the Swedish west coast (Jansson, 2008; Ungfors et al., in preparation), and can depress catches in specific areas or seasons so that fishers need to change area. Devices for decreasing the size of the entrance ring have been developed and are used seasonally.

Improvements for increasing the capture of *Nephrops* are ongoing as fishermen produce their own gears and gear manufacturers update their products. All are aiming to optimise the odour spread of the bait using thin mesh and minimal rope protection around the steel frames, while reducing the wear of the gear. Richardson (1996) found that blinding by high light intensities had no apparent effect on *Nephrops* responses to a baited creel and that catch rates were very low for blind and visually intact animals. Anecdotal knowledge of fishers indicates that the colour of the mesh may be of importance (Ungfors, A., unpublished data). The creel net is composed of a knotted polyamide diamond mesh. However, it is more likely that any colour effect is dependent on the contrast of the gear appearance, given the limited spectral sensitivity of *Nephrops* (see Chapter 4).

Creels are a passive gear in the sense that they are not mobile during the capture process. The creel, however, actively fishes by attracting *Nephrops* to potential food. Bait is sensed by the aesthetasc setae on the antennules, which are involved in chemo orientation (Hallberg and Skog, 2011), but other regions such as chemo-sensitive hairs on the pereopods as well as vision are involved (see Chapters 3 and 4). Bjordal (1986) filmed *Nephrops* creels in Norwegian fiords and found that only 6.1% (15 of 246) *Nephrops* entered the creel (Figure 7.8). The *Nephrops* approached the creel from downstream at angles up to 30° either side from the current. Of the *Nephrops* that approached the creel, 65% made physical contact. It appeared that it was hard for the *Nephrops* to find the entrance (they spent 1–40 min around the creel). Bjordal (1979) filmed various designs of creel and found that a trap with four entrances had better capture efficiency than one-entrance traps. Of 51 *Nephrops* that came into the creel vicinity, 6 entered, one of which escaped. Catching *Nephrops* by creels can be divided into different stages (Miller, 1990) where the attraction stage, which involves long-distance attraction by olfactory stimuli, and the gear stage, which operates in the proximity of the creel, are of importance. As larger male individuals spend a longer time out of their burrows (Briggs, 1988, 1995), they have a larger chance of encountering the odour plume of the bait and starting a search. Bjordal (1986) noted that smaller *Nephrops* were chased away from the creel by conspecifics or other species, or were frightened by the repellent gear, which resulted in a reduced time for finding the entrance. This may be why the creel is selective for larger individuals. A similar pattern of low

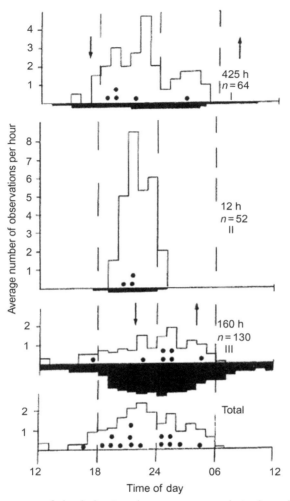

Figure 7.8 Summary of the behavioural responses to a baited creel. The capture efficiency for the creel is estimated based on UW-filming in Norway at three different locations (I Nyleia, II Nærøyfjorden, III Lysefjorden) and showed that 6.1% (15 of 246 approaches, summary below panel) entered the creel (Bjordal, 1986). Filled circles are catches of *Nephrops*, arrows indicate sunset and sunrise, and black field under the x-axis are the number of hours of observation (1–17 hrs).

capture efficiency (5%) was found for 300 h of filming around a creel in Western Scotland (Adey, 2007).

The CPUE (number or weight per trap) of crustacean increases with soak time, that is, the days the creels have been fishing at sea (Bennett and Brown, 1979; Miller, 1990).

In 2005–2006, an observer survey in the Swedish creel fishery (Jansson, 2008; more than 12,000 creels recorded for 26 vessels) found that on average 75% of the capture by weight consisted of the targeted *Nephrops*, 15% of under-sized *Nephrops* (under 40-mm CL) and 10% of fish such as cod and poor cod (*Trisopterus minutus*). This proportion is almost the opposite of the trawl fishery where the target *Nephrops* make up 30% in trawls using the Swedish 35-mm grid (and 70-mm square mesh) or only 15% in trawls without a grid (conventional 90-mm diamond mesh; Valentinsson and Ulmestrand, 2008; Ziegler, 2006). The creels capture larger individuals; the size at first catch is around 27 mm for the creels compared to at least 1 cm smaller CL for trawls. However, the Jansson (2008) creel study found the sex ratio to be equal to the trawl, 65 M:35 F.

Today, oily fishes such as herring or mackerel are mostly used as bait. In Sweden, the herring is salted before use (25 kg salt to 125 kg herring), either fresh or frozen, to allow for longer and easier use. In the Mediterranean, salted pilchard is used as bait (Morello et al., 2009). Smaller herring (50–100 g) are used whole, but larger ones are divided (Figure 7.7D). The herring bait is renewed at every hauling occasion, meaning every day or every second day, or occasionally at longer durations. Devices such as bait bags (Archdale and Kawamura, 2011) increase the length of time that the bait is attractive. This might be caused by the fact that the captured animal is unable to physically tear the bait apart and eat it, or that it reduces access for scavenging amphipods and isopods. In areas with a very high scavenger activity, such as the Adriatic Sea, up to 40% of bait can be consumed within 12 h and over 80% within 24 h (Morello et al., 2009). Bjordal (1979) evaluated different baits and found no statistical difference between the attractiveness of herring and mackerel, squid and trout pellet, or mackerel and trout pellet to *Nephrops*.

There have been several surveys and experiments examining which constituents of bait are the most attractive to crustaceans (Carr and Derby, 1986; Derby and Sorensen, 2008; Mackie, 1973). Some of the known and described feeding attractants in crustaceans are amines and certain amino acids (Levandowsky and Hodgson, 1965; Zimmer-Faust et al., 1984), with glycine and trimethylamine oxide being the most potent ones. Glycine is a free amino acid commonly found in invertebrate tissues and excretory products. Trimethylamine oxide is commonly found in plant and animal tissues. Other amino acids such as betaine have also been described to be an effective feeding stimulant (Hayden et al., 2007), and in some commercially available feeding attractants, a mix of up to nine different amino acids is used (Nunes et al., 2006). Other possible chemical attractants that could be used in bait would be sex pheromones, especially those that attract mating partners (Hardege et al., 2011), but besides the fact that they have not been tested for these purposes in the field yet, they would be seasonally restricted.

Presently, bait is one of the problems the creelers have to overcome, as the cost to buy it is around 5–10% of the value from the *Nephrops* first-hand sale. Around 1.1 kg of salted herring is used per kilogram of *Nephrops* landed (Ziegler, 2006). The herring is often caught by the fishermen themselves or bought from pelagic trawlers, and the baiting of creels with salted herring was estimated to be responsible for 10% of the total fuel use in the creel fishery (Ziegler, 2006). Artificial bait in pelleted, dried, or gelatinized form has been tried for other crustacean species such as brown crab (Dale et al., 2007), European lobster (Mackie et al., 1980), spiny lobster (*Panulirus interruptus*; Chanes-Miranda and Viana, 2000), and sand crab (*Ovalipes punctatus*; Archdale and Kawamura, 2011) to mention a few, and the preliminary results for *Nephrops* are promising (Ungfors et al., in preparation). Hancock (1974) reviewed the potential of using by-catch repellent bait ingredients to increase target captures. Baiting has also been suggested as potentially adding to the production of crustacean species, for example, to the American lobster in Maine (Grabowski et al., 2010; Saila et al., 2002) and the western rock lobster (*Panulirus cygnus*) in Western Australia (Waddington and Meeuwig, 2009). Bait has been calculated to contribute up to 13% of lobster food requirements over the whole ecosystem. This factor could contribute to the superior condition of *Nephrops* on creel bottoms compared to trawled ones as demonstrated (Eriksson, 2006).

Creel fishing has been shown to have no, or very little, disturbance effect on benthic fauna (Bergmann et al., 2002; Eno, et al., 2001). The survival of crustacean discard is also thought to be high (Bergmann and Moore, 2001), one reason why creeling has a light ecological footprint. Investigations into the stress of *Nephrops* experiencing reduced surface salinity during the trawling lift (or creeling) in the Kattegat have been carried out using exposure to 15 PSU water for 6 min and air exposure in different temperatures (5 or 15 °C; Harris and Ulmestrand, 2004), this resulting in increased mortality ranging between 25% and 42% compared to 0–8% for the control animals (full salinity treatment with or without air exposure). One negative aspect of creeling is the potential for ghost fishing. However, it has been found that the fishing potential for *Nephrops* creels was low after the bait had been consumed (Adey, et al., 2008). In addition, fishers state that it is seldom that they lose creels, and if they do, they can generally find them again.

The small size of *Nephrops* in the area and the intense activity of scavengers feeding on bait in the Central Adriatic Sea resulting in low capture rates has prevented the establishment of a creel fishery in this area (Morello et al., 2009; Panfili et al., 2007). In addition to this, the long distance to the fishing ground is not ideal for smaller creeling vessels (Morello et al., 2009).

3. FISHERY MANAGEMENT

There are several authorities in the North Atlantic that have jurisdiction over *Nephrops* fisheries. Norway, Iceland, and the Faroe Islands manage their fisheries independently, while the European Union manages *Nephrops* fisheries collectively under the TAC scheme within the Common Fisheries Policy (CFP). Input (i.e. effort control, closed seasons, minimum mesh size) and output measures (i.e. TAC quota, MLS, catch composition) are all used for Nephrops fishery management in a variety of combinations. It should be noted that the wording of 'total allowable catch' is generally misleading, as it is the landings that are the governing metric rather than the catch (which often contains individuals that cannot be landed due to size and/or sex limitations).

At its most basic, the purpose of fishery management should be to ensure that fishing activity does not induce permanent negative change to the stocks being targeted (or the habitat upon which they depend). The vast majority of commercially exploited stocks have at least one dispersive phase and, therefore, separation into distinct and meaningful stock units is often a subjective exercise. In the case of *Nephrops*, the post-settled individuals display a limited capacity for migration (Chapman and Rice, 1971) and the substrate requirements mean that there are fairly well-defined boundaries for the fishing grounds. Larval drift through successive generations may ensure that the genetic structure across a number of habitat patches is relatively uniform. Stock units can, therefore, be considered to be geographically large and it is likely that periodic local depletions are not necessarily disastrous. The time-scale for fishery management, including the economic and social viability of fisheries, is typically 1–5 years and is decreasing with the increasing use of technology and co-management. Therefore for management purposes, the spatial scale of 'stock' identity should be reduced to match (Kraak et al., 2012).

For the purposes of fishery management and reporting, the Northern Atlantic is divided into a number of geographic zones. The International Council for the Exploration of the Seas (ICES, the umbrella scientific organisation on the eastern side of the North Atlantic) divides the sea into a grid of rectangles with dimensions of 1° longitude and half a degree latitude.

These rectangles are then grouped into nested classes of sub-divisions and divisions for the purposes of stock assessment and management.[2]

1. Iceland: Iceland sits within ICES sub-division Va and has independent control of waters out to 200 nautical miles. All *Nephrops* stocks in this sub-division are found within these territorial waters. The Icelandic management system is unique in Northern European waters in that there is a total ban on the landing of females. In addition to this, there is a quota system in force, which integrates the 10 identified grounds. TAC quotas are determined annually by the Icelandic Fisheries Ministry and they generally (but not always) are in line with the TAC recommended by the Marine Institute. Reported landings typically exceed the TAC.

2. Norway: There is relatively little restriction placed upon Norwegian fishers fishing in Norwegian waters for *Nephrops*, an MLS being the only quoted management measure (ICES, 2012a).

3. European Union: At the time of writing, *Nephrops* are the only crustacean fisheries to fall within the TAC regime of the European CFP. Seven management areas under EU jurisdiction are in force, with an additional allowance for landings by EU vessels operating in the Norwegian sector:

   a. EC waters of IIa and IV (essentially the EU zone of the North Sea),
   b. IIIa; EC waters of IIIb (EU zone of the Skagerrak and Kattegat),
   c. VI; EC waters of Vb (effectively the West coast of Scotland),
   d. VII (Irish Sea, Celtic Sea, and West coast of Ireland),
   e. VIIIabde (Bay of Biscay),
   f. VIIIc (Northern Spain),
   g. IX, X, CECAF 34.1.1(1) (essentially the Iberian Peninsula).

For each of the management areas, a landings quota would be agreed upon between the Member States, which would then be divided among the countries according to a fixed allocation key. The allocation keys, referred to as Relative Stability and enshrined within the Treaties establishing the European Union, were determined on the basis of the historical (1994–1996) track record of landings per species, country, and management area and were introduced in 1999. The use of a fixed key for national allocations has both benefits and drawbacks. It would be unfeasible, and arguably impossible, to re-negotiate national allocations for each species under

---

[2] The counterpart to ICES covering the North-West Atlantic is NAFO, but there are no stocks reported in these waters.

the TAC scheme on an annual basis and it provides a degree of stability in the fleet composition (by restricting the ability of new countries to enter a fishery). National fleet composition and their markets are not static and as a consequence of the TAC setting decisions, there would follow a series of negotiations at the national level to trade their quota allocations for other species or even different commodities.

Given the limited dispersal abilities of *Nephrops*, the ICES sub-division scale or quota setting often encompasses multiple *Nephrops* stock units. Fisheries scientists therefore undertake stock assessment at a finer scale and need to balance the geographical ranges of the non-migratory, post-settled population with the potential for larval distribution and the scale at which reliable data are available. The finest geographic scale at which fisheries statistics (i.e. landings and effort) are routinely recorded in European waters is, therefore, the ICES rectangle for assessment purposes, *Nephrops* stocks are defined as a collection of rectangles and known as FUs. As of 2012, there were 34 FUs defined in the North-East Atlantic, with a further 8 in the Mediterranean. Of the 34 North Atlantic FUs, 29 lie within EU waters (see Table 7.1). The units are distributed as follows: 11 FUs within the North Sea, 3 FUs on the Scottish West coast, 3 in the Celtic Sea, 4 west of Ireland, 2 in the Irish Sea, 1 FU each in Iceland and Faroe Island, 4 in the Bay of Biscay and 5 at the Iberian Peninsula (Table 7.3).

Table 7.3 Functional units and stock status for Norway lobster (*Nephrops norvegicus*)

Functional units (FUs) or groupings		Stock status based on ICES assessment
1	Iceland	
2	Faroe Islands	
3	Skagerrak	Units now are considered to be one stock. F decreasing and below F_{MSY}. Bmsy undefined, but all indices suggest the stock is exploited sustainably
4	Kattegat	
5	Botney Gut	No reference points defined. Data-limited approach has been applied, indicating that stock is exploited at a sustainable level
6	Farn Deeps	F above F_{MSY} and B below B_{trig}
7	Fladen Ground	F below F_{MSY} and B above B_{trig} for time series. Abundance does appear to be declining

Table 7.3 Functional units and stock status for Norway lobster (*Nephrops norvegicus*)—cont'd

Functional units (FUs) or groupings		Stock status based on ICES assessment
8	Firth of Forth	F above F_{MSY} and has fluctuated just above the level for a period of time. B above B_{trig} for the time series
9	Moray Firth	F above F_{MSY} for 2011 after a period at or below. B above trigger throughout time series. Stock slightly declining
10	Noup	No reference points defined. Data-limited approach has been applied, indicating that stock has been exploited slightly above the recommended level
32	Norwegian Deep	Reference points are not defined for this stock. Data-limited approach has been applied. F appears to be below possible reference points and B is stable. Landings appear to be at a sustainable level
33	Off Horns Reef	Reference points not defined. Data-limited approach has been applied indicating that landings are at or around desirable levels. B appears to be increasing although F is unknown
34	Devil's Hole	Reference points not defined. Data-limited approach has been applied, indicating that landings may be slightly above desirable levels. F is unknown, and B appears to be declining
11	North Minch	Fishery exploited sustainably. B above trigger for time series and F below target
12	South Minch	Fishery exploited sustainably. B above trigger for the time series and F below target
13	Firth of Clyde and Sound of Jura	F above sustainable yields but B well above trigger for the time series. Sound of Jura—F below F_{MSY} and B undefined due to short, patchy time series
14	Irish Sea East	F below F_{MSY}. B_{trig} undefined due to short time series. Stock is being exploited sustainably
15	Irish Sea West	F above F_{MSY}, although it has fluctuated around it for the time series. B above trigger throughout time series

Continued

Table 7.3 Functional units and stock status for Norway lobster (*Nephrops norvegicus*)—cont'd

Functional units (FUs) or groupings		Stock status based on ICES assessment
16	Porcupine Bank	Advice revised in a year to account for first UWTV survey. F is thought to be below possible reference points, and B is increasing and is above the average of the time series
17	Aran Grounds	F below F_{MSY}. B_{trig} undefined, and B lowest in time series
19	Ireland SW and SE coast	F below F_{MSY}. B_{trig} undefined with no trend in B due to short time series
20 + 21	Celtic Sea—Labadie	Reference points not defined. Data-limited approach has been applied indicating that landings have been within sustainable levels in recent years. Trend indicates F is decreasing. Biomass is unknown, but LPUE has decreased in the last 2 years, suggesting declining abundance
22	Celtic Sea—The Smalls	F is below F_{MSY} after a period above. B_{trig} is not defined, but the trend indicates increasing B after a period of stability
23 + 24	Bay of Biscay	Data-limited approach has been applied. Qualitative evaluation suggests that F is above possible reference points with B increasing. Catch possibly slightly above sustainable levels (although precautionary)
25	North Galicia	No assessment for 2012. Qualitative evaluation suggests declining B with F unknown. Zero catch recommended. Recovery plan in place but not assessed by ICES
31	Cantabrian Sea	No assessment for 2012. Qualitative evaluation suggests declining B with F unknown. Zero catch recommended. Recovery plan in place but not assessed by ICES
26 + 27	West Galicia and North Portugal	No assessment for 2012. Qualitative evaluation suggests declining B with F unknown. Zero catch recommended. Recovery plan in place but not assessed by ICES

Table 7.3 Functional units and stock status for Norway lobster (*Nephrops norvegicus*)—cont'd

Functional units (FUs) or groupings		Stock status based on ICES assessment
28 + 29	South-west and South Portugal	Data-limited approach has been applied indicating that catch is at or about sustainable levels. Both F and B appear to be declining. Recovery plan in place but not assessed by ICES
30	Gulf of Cadiz	No assessment for 2012 although the data-limited approach has been applied and suggests that catch may be slightly above sustainable levels. Qualitative evaluation suggests declining B with F unknown. Recovery plan in place but not assessed by ICES

Data from ICES (2012b).

3.1. Management and policy making in the European Union

The ICES provides independent scientific advice to clients including the European Commission through the co-ordination of information from scientists from the 20 constituent nations. A schematic graph of the authorities and associated groups for *Nephrops* management within the EU is depicted in Figure 7.9. Members include those not within the European Union (e.g. Norway, Iceland, Faroes). The stock assessment advice for the major stocks of *Nephrops* is provided either on an annual or on a biannual basis with the most recent update provided in June 2012 (ICES, 2012b). The advice is timed to feed into the annual TAC and quota negotiations, with initial release in June and any required revisions between September and the beginning of December. In addition to this annual cycle, symposia and other workshops are organised throughout the calendar year.

Scientific advice submitted to the European Commission is subsequently reviewed by an internal group, the Scientific, Technical and Economic Committee for Fisheries (STECF). This body was established in 2002 (Council Regulation EC 2371/2002[3]), with an updated process in 2005 (Commission Decision 2005/629/EC[4]), and comprises 30–35 members who are experts in fisheries and marine biology from the member states. The annual

[3] Council Regulation (EC) No. 2371/2002 of 20 December 2002 on the conservation and sustainable exploitation of fisheries resources under the Common Fisheries Policy. Official Journal of the European Communities, L358/59.

[4] Commission Decision of 26 August 2005 establishing a Scientific Technical and Economic Committee for Fisheries (2005/629/EC). Official Journal of the European Union, L225/18.

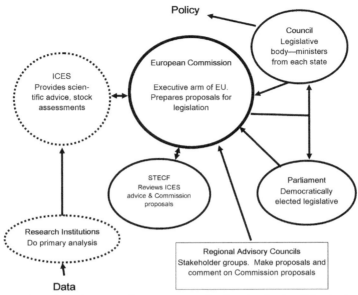

Figure 7.9 Flow chart depicting the process involved in developing fisheries legislation within the European Union.

report of STECF (e.g. STECF, 2012) containing recommendations for fishery management is passed on to the European Commission, which then generates proposals for the different geographic EU EEZ areas. Expert Working Groups with at least two of the STECF members can focus on different topics to prepare the recommendations. Both ICES and STECF comment on legislative proposals throughout the year, including long-term management plans (LTMPs).

The European Commission then takes the advice from ICES and STECF and makes proposals. The Commission is the executive arm of the European Union and has the sole responsibility of making proposals for legislation, a duty known as 'right of initiative'. The route that the legislation then takes is dependent on whether it is subject to a process known as co-decision or not. In the annual TAC and quota negotiations, normally held at the end of the calendar year and setting fishing opportunities for the following year, it is the Council of the European Union (formerly known as the Council of Ministers) that adopts the legislation. Other fisheries legislation, generally legislating for more than 1 year, must go through the process of co-decision where the European Parliament shares legislative power equally with the Council. If the two institutions cannot agree, the issue is passed on to a conciliation committee, comprising an equal number of members from both, to seek agreement. The process of co-decision, although more democratic, is somewhat unwieldy due to the number of people involved and the political nature of the Parliament. To date, the amount of time taken for legislation to pass through the process has been excessive, leading to some frustration among the stakeholders, particularly in relation to long-term management plans (LTMPs). That said, the opportunity for lobbying Members of the European Parliament has not been missed by some elements and some sectorally unpopular proposals have been withdrawn due to this.

The phraseology of ICES advice has evolved considerably in the last decade and there has not been consistent advice for a maximum biomass of landings. Phrases such as 'no increase in effort', 'reduce catches' and 'no increase in catches' have been used for FUs without a TV survey and it was left to the European Commission to convert this qualitative advice into the quantitative tonnages that the TAC system required. Even without these complications, the quota negotiations at the ministerial level means that changes in ICES advice do not necessarily translate into changes in TAC.

3.1.1 Regional Advisory Councils

Regional Advisory Councils (RACs) were established through a Council Decision as part of the CFP reforms in 2004 (European Union, 2004) and are, in the most part, geographically based. The RACs were intended to enable stakeholders to become more closely involved in the decision-making processes relating to fisheries by making recommendations and suggestions to the Commission and to national authorities. The RACs tend to form the main channel of consultation with stakeholders and are expected to have an expanded role under the 2013 CFP reform. The two RACs with the most significant interest in *Nephrops* are the North Sea RAC and the North-western Waters RAC.

3.1.2 North Sea Regional Advisory Council

The work of the North Sea Regional Advisory Council (NSRAC) is delivered by three working groups namely Demersal, Skagerrak and Kattegat, and Spatial Planning. These groups meet to discuss current and emerging topics and to develop advice and policy on behalf of the NSRAC membership. The working groups meet three–four times a year and enable a wide range of people to become involved in the NSRAC activities, including scientists, fishers, environmental specialists, economists and others. Each working group may be supported by a number of Focus Groups. Focus Groups are smaller groups, which are set up to address a specific topic. Focus Groups are flexible in their approach, drawing in representatives and experts from a number of sources. The *Nephrops* Focus Group (NFG) supports the Demersal Working Group and is developing an LTMP for the species.

3.1.3 Northwestern Waters Regional Advisory Council

The Northwestern Waters Regional Advisory Council (NWWRAC) also delivers policy advice through a number of working groups but on a sub-regional rather than thematic basis, namely, West of Scotland, West of Ireland and the Celtic Sea, the English Channel and the Irish Sea (NWWRAC, 2013). The working groups meet four times a year and are complemented by a Horizontal Working Group, covering issues of a more general nature, and subject-specific Focus Groups. The RAC does not currently have a specific NFG, but advice on the species is an active topic for informal groupings within the RAC and within the working groups.

3.1.4 The LTMP and the rebuilding plans

Although still under development, the NSRAC NFG has been working on the construction of an LTMP for the species since 2009. The group, comprising industry representatives from a number of North Sea nations and supported by CEFAS and Marine Scotland, agreed that the plan needed to be more than a harvest control rule and that it should take into account the differences in the national and sub-national fleets and the characteristics of their fishing grounds. The ability to protect discrete FUs on an individual basis is at the heart of the plan, but FU TACs or effort management are not considered to be the only means of delivery. In order to deliver management measures appropriate for the individual FUs, the plan consists of overarching objectives complemented by area-specific rebuilding of fishing plans that are designed to be implemented when required but can be revoked when any threat to the FU is deemed to have diminished sufficiently. The implementation of measures in each plan would be initiated at the point when the biomass or fishing mortality breaches a trigger that buffers the relevant reference points and promotes a flexible and proactive form of fisheries management.

The measures included in the plan were initially identified through the Focus Group but have since undergone significant stakeholder consultation to assess their acceptability, in terms of both efficiency and acceptability to the fleet. The latter point is important as fisher buy-in or, to a certain extent, ownership of measures increases the likelihood of compliance (Eggert and Ellegard, 2003; Gutiérrez et al., 2011), enhancing their efficacy and reducing the cost of enforcement. The first round of consultation was undertaken in the summer of 2010 and was general in its approach, taking account of a number of events throughout the North Sea coast of Scotland and North-East England. The results from the events were varied, depending on the fishing area and, more importantly, the characteristics of the consultees. As is usually the case (Johnson and Prime, 2009), the preferences of the more artisanal fishers varied significantly from those of the more nomadic or larger vessels. The second round, initiated in summer 2012 (Bailey et al., 2012), is in progress and has been more focussed on the specific management measures applicable to each area. These include, although are not limited to, gear restrictions, horsepower restrictions and the use of 'of which no more than' quotas.

3.2. Regulation of the *Nephrops* fishery in the European Union

The institutions and basic processes of the high-level policy making for the management of *Nephrops* in the European Union have been described in a previous section (3.1). Of the regulations that come from the process, there are a variety that are applied to *Nephrops* fisheries, some of which are specifically designed to manage the *Nephrops* fisheries and others that manage the stocks associated with them. Other regulations aim to set general rules for fisheries of specific types, often described by mesh size. These regulations are transposed into Member State legislation and are often complemented by bespoke domestic legislation, some examples of which are described here.

The basic conditions of any fishery within European waters are described within Council Regulation, 850/98 and its subsequent amendments.[5] Council Regulation, 850/98 describes fisheries by area of capture, target species and mesh size and defines the minimum quantity of the target species that should be retained by a vessel. Further, the regulation determines the species and the quantities of the species that can be landed alongside the target. These principles are intended to prevent larger fish species being targeted by smaller mesh fisheries and to ensure the protection of juveniles. This regulation should have been repealed and replaced during 2012, but the

---

[5] Council Regulation (EC) No. 850/98 for the conservation of fishery resources through technical measures for the protection of juveniles of marine organisms (europa.eu, ID number LEX-FAOC018268).

co-decision process has delayed agreement with the new regulation. The regulations also define specific characteristics of the fishing gear, such as twine thickness and positioning of SMPs, technical specifications that affect the selectivity of the gear in terms of both *Nephrops* and by-catch species. In many cases, Member States enhance these regulations with complementary use of domestic legislation, such as Statutory Instruments (UK and IRE). One example of such use is in Scotland where instruments, applicable to all UK vessels in Scottish waters and all Scottish vessels in UK waters, are used to legislate for a variety of measures including provisions increasing the selectivity of the gear (SSI 165/2009[6]) and legislating against multi-rigged gear (SSI 13/2007[7]).

Management and recovery plans also have the potential to affect the *Nephrops* fishery. Currently, there are no management plans applicable to the species, although the draft LTMP for North Sea *Nephrops* is discussed earlier in this chapter. There are, however, two recovery plans that impinge on the regulation of *Nephrops* fisheries. Council Regulation 2166/2005 amends 850/98 to legislate for the recovery of *Nephrops* in the Cantabrian Sea and the Western Iberian Peninsula.[8] The plan assumes a direct relationship between fishing effort and the level at which the TAC is set and aims to return the stocks to safe biological limits within 10 years by reducing the fishing mortality by 10% year on year. Effort restrictions, *per se*, are not in place but are effected by appropriate reductions in the available TAC. ICES have not evaluated the plan. The Cod Recovery Plan (Council Regulation 1342/2008, 2008[9]) applies to fisheries in European Union waters of ICES sub-divisions III, IV, VI and VIIa. The plan does not relate directly to *Nephrops* but is applied to fisheries that encounter cod as a by-catch and includes many of the *Nephrops* fisheries on that basis. The plan assumes a 1:1 relationship between fishing effort and cod mortality and, as a result, restricts the fishing effort available to sections of the fleet in a move to increase the biomass of cod to above B_{trig}. The regulation, replacing Council

[6] SSI 2009 No. 165, The Sea Fish (Specified Sea Areas; Regulation of Nets and Other Fishing Gear; Scotland) Amendment Order 2009. www.scotland.gov.uk

[7] SSI 2007 No. 13, The Prohibition of Fishing with Multiple Trawls (No. 2; Scotland) Amendment Order 2007. www.scotland.gov.uk

[8] Council Regulation (EC) No. 2166/2005 establishing measures for the recovery of the Southern hake and Norway lobster stocks in the Cantabrian Sea and Western Iberian Peninsula and amending Regulation (EC) No. 850/98 for the conservation of fishery resources through technical measures for the protection of juveniles of marine organisms. (europa.eu, ID Number LEX-FAOC061052)

[9] Council Regulation (EC) No. 1342/2008 establishing a long-term plan for cod stocks and the fisheries exploiting those stocks and repealing Regulation (EC) No. 423/2004. (europa.eu, ID number LEX-FAOC084627)

Regulation 423/2004,[10] made it possible for the Member States to manage the effort allocated to its fleets (as opposed to management at EU level), and it is this ability that has influenced the *Nephrops* fleets in III, IV, VI and VIIa, and their legislators more than anything else in recent history. The Swedish fleet in the Skagerrak and Kattegat are awarded for using a species–selective trawl equipped with a 35-mm grid in combination with a square mesh cod-end. These incentives, and also the potential trawl area, were reduced for Swedish fishers not using the improved trawl. In Sweden, around 100 over 10-m vessels have permits for trawling. There are 94 permits for trawling with the grid and 100 for trawling without it. Under 10-m vessels are currently exempt from CRP effort restrictions. In 2010, incentives became similar for the Danish fishers to use an SMP device in the cod-end (SELTRA 300; Madsen et al., 2012). Since 2007, Danish fishers have had a vessel quota system where each fisher is allocated an annual share of the Danish quota (ICES, 2012b; IIIa). The Scottish Conservation Credits Scheme requires additional selectivity measures in return for additional days at sea (Marine Scotland, 2012). The Scottish fleet in 2012 was required to operate highly selective gear for the entire year in return for a maximum of 200 days at sea. Many other Member States are operating similar schemes.

The selectivity measures outlined in the previous paragraph are mainly in response to the CRP, but, as a general rule, there is a move to discard reduction of other species as well and all commercial fisheries within the EU will be subject to a discard ban by 2019. This means that fisheries using relatively small mesh to catch a target species will have to make changes in their gears to significantly reduce the incidental by-catch of species that fishers cannot legally land or for which they cannot source quota.

As mentioned previously, *Nephrops* is currently the only crustacean subject to the TAC regime. The overall TAC is agreed on an annual basis, generally at the December Council of Ministers, and is not subject to co-decision. The TAC for each area is then distributed to Member States on the basis of Relative Stability. Member States have a variety of quota management mechanisms. Some, such as Denmark and the United Kingdom, use versions of individual quotas and allocate the quota to individual vessels through Producer Organisations. Others, such as the Republic of Ireland, use a pool system with monthly allocations that are the same for each vessel within a certain group.

---

[10] Council Regulation (EC) No. 423/2004 establishing measures for the recovery of cod stocks. (europa. eu, ID Number LEX-FAOC041519)

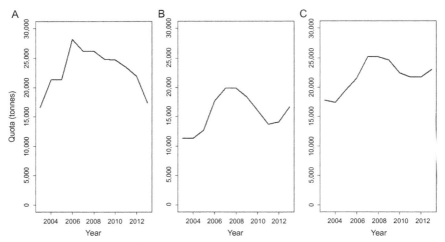

Figure 7.10 Overall TAC for Norway lobster *Nephrops* in ICES subdivision (A) IV, (B) VI and (C) VII. (For the colour version of this figure, the reader is referred to the online version of this chapter.)

The TAC for *Nephrops* can be variable and, as noted previously, is managed at an ICES sub-area level. This means that a significant shift in the abundance of one of the larger FUs can have a significant impact on the overall quota and, therefore, the management of the quota. Figure 7.10 shows the changes in TAC in Area IV for the years 1983–2013. As the figure shows, the Area IV TAC (encompassing all FUs) has declined in recent years, broadly following the declines in abundance, as measured by the UWTV surveys, that have been observed in FUs 7, 8 and 9 over the preceding four years. The largest FU within the North Sea is the Fladen area (FU 7), and the ICES advice for this unit drives much of the overall TAC set for the North Sea by the European Union. ICES advice for 2013, published in 2012, stated that catches from FU 7 should be no more than 10,000 tonnes, a reduction of 4100 tonnes from the 2012 advice, considerably impacting the fishing opportunities available to the North Sea fishermen. The overall TAC changes for the 10 year period are also illustrated for Areas VI and VII (Figure 7.10).

The issue of FU management of *Nephrops* has been touched upon in several parts of this chapter. The developments within this area will undoubtedly significantly influence the *Nephrops* fisheries in the EU as a whole. The only real example of individual FU management is in FU 16, the Porcupine Bank in Area VII. FU 16 is subject to additional regulation in the form of the use of a separate 'of which no more than' quota and a seasonal closure.

In 2010, the area was considered to be in decline and an annual closed area was introduced from 1 May to 31 July at the request of industry through NWWRAC. In 2011, a separate quota, an 'of which no more than' quota, was introduced to further protect the area. The individual TAC was unpopular but, combined with the closure, was deemed to have the potential to demonstrate a real move towards the proactive management of individual fisheries, provided it had the same principles as those underlying the work of the NSRAC LTMP, including flexibility. This type of regulation needs to be flexible and to be revoked when the stock no longer needs the level of protection. NWWRAC (2012) proposed a rule to enhance the efficacy of the closed area, associated with the available recruitment and abundance indices, to be applied to determine the length of the closure. As recruitment and abundance are fairly good, the closure was reduced to 1 month's duration for 2013. The RAC proposed a similar rule for the use of the 'of which no more than quota' which, although not adopted at the 2012 December Council of Ministers, is hoped to be implemented in the near future.

3.3. Mixed fishery issues

There are limits to the extent to which demersal trawl gear can be made to target specific species, and *Nephrops* trawls are no exception. A wide range of finfish species co-occur on the muddy substrates inhabited by *Nephrops*, and a wide range of commercially important species are observed in the catches including whiting, dab (*Limanda limanda*), haddock, plaice, cod, hake and other crustacean species. Discards of blue whiting and horse mackerel can be significant in the Bay of Biscay (ICES, 2012b). In South Portugal, the crustacean fleet also generates high discard rates, which can reach over 70% per weight (Monteiro et al., 2001). Most of the discarded species have little or no commercial value (e.g. the blue whiting and boarfish *Capros aper*). In addition to the commercially landed species, there are a large number of other species that have appeared in trawls targeting *Nephrops*: the English observer scheme has recorded 119 different species in *Nephrops* fisheries across the North and Irish Seas. Species subject to quota management will only be landed where the vessel has available quota, and where the sizes are above the MLS regulations. Non-quota species will only be landed where there is a sufficient market.

The mixed nature of these fisheries has been a mixed blessing, with some fishers regarding a whitefish by-catch as an economically vital part of their operations. In some sectors, however, this was offset by the imposition of by-catch limits and restrictions to the number of days at sea, as noted in pre-vious sections. One of the major influences on the activities of the trawl-based *Nephrops* fleets in the North and Irish Seas in the last decade has been the state of the cod fisheries in those regions. The problem was compounded by the effort regulations for different mesh size categories. *Nephrops* trawl fisheries predominantly use mesh sizes between 80 and 90 mm; however, this size mesh will also retain finfish down to quite small (and therefore young) sizes. There was a period between 2002 and 2004 during which effort was transferring out of the larger meshed (over 120 mm) trawl fisheries into the 80–99 mm categories as the 'days at sea' effort restriction was less prescriptive for this sector. As a measure for protecting the cod stock, this was obviously counter-productive, due to the higher catch rates of under-sized individuals by this smaller mesh size and the effort transfer was halted. In 2010, ICES started producing mixed fishery forecasts, in which the catch composition of the various fleet components were analysed and future catches predicted based upon historical catch patterns and the var-ious management objectives for the species complexes. For example, if the availability of cod quota was the most limiting factor and fisheries were required to stop all activities once the quota was exhausted, then the fleets catching the most cod would be required to cease activity ahead of the fleet sectors with little or no cod catch. Alternatively, if fishing activity was to continue until all quotas were exhausted, then the model could look at the potential for changes to discarding practice resulting from management restrictions on landings (ICES, 2012b; WGMIXFISH).

3.4. Minimum landing size

The MLS for *Nephrops* is, in general, 25-mm CL; or 85-mm total length, (TL), except for the sub-division VIa (FUs 11–13), VIIa (FUs 14–15), VIII (Bay of Biscay) and IX (Iberian Peninsula), which has an MLS of 20-mm CL (70-mm TL). Also, the MLS is different for Skagerrak and Kattegat, it being 40-mm CL (130-mm TL; Council regulation (EC), 850/98; Table 7.1, Figures 7.2 and 7.11). Variations in SOM between FUs may have important implications for fisheries management, particularly if mesh sizes are set to avoid the capture of immature individuals. There are different methods and techniques for investigating the maturity in female and male *Nephrops*. The histological approach in female *Nephrops* uses the presence/absence of spermatophores to confirm that individuals have mated, or the maturity stage of the ovaries (where stage 3 and above are considered as mature),

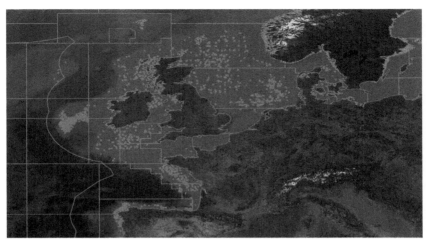

Figure 7.11 Minimum landing sizes (MLS) for Norway lobster (*Nephrops norvegicus*) can be checked with a specific gauge. The MLS for the regions are 25-mm CL (or over 85-mm total length) in the North Sea (IV, FUs 5–10), around Ireland (FUs 16–22) and the Norwegian Deep (FU 32), and 20-mm CL (or over 70-mm total length) on the West coast (VIa, FUs 11–13), the Irish Sea (VIIa, FUs 14–15) and the Bay of Biscay (VIII), and the Iberian Peninsula (IX); for Sweden, and Skagerrak and Kattegat (FUs 3 and 4), it is 40-mm CL (>13 cm total length). Photo of a Carapax gauge: Anette Ungfors.

which is an indication of the physiological maturation. Maturity diagnostics for male *Nephrops* investigate the presence/absence of spermatophores in the vas deferentia, showing the grade of physiological maturation. The morphological approach determines maturity in female *Nephrops* using the abdomen width as an indicator of maturity, while in male *Nephrops*, the appendix masculina or the claw propodus length are used (McQuaid et al., 2006).

The results of the ICES (2006) SOM analyses showed differences in the female size at maturity for different stocks, although there also appeared to be some variability within stocks, especially between years (Figure 7.12, Table 7.4). SOM values are often given as different proportions of the stock being mature based on different characters (see later), and often the length is given where 50% of the individuals (L_{50}) have reached a mature stage. In particular, the southern Iberian stocks show a much higher size at maturity (L_{50} 28–38 mm CL), re-emphasising the idea that the biological characteristics of these deep-water stocks more closely resemble those of the Mediterranean stocks rather than those of the other North Atlantic stocks. L_{50} for the North Sea, western waters and the Bay of Biscay (FUs 5–24) are of a similar size with no specific geographical pattern (L_{50} 21.7–25.5 mm CL),

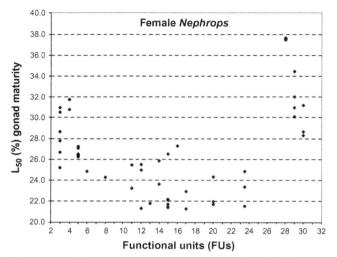

Figure 7.12 Overview of the size at onset of sexual maturity (SOM) for female Norway lobster (*Nephrops norvegicus*). The carapace length where 50% of the stock is mature (L$_{50}$), based on gonad maturity staging. In general, females from FUs 3–4 (Skagerrak and Kattegat) and FUs 25–30 (Iberian Peninsula) become mature at a larger size than females from the North Sea, Western Scotland, Celtic Sea, Irish Sea and around Ireland—Bay of Biscay. *Redrawn from ICES (2006).*

Table 7.4 Female Norway lobster (*Nephrops norvegicus*) size at onset of sexual maturity (SOM) based on the biological characters in use from different geographical areas

Area	Character	L$_{50}$ or inflexion point (mm CL)	References
Histological character			
Skagerrak	Ovary maturation stage (>III)	28.5	Eggert and Ulmestrand (1999)
Firth of Clyde	Ovary maturation stage (>III)	22.6–33.5	Tuck et al. (2000)
Irish Sea	Ovary maturation stage (>III)	22.9	McQuaid et al. (2006)
Skagerrak	Ovary maturation stage (>III)	30.5	ICES (2006)
Scotland		23.3–25.5	
England		24.8	
Ireland		21.7–24.3	
France		21.5–24.8	

Table 7.4 Female Norway lobster (*Nephrops norvegicus*) size at onset of sexual maturity (SOM) based on the biological characters in use from different geographical areas—cont'd

Area	Character	L_{50} or inflexion point (mm CL)	References
Portugal		37.6	
Spain		28.6–31.2	
Strait of Sicily	Ovary maturation stage (>III)	30.9	Bianchini et al. (1998)
Algarve	Ovary maturation stage (>III)	30	Relini et al. (1998)
Alboran		36	
Catalan		30	
Ligurian		32	
Tyrrhenian		32	
Adriatic		30	
Euboikos		33	
Irish Sea	Spermatheca	21.5	McQuaid et al. (2006)
Morphological character			
Irish Sea	Abdomen width	23.2–27.6	McQuaid et al. (2006)
Firth of Clyde	Abdomen width	21.4–34.6	Tuck et al. (2000)
Functional maturity			
Firth of Clyde	Ovigerous, smallest	22–32	Tuck et al. (2000)
Off Portuguese coast	Ovigerous, smallest	36	Ayza et al. (2011)

but the SOM at adjacent Skagerrak and Kattegat (FUs 3 and 4) is higher (L_{50} 30.8–31.7 mm CL). In the Irish Sea, the lowest estimate for SOM in females was at 21-mm CL, and in males at 15-mm CL, based on histological examinations (McQuaid et al., 2006). Morphology-based estimations showed a bigger variation in size at SOM; in females, it ranged from 23.2 to 27.6 mm CL. Differences in SOM for males, as estimated from the break point in the growth rate of the appendix masculina (situated on the second pleopod), are judged to be noisy and biased due to methodological problems, mainly caused by the problems in consistency in measurement of the appendix masculina, as even different hauls within same FUs gave the same variation as for the whole FU data set (ICES, 2006). The difference in break points for the males between the FUs is therefore not believed to be caused by genuine regional differences. The break point is most often found around 30-mm CL, which indicates the size of sexual maturity (functional maturity).

4. STOCK ASSESSMENT

There are many methods to assess or estimate crustacean population abundance (e.g. Smith and Addison, 2003). CPUE, length cohort analysis (LCA), virtual population analysis (VPA) and, more recently, UWTV surveys are commonly used methods of assessment for *Nephrops* population estimates. The former three are of limited reliability, in comparison with the UWTV, due to bias in fisheries dependent data and internal model assumptions. LCA/VPA analyses investigate the fishery mortality of males and females per year and the result from the analyses has traditionally been compared to CPUE trends from the commercial log books or survey data, to 'tune' the analyses. However, as fishermen visit the grounds with the highest abundance, for example, observed with UWTV on Farn Deep in the winter fishery 2002/2003 (Bell et al., 2005), declines in stocks from CPUE may be masked. Fishermen report the captures per statistical rectangle, $1°$ of longitude and $0.5°$ of latitude, ca. 56×65 km, and Bell et al. (2005) state that CPUE should be aggregated on a finer geographical scale of 10–20 km, to be able to detect density declines. Surveys are undertaken nationally to collect data (data collection regulation); ICES Fishmap shows the Fish trawl survey stations for which data on *Nephrops* have been collected, for all the ICES states in 1993–2012 (Figure 7.13).

4.1. VPA (including multi-species)

VPA uses catch data (fishery-dependent data collection) to determine the past stock size using mortality rates (Jennings et al., 2001). It uses fishing mortality (number landed) and natural mortality (estimated) to discern the number in the stock for the previous time period (i.e. the previous year/month) and survivors. This data can then be compared to real-time data across set number or repetitions (ICES, 2006; Sardá and Aguzzi, 2012)

Figure 7.13 ICES Fishmap showing *Nephrops* surveys (fish trawl surveys) for all years 1993–2012. ICES FishMap. 'Norway lobster, *Nephrops norvegicus*'. http://www.ices.dk/ marineworld/fishmap/ices/ (accessed 28 December 2012).

and modified to analyse other factors such as reproduction potential and recruitment (Relini et al., 1998).

VPA uses age-based data for population dynamic modelling to describe the influence of fishing pressure upon the stock size. However, as with crustaceans generally, identifying the age of *Nephrops* is problematic because there are no easy or relatively cheap methods to identify age, any calcified markings being lost with the old outer exoskeleton (Ulmestrand and Eggert, 2001). To overcome this problem, the von Bertalanffy growth parameters are used to convert length data into age classes, referred to as 'slicing' of the data (ICES, 2004a,b). However, this method has come under scrutiny because of the lack of assumed variability in growth, and other external influences such as sampling and capture methodology. Thus, there is a strong reason for scientists to use VPA-derived models with caution (ICES, 2006; Sardá and Aguzzi, 2012; Tuck et al., 1997; Ulmestrand and Eggert, 2001). Most parameters can be found in the literature, such as regional fishing effort and technology, natural mortality rates, for example, from stomach analysis, SOM and size–weight relationships (Sardá and Aguzzi, 2012; Sparre, 1991; ICES, 2012a). The demand for developments within the VPA models for estimating growth from age–size and tagging studies is increasing to reduce the bias set in using estimated or fitted growth functions (Walters and Martell, 2004). There are, however, several methods that have been used for determining the age of individuals. Indirect age measurement using pigment aggregation in the eyestalk or neural areas over time has been evaluated for different species, for example, the European lobster (O'Donovan and Tully, 1996; Sheehy et al., 1999) and the brown crab (Sheehy and Prior, 2008), as well as for *Nephrops* (Belchier et al., 1994). This technique requires calibration for each species and environment using specimens of known age (Vogt, 2012). Lipofuscin analyses using spectrophotometric methods have also been evaluated for, for example, the blue crab *Callinectes sapidus* (Ju et al., 2001). Recently, a new direct technique has been presented where cuticle growth bands are observed on the eyestalk or gastric mill stones for commercial crustaceans such as the American lobster and snow crab (Kilada et al., 2012).

There are, however, a number of alternative models that can incorporate data derived from *Nephrops* fisheries without the need for aging techniques. These include Stock Synthesis and GADGET modelling frameworks that allow for *Nephrops*-specific variation in capture. This variation could be in the form of density-dependent growth and maturity rates, and reduced female emergence patterns that typically reduce the availability of other models or assessments to be used. There are several VPA methods used for *Nephrops* population assessments that differ in the way the data are presented and calculated. ADAPT and XSA are a few methods, which use catch data affiliated with survey data, that are becoming increasingly popular.

Multi-species VPA (MSVPA) differs from the single population model as variation in mortality and growth can be influenced in each time period due to interactions with other species (Magnusson, 1995; FAO). In standard VPA, natural mortality (M) is considered constant, whereas MSVPA assumes that M is iterative for each time period and age. Predation is also included and influences the mortality within the model. There have been few attempts with this method as it seems both complex and time consuming regarding the broadness of scope in the model in determining the predator–prey relationship within the fishery (Molina and Livingston, 2004).

Overall, it seems that VPA and Length Cohort Analysis provide the most realistic and reliable estimate of populations as they use direct data from the individuals caught (Sardá and Aguzzi, 2012). A note of caution with these approaches is that absolute biomass estimates are dependent upon the landings being reliably reported. Prior to 2006, there was widespread under-reporting of a large number of species including *Nephrops*. The introduction of 'Buyers and Sellers' legislation in the United Kingdom appears to have significantly improved the reporting of landings. Under this legislation, it is required that the purchases of fish be registered. Most assessment studies are now using VPA/LCA in conjunction with other assessment methods, which have been similar in their findings, such as CPUE and UWTV monitoring. The Western Irish Sea FU is interesting in terms of *Nephrops* population assessment as it has long-term capture data and survey data and is developing techniques for UWTV that are suitable for exploring several assessment methods in an integrated approach for advising management.

4.2. UWTV surveys

Fishery-independent abundance-estimating methods are therefore important as a tool for managers. UWTV surveys for counting *Nephrops* burrows as a fishery-independent assessment have been conducted for research purposes since the 1980s (Bailey et al., 1993; Chapman, 1985), as assessment guidance since 1994 for Scottish waters, refined in EU projects (Marrs et al., 1996; Tuck et al., 1997), and for absolute indices of density since 2009 (ICES, 2012b). Sarda and Aguzzi (2012) review the burrow counting method as an alternative to traditional assessment methods, such as VPA modelling. This method is independent of the fishery and therefore reduces the associated bias; however, the method is not without analytical bias (ICES, 2009, 2012b), as discussed later. Decision on the use of fishery-independent methods such as UWTV was made following difficulties associated with assessing non-aged species and bias in reporting of commercial data affecting VPA/LCA and CPUE trend analyses.

The seabed is monitored using TV cameras fastened on a sledge towed after a vessel at a speed of 0.5–0.7 knot (e.g. Campbell et al., 2009b). The area covered is calculated from the known width of the visual frame of the camera corrected for the distance above bottom (recently laser points have been used as frame marks) and the distance covered over the filming based either on vessel GPS tracks or odometer. Burrow systems (rather than individual entrances) are counted. This means that the individual counter has to assess how many entrances are likely to be associated with a specific burrow using visual cues such as burrow entrance shape, orientation, and distance from other entrances. It is also important for the counter to ignore the burrows associated with other burrowing fauna. The assumption behind the method is that each burrow system is inhabited by one adult (larger than 17 mm) *Nephrops*, that is, occupancy is assumed to be 100%. Burrows not occupied are thought to infill quickly and it is suspected that unoccupied burrows are not counted during surveys.

Uncertainties concerning the resulting abundance have been debated and are summarised (Uncertainties table, ICES, 2007), and burrow occupancy and burrow system morphology are an important focus of research. Improvements in methodology since the WKNEPHTV 2007 (ICES, 2007) are given in ICES (2009, 2012c) and methodological modellings to answer the critics have been made (Campbell et al., 2009a). Edge-effect uncertainty may overestimate the population as burrow systems not totally within frames are counted and extrapolation to a total substrate area yields a higher density than in reality and is corrected for in estimations in the Adriatic Sea (Morello et al., 2007). Investigations on whether the burrow system is inhabited by one or more individuals have mostly been performed in shallow lochs accessible by SCUBA and need to be done in other habitats (depth, areas). The time taken for the collapse of an uninhabited burrow is dependent on the bottom (from days to weeks to years) where trawling frequency, current and bioturbation of other species have an impact, and a better idea of collapse time is needed. The shape and extension of a burrow also needs more attention, for example, using ROVs to inject coloured dye into the burrow to learn about the burrow system. The identification of the species inhabiting a burrow (shape of the entrance and other signs) can be a problem, especially at the edges of the frame, where light and quality of picture are not so good. Other burrowing species inhabiting soft areas that can lead to overestimation of the density are the thalassinid shrimp (*Calocaris macandreae*), angular crab (*Goneplax rhomboides*) and the fish Fries' goby (*Lesueurigobius friesii*). However, experienced readers can discriminate between burrows and a conservative attitude is taken, that is, not including those burrows where there are reasonable uncertainties.

There are two main types of design for investigating burrow densities: the use of a randomised grid and the pseudo-randomised stratified-type. One analytical issue of UWTV is how many stations per strata are necessary to give a fair picture of the real abundance with a given precision, that is, a variance in burrow density needs to be stated and aimed for (ICES, 2007). For the randomised method where a station can be placed anywhere within the strata, the co-efficient of variation (CV, relative standard error) is used to precisely determine the average burrow density and is decreased with the number of stations. On the other hand, variation of the grid design method may be estimated from probability calculations and comparison with earlier data from the strata (Thompson, 2002). This gives an indication of how sampling effort can be adjusted in relation to the relative error. This serves as a proxy for the relative standard errors used in the confidence intervals calculation. The CV from randomised strata locations is mostly below 20%, whereas grid locations have an estimate of variation of below 5%

(ICES, 2012b). The Study Group on *Nephrops* Surveys (SGNEPS) has an important role, that of international coordination of UWTV efforts, and is focussing on planning, protocols, quality control, design and issues regarding survey development (ICES, 2012c). SGNEPS recommend that a CV (or relative standard error) of <20% is an acceptable precision level for UWTV survey estimates of abundance. The lowest survey densities are on the Fladen ground with 2.5 stations per 1000 km², while the highest densities are on the Aran Grounds with 79.9 stations per 1000 km². Currently, a randomised fixed grid design is used for all UWTV surveys operated under the Marine Institute, except for the Aran Grounds. For some grounds, such as the Aran Grounds, fewer stations using longer distances between stations are recommended.

4.2.1 Reference point determination
The ability to determine a standing biomass and a harvest rate is only part of establishing how a fishery is performing. It is important to understand what the exploitation limits (in terms of long-term sustainability) of any harvesting regime are. The evolution of the wording of ICES advice over the last decade has been in response to changes in philosophy, moving from MBAL (minimum biological acceptable limits), to the Precautionary Approach and, most recently to MSY (maximum sustainable yield), the latter in response to the European Union's 'Marine Strategy Framework Directive' (MSFD, EC regulation 2008/56/EC[11]). Within the MSFD, there are 11 separate descriptors of what Good Environmental Status (GES) should be (including food-webs, marine litter, hydrodynamics and contaminants), and there is a specific section for commercially exploited fish and shellfish species. Under this, descriptor stocks shall (1) be exploited sustainably consistent with high long-term yields, (2) have full reproductive capacity in order to maintain stock biomass, and (3) the proportion of older and larger fish/shellfish should be maintained (or increased). The time scale for this commitment is that GES should be achieved by 2016 or 2020 at the latest. GES for commercial species has been interpreted by ICES as a requirement for stocks to be exploited at MSY. Analytical determination of MSY requires a parameterised functional relationship between the size of spawning stock and the number of recruits, which, for the majority of *Nephrops* stocks, is not possible given the lack of a fully age-based assessment. Proxies for

---

[11] Directive 2008/56/EC of the European Parliament and of the Council of 17 June 2008. Establishing a framework for community action in the field of marine environmental policy (Marine Strategy Framework Directive) Official Journal of the European Union L164/19.

harvesting strategies, which are likely to generate MSY, are therefore used and the three candidates for F_{MSY} are $F_{0.1}$, $F_{35\%SPR}$, and F_{max}. These proxies are determined using processes similar to the Jones Length Cohort Analysis (Jones, 1984). As LCA assumes that the input length frequency comes from a population at equilibrium, a 3-year average length frequency is used. Anything that might change the shape of the length frequency (e.g. gear selection pattern, discarding patterns, stock–recruitment changes) will therefore induce a change in the associated reference points; so, regular updating of these reference points is required.

There may be strong differences in relative exploitation rates between the sexes in many stocks. To account for this, values for the candidates have been determined for males, females, and the two sexes combined. The F_{MSY} candidate considered to be most appropriate has been selected for each FU independently according to the perception of stock resilience, factors affecting recruitment, population density, knowledge of biological parameters, and the nature of the fishery (relative exploitation of the sexes and historical harvest rate vs. stock status). A formalised decision-making framework was developed for the selection of preliminary stock-specific F_{MSY} proxies (ICES, 2012a). These proxies may be modified following further data exploration and analysis. The combined-sex F_{MSY} proxy is considered appropriate if the resulting percentage of virgin spawner per recruit for males or females does not fall below 20%. When this does happen, a more conservative sex-specific F_{MSY} proxy is selected instead of the combined proxy.

ACKNOWLEDGEMENTS

The authors thank Bill Wiseman, Billy Hughes, Frank Stride and others from the Scottish *Nephrops* industry for their industry insight; Eibhlin O'Sullivan and Judith Farrell for their up-to-date commentary on the Celtic and Irish Seas; Daniel Valentinsson, who contributed information on the Swedish management and regulations; all the helpful creel fishers, Ingemar G., Patrik I., Tony & Kerstin, within the EU project *Nephrops* for sharing their knowledge from the daily operations at sea and Ana Leocádio, who contributed information on the Iberian fisheries.

REFERENCES

Abelló, P., Abella, A., Adamidou, A., Jukic-Peladic, S., Maiorano, P., Spedicato, M.T., 2002. Geographical patterns in abundance and population structure of *Nephrops norvegicus* and *Parapenaeus longirostris* (Crustacea: Decapoda) along the European Mediterranean coasts. Sci. Mar. 66 (Suppl. 2), 125–141.

Adey, J.M., 2007. Aspects of the sustainability of creel fishing for Norway lobster, Nephrops norvegicus (L.), on the west coast of Scotland. PhD thesis, University of Glasgow, pp. 488.

Adey, J.M., Smith, I.P., Hawkins, A.D., Atkinson, R.J.A., Tuck, I.D., Taylor, A.C., 2008. 'Ghost fishing' of target and non-target species by Norway lobster *Nephrops norvegicus* creels. Mar. Ecol. Prog. Ser. 366, 119–127.

Aguzzi, J., Sarda, F., 2008. A history of recent advancements on *Nephrops norvegicus* behavioral and physiological rhythms. Rev. Fish Biol. Fisheries 18, 235–248.

Aguzzi, J., Sardá, F., Abello, P., Company, J.B., Rotllant, G., 2003. Diel and seasonal patterns of *Nephrops norvegicus* (Decapoda: Nephropidae) catchability in the western Mediterranean. Mar. Ecol. Prog. Ser. 258, 201–211.

Aguzzi, J., Allué, R., Sardà, F., 2005. Seasonal dynamics in *Nephrops norvegicus* (Decapoda: Nephropidae) catches off the Catalan Coasts (Western Mediterranean). Fish. Res. 69, 293–300.

Aguzzi, J., Company, J.B., Sardà, F., 2007. The activity rhythm of berried and unberried females of *Nephrops norvegicus* (Crustacea, Decapoda). Crustaceana 80 (9), 1121–1134.

ANIFPO, 2007. A pilot pot fishery for *Nephrops norvegicus* off the Northern Ireland coast, 15 pp.

Archdale, M.V., Kawamura, G., 2011. Evaluation of artificial and natural baits for the pot fishery of the sand crab *Ovalipes punctatus* (De Haan, 1833). Fish. Res. 111, 159–163.

Arnold, K.K., Findlay, H.S., Spicer, J.I., Daniels, C.L., Boothroyd, D., 2009. Effect of CO_2-related acidification on aspects of the larval development of the European lobster, *Homarus gammarus* (L.). Biogeosciences 6 (8), 1747–1754.

Ayza, O., Tuset, V.M., González, J.A., 2011. Estimation of size at onset of sexual maturity and growth parameters in Norway lobster (*Nephrops norvegicus*) off the Portuguese coast. Fish. Res. 108, 205–208.

Bailey, N., Chapman, C.J., Kinnear, J., Bova, D., Weetman, A., 1993. Estimation of *Nephrops* stock biomass on the Fladen ground by TV survey. ICES Document CM 1993/K: 34, 9 pp.

Bailey, M.C., Polunin, N.V.C., Hawkins, A.D., 2012. A sustainable fishing plan for the Farne Deeps *Nephrops* fishery. Rep. Mar. Manag. Org. 85 pp.

Ball, B.J., Fox, G., Munday, B.W., 2000. Long- and short-term consequences of a *Nephrops* trawl fishery on the benthos and environment of the Irish Sea. ICES J. Mar. Sci. 57, 1315–1320.

Belchier, M., Shelton, P.M.J., Chapman, C.J., 1994. The indication and measurement of fluorescent age-pigment abundance in the brain of a crustacean (*Nephrops norvegicus*) by confocal microscopy. Comp. Biochem. Physiol. 108B, 157–164.

Bell, M.C., Elson, J.M., Addison, J.T., 2005. The effects of spatial targeting of fishing effort on the distribution of the Norway lobster, *Nephrops norvegicus*, on the Farn Deeps grounds, northeast England. N. Z. J. Mar. Freshw. Res. 39, 1023–1037.

Bell, M., Redant, F., Tuck, I., 2006. Nephrops species. In: Phillips, B. (Ed.), Lobsters: Biology, Management, Aquaculture and Fisheries, Blackwell Publishing, Oxford Chapter 13.

Bennett, D.B., Brown, C.G., 1979. The problems of pot immersion time in recording and analysing catch-effort data from a trap fishery. Rapp. P.-v. Reun. Cons. int. Explor. Mer. 175, 186–189.

Bennett. D., Hough, A., 2007. Re-certification report for loch Torridon *Nephrops* creel fishery client: Torridon *Nephrops* Management Group, Moody Marine.

Bergmann, M., Moore, P.G., 2001. Survival of decapod crustaceans discarded in the *Nephrops* fishery of the Clyde Sea area, Scotland. ICES J. Mar. Sci. 58, 163–171.

Bergmann, M., Wieczorek, S.K., Moore, P.G., Atkinson, R.J.A., 2002. Discard composition of the *Nephrops* fishery in the Clyde Sea area, Scotland. Fish. Res. 57, 169–183.

Bianchini, M.L., Di Stefano, L., Ragonese, S., 1998. Size and age at onset of sexual maturity of female Norway lobster *Nephrops norvegicus* L. (Crustacea: Nephropidae) in the Strait of Sicily (Central Mediterranean Sea). Sci. Mar. 62 (1–2), 151–159.

Bjordal, Å., 1979. Factors effecting creel capture of Norway lobster (*Nephrops norvegicus*) and Northern shrimp (*Pandalus borealis*), investigations by fishery and baiting experiments (in Norwegian). Masters thesis, Department of Fisheries, University of Bergen, 97 pp.

Bjordal, Å., 1986. The behaviour of Norway lobster towards baited creel and size selection of creels and trawls. Fiskeridirektoratets Skrifter Serie Havundersokelser (Report on Norwegian fishery and marine investigations) 18, 131–137.

Brander, K.M., Bennett, D.B., 1989. Norway lobsters in the Irish Sea: modeling one component of a multispecies resource. In: Caddy, J.F. (Ed.), Marine Invertebrate Fisheries: Their Assessment and Management, John Wiley, New York, pp. 183–204.

Briggs, R.P., 1988. A preliminary analysis of maturity data for northwest Irish Sea Nephrops fishery. ICES Document CM 1988/K:21.

Briggs, R.P., 1995. Variability in Northwest Irish Sea *Nephrops*. Fish. Res. 23, 175–187.

Briggs, R.P., 2010. A novel escape panel for trawl nets used in the Irish Sea *Nephrops* fishery. Fish. Res. 105, 118–124.

Briggs, R.P., McAliskey, M., 2002. The prevalence of *Hematodinium* in *Nephrops norvegicus* from the western Irish Sea. J. Mar. Biol. Ass. U.K. 82, 427–433.

Campbell, N., Dobby, H., Bailey, N., 2009a. Investigating and mitigating uncertainties in the assessment of Scottish *Nephrops norvegicus* populations using underwater television data. ICES J. Mar. Sci. 66, 646–655.

Campbell, N., Allan, L., Weetman, A., Dobby, H., 2009b. Investigating the link between *Nephrops norvegicus* burrow density and sediment composition in Scottish waters. ICES J. Mar. Sci. 66, 2052–2059.

Cardinale, M., Raetz, H.J., Charef, A., 2011. Assessment of Mediterranean Sea stocks—part 2. EU: Report EUR 25053 EN.

Cardinale, M., Osio, G.C., Charef, A., 2012. Report of the scientific, technical and economic committee for fisheries on assessment of Mediterranean Sea stocks—part 1. EU: Report EUR 25602 EN.

Carr, W.E.S., Derby, C.D., 1986. Chemically stimulated feeding behavior in marine animals: the importance of chemical mixtures and the involvement of mixture interactions. J. Chem. Ecol. 12, 989–1011.

Catchpole, T.L., Revill, A.S., 2008. Gear technology in *Nephrops* trawl fisheries. Rev. Fish Biol. Fisheries 18 (1), 17–31.

Catchpole, T.L., Frid, C.L.J., Gray, T.S., 2005. Discards in North Sea fisheries: causes, consequences and solutions. Mar. Policy 29, 421–430.

Catchpole, T.L., Frid, C.L.J., Gray, T.S., 2006a. Resolving the discard problem—a case study of the English *Nephrops* fishery. Mar. Policy 30, 821–831.

Catchpole, T.L., Revill, A.S., Dunlin, G., 2006b. An assessment of the Swedish grid and square-mesh codend in the English (Farn Deeps) *Nephrops* fishery. Fish. Res. 81, 118–125.

Catchpole, T.L., van Keeken, O., Gray, T., Piet, G., 2008. The discard problem—a comparative analysis of two fisheries: The English *Nephrops* fishery and the Dutch beam trawl fishery. Ocean Coast Manag 51, 772–778.

Chanes-Miranda, L., Viana, M.T., 2000. Development of artificial lobster baits using fish silage from tuna by-products. J. Shellfish Res. 19 (1), 259–263.

Chapman, C.J., 1985. Observing Norway lobster, *Nephrops norvegicus* (L) by towed sledge fitted with photographic and television cameras. In: George, J.D., Lythgoe, G.I., Lythgoe, J.N. (Eds.), Underwater Photography and Television for Scientists, Clarendon Press, Oxford, UK, pp. 100–108.

Chapman, C.J., Rice, A.L., 1971. Some direct observations on the ecology and behaviour of the Norway Lobster *Nephrops norvegicus* using different methods. Mar. Biol. 10 (4), 321–329.

Cheung, W.W.L., Pinnegar, J., Merino, G., Jones, M.C., Barange, M., 2012. Review of climate change impacts on marine fisheries in the UK and Ireland. Aquat. Conserv. Mar. Freshw. Ecosyst. 22, 368–388.

Combes, J., Lart, W., 2007. Clyde environment and fisheries review and sustainable supply chain project report. Clyde Fisheries Development Project.

Conan, G.Y., Comeau, M., Gosset, C., Robichaud, G., Gara·icoechea, C., 1994. The Big-ouden *Nephrops* trawl, and the Devismes trawl, two otter trawls efficiently catching benthic stages of Snow Crab (*Chionoecetes opilio*), and American Lobster (*Homarus americanus*). Can. Tech. Rep. Fish. Aquat. Sci. 27, 1992.

COUNCIL REGULATION (EC) No 850/98 for the conservation of fishery resources through technical measures for the protection of juveniles of marine organisms, pp. 55.

Dale, T., Siikavuopio, S.I., Aas, K., 2007. Development of formulated bait for edible crab (*Cancer pagurus* L.), using by-products from the fisheries and aquaculture industry. J. Shellfish Res. 26, 597–602.

De Buen, O., 1916. El Instituto español de oceanografía y sus primera campañas. Trabajos de oceanografía e biologia marina.

Derby, C.D., Sorensen, P.W., 2008. Neural processing, perception, and behavioural responses to natural chemical stimuli by fish and crustaceans. J. Chem. Ecol. 34, 898–914.

Drewery, J., Bova, D., Kynoch, R.J., Edridge, A., Fryer, R.J., O'Neill, F.G., 2010. The selectivity of the Swedish grid and 120 mm square mesh panels in the Scottish Nephrops trawl fishery. Fish. Res. 106, 454–459.

Eggert, H., Ellegard, A., 2003. Fishery control and regulation compliance: a case for co-management in Swedish commercial fisheries. Mar. Policy 27, 525–533.

Eggert, H., Ulmestrand, M., 1999. A bioeconomic analysis of the Swedish fishery for Norway lobster (*Nephrops norvegicus*). Mar. Resource Econ. 14, 225–244.

Eiriksson, H., 1999. Spatial variabilities of CPUE and mean size as possible criteria for unit stock demarcations in analytical assessments of *Nephrops* at Iceland. Rit Fiskideildar 16, 239–245.

Engelhard, G.H., Pinnegar, J.K., 2010. *Nephrops*. In: Rijnsdorp, A.D., Peck, M.A., Engelhard, G.H., Möllmann, C., Pinnegar, J.K. (Eds.), Resolving Climate Impacts on Fish Stocks. Copenhagen: ICES Cooperative Research Report 301, pp. 203–207.

Eno, N.C., MacDonald, S.D., Kinnear, J.A.M., Amos, C.S., Chapman, C.J., Clark, R.A., Bunker, F.P.D., Munro, C., 2001. Effects of crustacean traps on benthic fauna. ICES J. Mar. Sci. 58, 11–20.

Eriksson, S., 2006. Differences in the condition of Norway lobsters (*Nephrops norvegicus* (L.)) from trawled and creeled fishing areas. Mar. Biol. Res. 2, 52–58.

European Union, 2004. Council Decision of 19 July 2004 on establishing Regional Advisory Councils under the Common Fisheries Policy (2004/585/EC). Off. J. Eur. Union 256, 17–22.

FAO, 2010. Landing data for Nephrops norvegicus in 1955–2010 using FAO programme and database FishStatJ. http://www.fao.org/fishery/statistics/software/fishstatj/en.

Fariña, A.C., 1996. Megafauna de la plataforma continental y talud superior de Galicia. Biología de la cigala Nephrops norvegicus. Tesis Doctoral. Universidad da Coruña.

Farmer, A.D.S., 1975. Synopsis of biological data on Norway lobster *Nephrops norvegicus* (Linneo 1758). FAO Fish. Synop. 112, 1–97.

Field, R.H., Chapman, C.J., Taylor, A.C., Neil, D.M., Vickerman, K., 1992. Infection of the Norway lobster *Nephrops norvegicus* by a *Hematodinium*-like species of dinoflagellate on the west coast of Scotland. Dis. Aquat. Organ. 13, 1–15.

Figueiredo, M.J., Viriato, A., 1989. Localização e reconhecimento da topografia submarina dos principais pesqueiros de lagostins ao longo da costa Portuguesa, efectuados a bordo dos N/E "Noruega" e "Mestre Costeiro", em 1983/87. INIP, Relatórios Técnicos e Cient'ificos No. 4:37 pp.

Froese, R., Proelss, A., 2012. Evaluation and legal assessment of certified seafood. Mar. Policy 36 (6), 1284–1289.

González-Herraiz, I., Torres, M.A., Fariña, A.C., Freire, J., Cancelo, J.R., 2009. The NAO index and the long-term variability of *Nephrops norvegicus* population and fishery off the West of Ireland. Fish. Res. 98, 1–7.

Grabowski, J.H., Clescerci, E.J., Baukus, A.J., Gaudette, J., Weber, M., Yund, P.O., 2010. Use of herring bait to farm lobsters in the Gulf of Maine. PLoS One 5 (4), 1–11.

Guijarro, B., 2010. *GSA05* Norway lobster. Balearic Islands. Report of the SCSA working group on stock assessment of demersal species. Istanbul: GFCM, pp. 56–57.

Gutiérrez, N.L., Hilborn, R., Defeo, O., 2011. Leadership, social capital and incentives promote successful fisheries. Nature 470 (7334), 386–389.

Hallberg, E., Skog, M., 2011. Chemosensory sensilla in crustaceans. Chapter 6 in Chemical Communication in Crustaceans. In: Breithaupt, T., Thiel, M. (Eds.), Springer, pp. 103–121, 565 pp.

Hancock, D.A., 1974. Attraction and avoidance in marine invertebrates—their possible role in developing an artificial bait. J. Cons. Int. Explor. Mer. 35, 328–331.

Hardege, J.D., Bartels-Hardege, H.D., Fletcher, N., Terschak, J.A., Harley, M., Smith, M.A., Davidson, L., Hayden, D., Mueller, C.T., Lorch, M., Welham, K., Walter, T., Bublitz, R., 2011. Identification of a female sex pheromone in *Carcinus maenas*. Mar. Ecol. Prog. Ser. 436, 177–189.

Harris, R.R., Ulmestrand, M., 2004. Discarding Norway lobster (*Nephrops norvegicus* L.) through low salinity layers—mortality and damage seen in simulation experiments. ICES J. Mar. Sci. 61, 127–139.

Hayden, D., Jennings, A., Müller, C., Pascoe, D., Bublitz, R., Webb, H., Breithaupt, T., Watkins, L., Hardege, J.D., 2007. Sex-specific mediation of foraging in the shore crab, *Carcinus maenas*. Horm. Behav. 52, 162–168.

Heath, M.R., Neat, F.C., Pinnegar, J.K., Reid, D.G., Sims, D.W., Wright, P.J., 2012. Review of climate change impacts on marine fish and shellfish around the UK and Ireland. Aquat. Conserv. Mar. Freshw. Ecosyst. 22 (3), 337–367.

Hornborg, S., Nilsson, P., Valentinsson, D., Ziegler, F., 2012. Integrated environmental assessment of fisheries management: Swedish *Nephrops* trawl fisheries evaluated using a life cycle approach. Mar. Policy 36, 1193–1201.

ICES, 2004a. The Nephrops fisheries of the Northeast Atlantic and Mediterranean—a review and assessment of fishing gear design. ICES Cooperative Research Report, No. 270, 40 pp.

ICES, 2004b. ICES Advisory Committee on Fishery Management, report of the Working Group on Nephrops Stocks. ICES CM 2004/ACFM:19.

ICES, 2006. Report of the Workshop on Nephrops Stocks (WKNEPH), 24–27 January 2006, ICES Headquarters. ICES CM 2006/ACFM:12, 85 pp.

ICES, 2007. Workshop on the use of UWTV surveys for determining abundance in Nephrops stocks throughout European waters, 17–21 April 2007, Heraklion, Crete, Greece. ICES CM 2007/ACFM:14, 198 pp.

ICES, 2009. Report of the Benchmark workshop on Nephrops (WKNEPH), 2–6 March 2009, Aberdeen, UK. ICES CM 2009/ACOM:33, 156 pp.

ICES, 2012a. Report of the working group on the Assessment of Demersal Stocks in the North Sea and Skagerrak (WGNSSK), 27 April–3 May 2012, ICES Headquarters, Copenhagen. ICES CM 2012/ACOM:13, 1346 pp.

ICES, 2012b. Report of the ICES Advisory Committee, 2012, ICES Advice, 2012.

ICES, 2012c. Report of the Study Group on Nephrops Surveys (SGNEPS), 6–8 March 2012, Acona, Italy. ICES CM 2012/SSGESST:19, 36 pp.

ICES FishMap, Norway lobster, *Nephrops norvegicus*. Available through: http://www.ices.dk/marineworld/fishmap/ices/ (accessed 28.12.12).

Ingólfsson, Ó.A., 2011. The effect of forced mesh opening in the upper panel of a Nephrops trawl on size selection of *Nephrops*, haddock and whiting. Fish. Res. 108, 218–222.

Jansson, T., 2008. Discards in the Swedish creel fishery for *Nephrops norvegicus*. MSc. Gothenburg University

Jennings, S., Kaiser, M.J., Reynolds, J.D., 2001. Marine Fisheries Ecology, Blackwell Science Ltd, Oxford, UK.

Johnson, M.L., Prime, M., 2009. Smalls have their say. Samudra 54, 34–38.

Jones, R., 1984. Assessing the effects of changes in exploitation pattern using length composition data (with notes on VPA and cohort analysis). FAO Fisheries Technical Paper No. 256.

Ju, S.J., Secor, D.H., Harvey, H.R., 2001. Growth rate variability and lipofuscin accumulation rates in the blue crab *Callinectes sapidus*. Mar. Ecol. Prog. Ser. 224, 197–205.

Jury, S.H., Howell, H., O'Grady, D.F., Watson III, W.H., 2001. Lobster trap video: in situ video surveillance of the behaviour of *Homarus americanus* in and around traps. Mar. Freshw. Res. 52, 1125–1132.

Kilada, R., Sainte-Marie, B., Rochette, R., Davis, N., Vanier, C., Campana, S., 2012. Direct determination of age in shrimps, crabs, and lobsters. Can. J. Fish. Aquat. Sci. 69, 1728–1733.

Kraak, S.B.M., Reid, D.G., Gerritsen, H.D., Kelly, C.J., Fitzpatrick, M., Codling, E.A., Rogan, E., 2012. 21st century fisheries management: a spatio-temporally explicit tariff-based approach combining multiple drivers and incentivising responsible fishing. ICES J. Mar. Sci. 69 (4), 590–601.

Krag, L.A., Frandsen, R.P., Madsen, N., 2008. Evaluation of a simple means to reduce discard in the Kattegat-Skagerrak Nephrops (*Nephrops norvegicus*) fishery: Commercial testing of different codends and square-mesh panels. Fish. Res. 91, 175–186.

Leocádio, A.M., Whitmarsh, D., Castro, M., 2012. Comparing trawl and creel fishing for Norway lobster (*Nephrops norvegicus*): biological and economic considerations. PLoS One 7 (7), e39567.

Levandowsky, M., Hodgson, E.S., 1965. Amino acid and amine receptors of lobsters. Comp. Biochem. Physiol. 16, 159–161.

Lovewell, S.R., Howard, A.E., Bennett, D.B., 1988. The effectiveness of parlour pots for catching lobsters (Homarus gammarus (L.)) and crabs (Cancer pagurus L.). J. Cons. Int. Explor. Mer 44, 247–252.

Mackie, A.M., 1973. The chemical basis of food detection in the lobster *Homarus gammarus*. Mar. Biol. 21, 103–108.

Mackie, A.M., Grant, P.T., Shelton, R.G.J., Hepper, B.T., Walne, P.R., 1980. The relative efficiencies of natural and artificial baits for the lobster, *Homarus gammarus*: laboratory and field trials. J. Cons. Int. Explor. Mer 39 (2), 123–129.

Madsen, N., Valentinsson, D., 2010. Use of selective devices in trawls to support recovery of the Kattegat cod stock: a review of experiments and experience. ICES J. Mar. Sci. 67, 2042–2050.

Madsen, N., Holst, R., Frandsen, R.P., Krag, L.A., 2012. Improving the effectiveness of escape windows in directed Norway lobster *Nephrops norvegicus* trawl fisheries. Fisheries Sci. 78 (5), 965–975.

Magnusson, K., 1995. An overview of the multispecies VPA - theory and applications. Rev. Fish Biol. Fisheries 5, 195–212.

Main, J., Sangster, G.I., 1981. The Behaviour of the Norway Lobster, *Nephrops norvegicus* (L.), During Trawling, Marine Laboratory, Aberdeen.

Marano, G., Ungaro, N., Marano, C.A., Marsan, R., 1998. La ricerca sulle risorse demersali del bacino Adriatico sud-occidentale (anni 1985-97): sintesi dei risultati. Biol. Mar. Medit. 5 (3), 109–119.

Marine Institute, 2012. Annual Review of Fish Stocks in 2012 with Management Advice for 2013. Marine Institute, Galway.

Marine Scotland, 2012. Conservation credits scheme rules (amended 17 October).

Marrs, S.J., Atkinson, R.J.A., Smith, C.J., Hills, J.M., 1996. Calibration of the towed under-water TV technique for use in stock assessment of Nephrops norvegicus. Final Report to the European Commission. Brussels: European Commission

Mason, E., Atkinson, R.J.A., Hough, A., 2002. Loch Torridon Nephrops Creel Fishery Cer-tification Report, Moody Marine Ltd., Birkenhead.

McQuaid, N., Briggs, R.P., Roberts, D., 2006. Estimation of the size of onset of sexual maturity in Nephrops norvegicus (L.). Fish. Res. 81, 26–36.

Miller, R.J., 1990. Effectiveness of crab and lobster traps. Can. J. Fish. Aquat. Sci. 47, 1228–1251.

Molina, J.J., Livingston, P., 2004. Sensitivity analysis of the multispecies virtual population analysis model parameterized for a system of trophically-linked species from the eastern Bering Sea. Ciencias Marinas 30 (2), 1–12.

Monteiro, P., Araújo, P., Erzini, K., Castro, M., 2001. Discards of the Algarve (southern Portugal) crustacean trawl fishery. Hydrobiologia 449, 267–277.

Morello, E.B., Froglia, C., Atkinson, R.J.A., 2007. Underwater television as a fishery-independent method for stock assessment of Norway lobster (Nephrops norvegicus) in the central Adriatic Sea (Italy). ICES J. Mar. Sci. 64, 1116–1123.

Morello, E.B., Antolini, B., Gramitto, M.E., Atkinson, R.J.A., Froglia, C., 2009. The fishery for Nephrops norvegicus (Linnaeus, 1758) in the central Adriatic Sea (Italy): preliminary observations comparing bottom trawl and baited creels. Fish. Res. 95, 325–331.

Nunes, A.J.P., Sá, M.V.C., Andriola-Neto, F.F., Lemos, D., 2006. Behavioral response to selected feed attractants and stimulants in Pacific white shrimp, Litopenaeus vannamei. Aquaculture 260 (1–4), 244–254.

NWWRAC, 2012. NWWRAC position paper re Management Measures for the Nephrops Stock in the Porcupine Bank (FU 16) (2012). http://www.nwwrac.org/admin/publication/upload/NWWRAC_Advice_Supporting_EAPO_Position_NEPHROPS_Mgment_FU16_14Dec2012_EN.pdf.

NWWRAC, 2013. http://www.nwwrac.org/About_NWWRAC/About_Us_ENG/Navigation.html.

O'Donovan, V., Tully, O., 1996. Lipofuscin (age pigment) as an index of crustacean age: correlation with age, temperature and body size in cultured juvenile Homarus gammarus (L.). J. Exp. Mar. Biol. Ecol. 207 (1–2), 1–14.

Panfili, M., Morello, E.B., Froglia, C., 2007. The impact of scavengers on the creel fishery for Nephrops norvegicus in the central Adriatic sea. Ra Comm. int. Mer M.dit. 38, 571.

Ramos, F., Sobrino, I., Jiménez, M.P., 1996. Cartografía de especies y caladeros. Golfo de Cádiz. Inf. Téc. 45/96 Consejería de Agricultura y Pesca. Junta de Andalucía.

Relini, L.O., Zamboni, A., Fiorentino, F., Massi, D., 1998. Reproductive patterns in Norway lobster Nephrops norvegicus (L.), (Crustacea Decapoda Nephropidae) of different Mediterranean areas. Scientia Marina 62 (Suppl. 1), 25–41.

Richardson, G., 1996. The effect of light-induced eye damage on the behaviour of Nephrops norvegicus, Ph.D., University of Leicester.

Ridgway, I.D., Taylor, A.C., Atkinson, R.J.A., Chang, E.S., Neil, D.M., 2006. Impact of capture method and trawl duration on the health status of the Norway lobster, Nephrops norvegicus. J. Exp. Mar. Biol. Ecol. 339, 135–147.

Saila, S.B., Nixon, S.W., Oviatt, C.A., 2002. Does lobster trap bait influence the Maine inshore trap fishery? N. Am. J. Fish. Manage. 22, 602–605.

Sandberg, M.G., Gjermundsen, A., Hempel, E., Olafsen, T., Curtis, H., Martin, A., 2004. Seafood Industry Value Chain Analysis—Cod, Haddock and Nephrops (Landed in UK), KPMG and Sea Fish Industry Authority, Edinburgh.

Sangster, G.I., Breen, M., 1998. Gear performance and catch comparison trials between a single trawl and a twin rigged gear. Fish. Res. 36, 15–26.

Sarda, F., Aguzzi, J., 2012. A review of burrow counting as an alternative to other typical methods of assessment of Norway lobster populations. Rev. Fish Biol. Fisheries 22 (2), 409–422.

Sheehy, M.R.J., Prior, A.E., 2008. Progress on an old question for stock assessment of the edible crab *Cancer pagurus*. Mar. Ecol. Prog. Ser. 353, 191–202.

Sheehy, M.R.J., Bannister, R.C.A., Wickins, J.F., Shelton, P.M.J., 1999. New perspectives on the growth and longevity of the European lobster (*Homarus gammarus*). Can. J. Fish. Aquat. Sci. 56 (10), 1904–1915.

Smith, M.T., Addison, J.T., 2003. Methods for stock assessment of crustacean fisheries. Fish. Res. 65, 231–256.

Sparre, P., 1991. Introduction to multispecies virtual population analysis. ICES Mar. Sci. Symp. 193, 12–21.

STECF, 2012. Review of Scientific Advice for 2013—Consolidated Advice on Fish Stocks of Interest to the European Union, STECF, Brussels.

SwAM The Swedish Agency for Marine and Water Management, 2004. Havs-och vattenmyndighetens föreskrifter om fiske i Skagerrak, Kattegatt och Östersjön (FIFS 2004:36).

Swedish National Board of Fisheries, 2006. The effects of a moved out trawl border on fish and benthic fauna. Analyses of an increased use of passive gears inside the trawl limit and the economic consequences for the fishing industry, Gothenburg.

Swedish National Board of Fisheries, 2010. Small scale coastal fisheries in Sweden, Gothenburg.

The Prohibition of Fishing with Multiple Trawls (No.2), 2007. SSI13/2007 Scotland: TSO, The Stationary Office.

The Sea Fish (Specified Sea Areas) (Regulation of Nets and Other Fishing Gear), 2009. SSI 165/2009 Scotland: TSO, The Stationary Office.

Thompson, S.K., 2002. Sampling, second ed. John Wiley & Sons, Inc., New York.

Tuck, I.D., Chapman, C.J., Atkinson, R.J.A., Bailey, N., Smith, R.S.M., 1997. A comparison of methods for stock assessment of the Norway lobster, *Nephrops norvegicus*, in the Firth of Clyde. Fish. Res. 32, 89–100.

Tuck, I.D., Hall, S.J., Robertson, M.R., Armstrong, E., Basford, D.J., 1998. Effects of physical trawling disturbance in a previously unfished sheltered Scottish sea loch. Mar. Ecol. Prog. Ser. 162, 227–242.

Tuck, I.D., Atkinson, R.J.A., Chapman, C.J., 2000. Population biology of the Norway lobster, *Nephrops norvegicus* (L.) in the Firth of Clyde, Scotland II: fecundity and size at onset of sexual maturity. ICES J. Mar. Sci. 57, 1227–1239.

Ulmestrand, M., Eggert, H., 2001. Growth of Norway lobster, *Nephrops norvegicus* (Linnaeus 1758), in Skagerrak, estimated from tagging experiments and length frequency data. ICES J. Mar. Sci. 58, 1326–1334.

Ungfors, A., Krång., A-S., Cuellar, E., Eriksson, S. (In prep). Potential novel baits for the Norway lobster (*Nephrops norvegicus*) creel fisheries investigated in Skagerrak and Kattegat.

Valentinsson, D., Ulmestrand, M., 2008. Species-selective *Nephrops* trawling: Swedish grid experiments. Fish. Res. 90, 109–117.

Vergnon, R., Blanchard, F., 2006. Evaluation of trawling disturbance on macrobenthic invertebrate communities in the Bay of Biscay, France: Abundance Biomass Comparison (ABC method). Aquat. Living Res. 19, 219–228.

Vogt, G., 2012. Ageing and longevity in the Decapoda. (Crustacea): a review. Zoologischer Anzeiger 251, 1–25.

Vrgoč, N., Arnen, E., Jukić-Peladić, S., Krstulović Šifner, S., Mannini, P., Marčeta, B., Osmani, K., Piccinetti, C., Ungaro, N., 2004. Review of current knowledge on shared demersal stocks of the Adriatic Sea. FAO-MiPAF Scientific Cooperation to Support

Responsible Fisheries in the Adriatic Sea. GCP/RER/010/ITA/TD. Adria. Med. Tech. 12, 1–9.

Waddington, K.I., Meeuwig, J., 2009. Contribution of bait to lobster production in an oligotrophic marine ecosystem as determined using a mass balance model. Fish. Res. 99, 1–6.

Walters, C.J., Martell, J.D., 2004. Fisheries Ecology and Management. Princeton University Press, United Kingdom, 448 pp.

Watson, W., Jury, S.H., 2013. The relationship between American lobster catch, entry rate into traps and density. Mar. Biol. Res. 9 (1), 59–68.

Wieczorek, S.K., Campagnuolo, S., Moore, P.G., Froglia, C., Atkinson, R.J.A., Gramitto, E.M., Bailey, N. 1999. The composition and fate of discards from *Nephrops* trawling in Scottish and Italian waters. Final Report to the European Commission. Reference 96/092, Study Project in support of the Common Fisheries Policy (XIV/96/C75), 323 pp.

Ziegler, F., 2006. Environmental life cycle assessment of Norway lobster (*Nephrops norvegicus*) fished by creels, conventional and species-selective trawls along the Swedish west coast. A data report. Borås: The Swedish Institute for Food and Biotechnology.

Ziegler, F., Valentinsson, D., 2008. Environmental life cycle assessment of Norway lobster (*Nephrops norvegicus*) caught along the Swedish west coast by creels and conventional trawls—LCA methodology with case study. Int. J. Life Cycle Assess. 13, 487–497.

Zimmer-Faust, R.K., Tyre, J.E., Michel, W.C., Case, J.F., 1984. Chemical mediation of appetitive feeding in a marine decapod crustacean: the importance of suppression and synergism. Biol. Bull. 167, 339–353.

Zuur, A.F., Tuck, I.D., Bailey, N., 2003. Dynamic factor analysis to estimate common trends in fisheries time series. Can. J. Fish. Aquat. Sci. 60, 542–552.

SUBJECT INDEX

Note: Page numbers followed by "*f*" indicate figures, and "*t*" indicate tables.

TAXONOMIC INDEX

Note: Page numbers followed by "*f*" indicate figures, and "*t*" indicate tables.